Learning with LabVIEW 7™ Express

Robert H. Bishop

The University of Texas at Austin

PEARSON

Prentice Hall

Upper Saddle River, NJ 07458

Library of Congress Cataloging-in-Publication Data on file.

Vice President and Editorial Director, ECS: *Marcia J. Horton*
Acquisitions Editor: *Laura Fischer*
Editorial Assistant: *Andrea Messineo*
Vice President and Director of Production and Manufacturing, ESM: *David W. Riccardi*
Executive Managing Editor: *Vince O'Brien*
Managing Editor: *David A. George*
Production Editor: *Craig Little*
Director of Creative Services: *Paul Belfanti*
Art Director and Cover Manager: *Paul Gourhan*
Cover Design: *National Instruments, Inc.*
Art Editor: *Xiaohong Zhu*
Manufacturing Manager: *Trudy Pisciotti*
Manufacturing Buyer: *Lisa McDowell*
Marketing Manager: *Holly Stark*

© 2004 by Pearson Education, Inc.
Pearson Prentice Hall
Pearson Education, Inc.
Upper Saddle River, NJ 07458

Printed in the United States of America

10 9 8 7 6 5 4 3 2 1

ISBN 0-13-117605-6

Pearson Education Ltd., *London*
Pearson Education Australia Pty., *Sydney*
Pearson Education Singapore, Pte. Ltd.
Pearson Education North Asia Ltd., *Hong Kong*
Pearson Education Canada, Inc., *Toronto*
Pearson Educación de Mexico, S.A. de C.V.
Pearson Education—Japan, *Tokyo*
Pearson Education Malaysia, Pte. Ltd.
Pearson Education Inc., *Upper Saddle River, New Jersey*

*To my parents, W. Robert Bishop
and Anna Maria DiPietro Bishop*

CONTENTS

PREFACE

Learning with LabVIEW 7 ™ Express is the textbook that accompanies the *Lab-VIEW 7 Express Student Edition* from National Instruments, Inc. This textbook, as well as the LabVIEW software, has undergone a significant revision from the previous edition to include all the latest features and technologies of LabVIEW 7 Express. The new Express technology makes learning and using LabVIEW even easier than before, allowing you to create complex and powerful applications in minimal time. As you read through the book and work through the examples, we hope you will agree that this book is more of a personal tour guide than a software manual.

LabVIEW is the leading graphical development environment for science and engineering with built-in functionality for simulation, data acquisition, instrument control, measurement analysis, and data presentation. The *LabVIEW 7 Express Student Edition* delivers all the capabilities of the professional version of LabVIEW, widely considered the industry standard for test, measurement, automation, and control. With LabVIEW, students can design graphical programming solutions to their classroom problems and laboratory experiments—an ideal tool for science and engineering applications—that is also fun to use! The *Lab-VIEW 7 Express Student Edition* affords students the opportunity for self-paced learning and independent project development.

The goal of this book is to help students learn to use LabVIEW on their own. With that goal in mind, this book is very art-intensive with over 400 figures in all. That means that there are numerous screen captures in each section taken from a typical LabVIEW session. The figures contain additional labels and pointers added to the LabVIEW screen captures to help students understand what they are seeing on their computer screens as they follow along in the book.

The most effective way to use *Learning with LabVIEW* is to have a concurrent LabVIEW session in progress on your computer and to follow along with the steps in the book. A directory of virtual instruments has been developed by the author exclusively for use by students using *Learning with LabVIEW* and is available on **www.prenhall.com/bishop**. These virtual instruments complement the material in the book. In most situations, the students are asked to develop the virtual instrument themselves following instructions given in the book, and then comparing their solutions with the solutions provided by the author to obtain immediate feedback. In other cases, students are asked to run a specified virtual instrument as a way to demonstrate an important LabVIEW concept.

THE *LABVIEW 7 EXPRESS STUDENT EDITION*

With the renewed emphasis in higher education on hands-on laboratory experience, many educational institutions are beginning to improve their laboratory facilities with the goal of increasing student exposure to practical problems. Educators continue to be pushed by industry to produce college graduates with experience in acquiring and analyzing data, constructing computer-based simulations of physical systems, and multi-purpose computer programming. To that end, LabVIEW offers a powerful, efficient, and easy-to-use development environment that provides educators the tools to teach their students a wide range of topics with just one open, industry-standard tool. Using LabVIEW you can create your own virtual instruments (VIs). An effective way to improve laboratory instrumentation is by modifying and improving a LabVIEW computer program (that is, a virtual instrument that emulates a standard instrument), rather than to retrofit a laboratory with new hardware equipment. In LabVIEW, the software is the instrument.

THE *LABVIEW 7 EXPRESS STUDENT EDITION*

The *LabVIEW 7 Express Student Edition* software package is a powerful and flexible instrumentation, analysis, and control software platform for PCs running Microsoft Windows or Apple Macintosh OS X. The student edition is designed to give students early exposure to the many uses of graphical programming. LabVIEW not only helps reinforce basic scientific, mathematical, and engineering principles, but it encourages students to explore advanced topics as well. Students can run LabVIEW programs designed to teach a specific topic, or they can use their skills to develop their own applications. LabVIEW provides a real-world, hands-on experience that complements the entire learning process.

WHAT'S NEW WITH THE LABVIEW 7 EXPRESS STUDENT EDITION?

The demand for LabVIEW in colleges and universities has led to the development of *LabVIEW 7 Express Student Edition* based on the industry version of LabVIEW 7 Express. This is a new and significant software revision of *LabVIEW Student Edition 6.0* that delivers all of the graphical programming capabilities of the professional edition. With the student edition, students can design graphical programming solutions for their classroom problems and laboratory experiments on their personal computers. The *LabVIEW 7 Express Student Edition* features include the following:

- New Express VIs that bring interactive, configuration-based application design for acquiring, analyzing, and presenting data.

- New interactive measurement assistants and a redesigned NI-DAQ driver make creating data acquisition and instrument control applications easier than ever.

- Full LabVIEW advanced analysis capability, including 13 new analysis Express VIs and over 400 native analysis and signal processing functions.

- Full compatibility with all National Instruments data acquisition and instrument control hardware.

- Support for all data types used in the LabVIEW Full Development System.

ORGANIZATION OF LEARNING WITH LABVIEW

This textbook serves as a LabVIEW resource for students. The pace of instruction is intended for both undergraduate and graduate students. The book is comprised of 12 chapters and should be read sequentially when first learning LabVIEW. For more experienced students, the book can be used as a reference book by using the index to find the desired topics. The 12 chapters are as follows:

CHAPTER 1: LabVIEW Basics—This chapter introduces the LabVIEW environment and helps orient students when they open a virtual instrument. Concepts such as windows, toolbars, menus, and palettes are discussed.

CHAPTER 2: Virtual Instruments—The components of a virtual instrument are introduced in this chapter: front panel, block diagram, and icon/connector pair. This chapter also introduces the concept of controls (inputs) and indicators (outputs) and how to wire objects together in the block diagram. Express VIs are introduced in the chapter.

CHAPTER 3: Editing and Debugging Virtual Instruments—Resizing, coloring, and labeling objects are just some of the editing techniques introduced in this chapter. Students can find errors using execution highlighting, probes, single-stepping, and breakpoints, just to name a few of the available debugging tools.

CHAPTER 4: Sub VIs—This chapter emphasizes the importance of reusing code and illustrates how to create a VI icon/connector. It also shows parallels between LabVIEW and text-based programming languages.

CHAPTER 5: Structures—This chapter presents loops, case structures, and flat and stacked sequence structures that govern the execution flow in a VI.

The Formula Node is introduced as a way to implement complex mathematical equations. The MATLAB Node is also introduced as a way to run m-files from within the LabVIEW environment.

CHAPTER 6: Arrays and Clusters—This chapter shows how data can be grouped, either with elements of the same type (arrays) or elements of a different type (clusters). This chapter also illustrates how to create and manipulate arrays and clusters.

CHAPTER 7: Charts and Graphs—This chapter shows how to display and customize the appearance of single and multiple charts and graphs.

CHAPTER 8: Data Acquisition—The basic characteristics of analog and digital signals are discussed in this chapter, as well as the factors students need to consider when acquiring and generating these signals. This chapter introduces students to the Measurement and Automation Explorer (MAX) and the DAQ Assistant.

CHAPTER 9: Strings and File I/O—This chapter shows how to create and manipulate strings on the front panel and block diagram. This chapter also explains how to write data to and read data from files.

CHAPTER 10: Instrument Control—The components of an instrument control system using a GPIB or serial interface are presented in this chapter. Students are introduced to the notion of instrument drivers and of using the Measurement and Automation Explorer (MAX) to detect and install instrument drivers. The Instrument I/O Assistant is introduced.

CHAPTER 11: Analysis—LabVIEW can be used in a variety of ways to support analysis of signals and systems. Several important analysis topics are discussed in this chapter, including how to use LabVIEW for signal generation, signal processing, linear algebra, curve fitting, formula display on the front panel, differential equations, fording roots (zero finder), and integration and differentiation.

CHAPTER 12: Other LabVIEW Applications—The concluding chapter discusses briefly other features of LabVIEW, including event-driven programming, remote Front Panels, and property nodes.

The important pedagogical elements in each chapter include the following:

1. A brief table of contents and a short preview of what to expect in the chapter.

2. A list of chapter goals to help focus the chapter discussions.

3. Margin icons that focus attention on a helpful hint or on a cautionary note.

4. An end-of-chapter summary and list of key terms.

KEY TERMS

5. Sections entitled **Building Blocks** near the end of each chapter present the continuous development and modification of a virtual instrument for measuring volume. The student is expected to construct the VIs based on the instructions given in the sections. The same VI is used as the starting point and then improved in each subsequent chapter as a means for the student to practice with the newly introduced chapter concepts. The VI for measuring volume was not used in Building Block sections whenever it proved more effective to use a different instrument to illustrate the important chapter concepts.

BUILDING BLOCK

6. Many worked examples are included in each chapter. In most cases, students construct the VIs discussed in the examples by following a series of instructions given in the text. In the early chapters, the instructions for building the VIs are quite specific, but in the later chapters, students are expected to construct the VIs without precise step-by-step instructions. Of course, in all chapters, working versions of the VIs are provided for all examples in the Learning directory included as part of the *LabVIEW 7 Express Student Edition*. Here is a sample of the worked examples:

   ▪ Temperature system demonstration.

- Solving a set of linear differential equations.
- Building your first virtual instrument.
- Computing a baseball batting average.
- Computing and graphing the time value of money.
- Studying chaos using the logistic difference equation.
- Acquiring data.
- Writing ASCII data to a file.

7. A section entitled *Relaxed Reading* that describes how LabVIEW is being utilized to solve interesting real-world problems. The material is intended to give students a break from the technical aspects of learning LabVIEW and to stimulate thinking about how LabVIEW can be used in various other situations.

8. End-of-chapter exercises and problems reinforce the main topics of the chapter and provide practice with LabVIEW.

ORIGINAL SOURCE MATERIALS

Learning with LabVIEW 7 Express was developed with the aid of several important references. Each reference is a manual published by National Instruments and one manual is actually used in a hands-on classroom environment taught by National Instruments instructors in their own classroom. Specifically, the following materials were used as primary sources:

- *LabVIEW Basics I*, Course Software Version 7.0, Copyright © 2003.
- *LabVIEW User Manual*, Copyright © 2003.

By design, there is a strong correlation between the material contained in the NI LabVIEW manuals and the material presented in this book. The information contained in the manuals was reduced in scope and refined to make it more accessible to students learning LabVIEW on their own.

LABVIEW 7 EXPRESS STUDENT EDITION SOFTWARE

It is assumed that the reader has a working knowledge of either the Windows or the Mac OS X operating system. If your computer experience is limited, you may first want to spend some time familiarizing yourself with your computer in order to understand the operation of your Mac or PC. You should know how to access pull-down menus, open and save files, install software from a CD, and use a mouse. You will find previous computer programming experience helpful—but not necessary.

A set of virtual instruments has been developed by the author for this book. You will need to obtain the Learning directory from the companion website to this book at Prentice Hall:

http://www.prenhall.com/bishop

For more information, you may also want to visit the National Instruments website at

http://www.ni.com/labviewse

All of the VI examples in this book were tested by the author on a Dell Dimensions XPS D300 running Windows XP Professional. Obviously, it is not possible to verify each VI on all the available Windows and Macintosh platforms that are compatible with LabVIEW so if you encounter platform-specific difficulties, please let us know.

If you would like information on upgrading to the LabVIEW 7 Express Professional Version, please write to

National Instruments
att.: Academic Sales
11500 North Mopac Expressway
Austin, TX 78759

or visit the National Instruments website: **http://www.ni.com**

LIMITED WARRANTY

The software and the documentation are provided "as is," without warranty of any kind, and no other warranties, either expressed or implied, are made with respect to the software. National Instruments does not warrant, guarantee, or make any representations regarding the use, or the results of the use, of the software or the documentation in terms of correctness, accuracy, reliability, or otherwise and does not warrant that the operation of the software will be uninterrupted or error-free. This software is not designed with components and testing for a level of reliability suitable for use in the diagnosis and treatment of humans or as critical components in any life-support systems whose failure to perform can reasonably be expected to cause significant injury to a human. National Instruments expressly disclaims any warranties not stated herein. Neither National Instruments nor Pearson Education shall be liable for any direct or indirect damages. The entire liability of National Instruments and its dealers, distributors, agents, or employees are set forth above. To the maximum extent permitted by applicable law, in no event shall National Instruments or its suppliers be liable for any damages, including any special, direct, indirect, incidental, exemplary, or consequential damages, expenses, lost profits, lost savings, business interruption, lost business information, or any other damages arising out of the use, or inability to use, the software or the documentation even if National Instruments has been advised of the possibility of such damages.

ACKNOWLEDGMENTS

Thanks to all the folks at National Instruments for their assistance and input during the development of *Learning with LabVIEW 7 Express*. A very special thanks to Jim Cahow of NI for providing day-to-day support during the final months of the project. Finally, I wish to express my appreciation to Lynda Bishop for assisting me with the manuscript preparation, for providing valuable comments on the text, and for handling my personal day-to-day activities associated with the entire production.

KEEP IN TOUCH!

The author and the staff at Pearson Prentice Hall and at National Instruments would like to establish an open line of communication with the users of the *LabVIEW 7 Express Student Edition*. We encourage students to e-mail the author with comments and suggestions for this and future editions.

Keep in touch!

ROBERT H. BISHOP
rhbishop@mail.utexas.edu
JIM CAHOW
National Instruments Academic Resources Manager
jim.cahow@ni.com

LabVIEW Basics

Welcome to the *Student Edition of LabVIEW*! **LabVIEW** is a powerful and complex programming environment. Once you have mastered the various concepts introduced in this book you will have the ability to develop applications in a graphical programming language and to develop virtual instruments for data acquisition, signal analysis, and instrument control. This introductory chapter provides a basic overview of LabVIEW and its components.

GOALS

1. Installation of the *Student Edition of LabVIEW*.

2. Familiarization with the basic components of LabVIEW.

3. Introduction to front panels and block diagrams, short cut and pull-down menus, palettes, VI libraries, and on-line help.

1.1   SYSTEM CONFIGURATION REQUIREMENTS

The *LabVIEW Student Edition* is distributed on CD-ROM and contains versions for Windows XP/2000/NT/ME/98 and Mac OS X.

Windows XP/2000/NT/ME/98

Processor:	Pentium III/Celeron 600 MHz or equivalent
RAM:	128 MB
Screen Resolution:	800 × 600 pixels
Operating System:	Windows XP/2000/NT/ME/98
Disk Space:	130MB

Macintosh OS X

Processor:	G3 or better
RAM:	128 MB
Screen Resolution:	800 × 600 pixels
Operating System:	Mac OS X (10.2 or later)
Disk Space:	380MB

1.2   INSTALLING THE *STUDENT EDITION OF LABVIEW*

Windows

1. Once you have inserted the CD in the CD-ROM drive, the LabVIEW 7 Express autorun screen should appear. Click on Install LabVIEW to begin installation.

   If the autorun screen does not appear, click on the My Computer icon of your computer and then double click on the CD-ROM icon. Next, double click on setup.exe to begin the installation.

2. This will begin the initialization of the installer. This process may take a few minutes to complete. When initialization is complete the software license agreements will be displayed. Upon reviewing of the software license agreements, click the **I accept the License Agreement(s)** button and click the **Next** button to continue.

3. The subsequent dialog box allows you to select the installation directory. If you do not have a preference, keep the default directory. Otherwise, click the

Browse button to select an alternate directory. When ready, click the **Next** button to continue.

4. The following screen will list LabVIEW 7.0 Student Edition under Features to be Changed or Added. Click the **Next** button to begin installation.

5. The installation should start installing files to your hard drive. Click on the **Finish** button when the installation has completed.

6. The installer may prompt you to reboot the system after finishing installation. If so, click **Yes** to restart your computer before using LabVIEW.

Macintosh

1. Once you have inserted the CD in the CD-ROM drive, double click on the CD icon labeled **LabVIEW 7.0 SE Mac OS X** on your desktop.

2. Double click on the **LabVIEW 7.0 Student Ed** icon in the folder that pops up.

3. At this point, Mac OS X will prompt you to supply the administrator password to install the LabVIEW Student Edition software. Type in your password and click **OK**.

4. The LabVIEW installer will begin to initialize. When complete, click on **Continue** to proceed to the next window.

5. The software license will now be displayed. Upon reviewing the license agreement, click the **Accept** button to continue.

6. The LabVIEW 7.0 Readme for Mac OS X will now be displayed. Once you have finished reading click the **Continue** button.

7. This dialog box will begin the installation. In the upper left corner there is a drop down menu with two options: **Easy Install** and **Uninstall**. Make sure that **Easy Install** is selected and click **Install** to begin installing the LabVIEW Student Edition.

8. When finished, you should get a dialog box claiming that the installation was successful.

1.3   THE LABVIEW ENVIRONMENT

LabVIEW is short for **Lab**oratory **V**irtual **I**nstrument **E**ngineering **W**orkbench. It is a powerful and flexible instrumentation and analysis software development application created by the folks at National Instruments—a company that creates hardware and software products that leverage computer technology to help engineers and scientists take measurements, control processes, and analyze and store data. National Instruments was founded over twenty-five years ago in Austin, Texas by James Truchard (known as Dr. T), Jeffrey Kodosky, and William Nowlin. At the time, all three men were working on sonar applications for the U.S. Navy at the Applied Research Laboratories at The University of Texas at Austin. Searching for a way to connect test equipment to DEC PDP-11 computers, Dr. T decided to develop an interface bus. He recruited Jeff and Bill to join him in his endeavor, and together they successfully developed LabVIEW and the notion of a "virtual instrument." In the process they managed to infuse their new company—National Instruments—with an entrepreneurial spirit that still pervades the company today.

Engineers and scientists in research, development, production, test, and service industries as diverse as automotive, semiconductor, aerospace, electronics, chemical, telecommunications, and pharmaceutical have used and continue to use LabVIEW to support their work. LabVIEW is a major player in the area of testing and measurements, industrial automation, and data analysis. For example, scientists at NASA's Jet Propulsion Laboratory used LabVIEW to analyze and display Mars Pathfinder Sojourner rover engineering data, including the position and temperature of the rover, how much power remained in the rover's battery, and generally to monitor Sojourner's overall health. This book is intended to help you learn to use LabVIEW as a programming tool and to serve as an introduction to data acquisition, instrument control, and data analysis.

LabVIEW programs are called **Virtual Instruments**, or VIs for short. LabVIEW is different from text-based programming languages (such as Fortran and C) in that LabVIEW uses a graphical programming language, known as the G programming language, to create programs relying on graphic symbols to describe programming actions. LabVIEW uses a terminology familiar to scientists and engineers, and the graphical icons used to construct the G programs are easily identified by visual inspection. You can learn LabVIEW even if you have little programming experience, but you will find knowledge of programming fundamentals helpful. If you have never programmed before (or maybe you have programming experience but have forgotten a few things) you may want to review the basic concepts of programming before diving into the G programming language.

LabVIEW provides an extensive library of virtual instruments and functions to help you in your programming. An important new development introduced in LabVIEW 7.0 is the so-called Express VIs. The Express VIs allow users to program common measurement tasks while requiring minimal wiring because the VI configuration is accomplished with dialog boxes. As we soon learn in this chapter, Express VIs are located on the **Functions** palette, and they appear on the **Functions** palette with white backgrounds surrounded by a blue border. LabVIEW also contains application-specific libraries for data acquisition (discussed in Chapter 8), file input/output (discussed in Chapter 9), GPIB and serial instrument control (discussed in Chapter 10), and data analysis (discussed in Chapter 11). It includes conventional program debugging tools with which you can set breakpoints, single-step through the program, and animate the execution so you can observe the flow of data. Editing and debugging VIs is the topic of Chapter 3.

LabVIEW has a good set of VIs for data presentation on various types of charts and graphs. Chapter 7 discusses the process of presenting data on charts and graphs.

The LabVIEW system consists of the LabVIEW application executable files and many associated files and folders. LabVIEW uses files and directories to store information necessary to create your VIs. Some of the more important files and directories are:

1. The LabVIEW executable. Use this to launch LabVIEW.

2. The vi.lib directory. This directory contains libraries of VIs such as data acquisition, instrument control, and analysis VIs; it must be in the same directory as the LabVIEW executable. Do not change the name of the vi.lib directory, because LabVIEW looks for this directory when it launches. If you change the name, you cannot use many of the controls and library functions.

3. The examples directory. This directory contains many sample VIs that demonstrate the functionality of LabVIEW.

4. The user.lib directory. This directory is where you can save VIs you have created, and they will appear in the LabVIEW **Functions** palette.

5. The instr.lib directory. This directory is where your instrument driver libraries are placed if you want them to appear in the **Functions palette** palette.

6. The Learning directory. This file contains a library of VIs that you will use with the *Learning with LabVIEW* book.

 The files in the Learning directory must be downloaded from the site http://www.ni.com/labviewse—the Learning directory VIs are not found on the installation CD! You can access the National Instruments site through the Internet using any standard web browser.

1.4   THE STARTUP SCREEN

When you launch LabVIEW by double-clicking on its icon, the startup screen appears as in Figure 1.1. The startup screen contains a navigation dialog box that includes introductory material and common commands. The dialog box includes a menu with standard items such as **File**, **Edit**, **Tools**, and **Help**. The dialog box also presents a set of buttons for creating and opening VIs, configuring data acquisition devices, and finding helpful information:

- Click the **New** button to create a new VI. Click the arrow on the **New** button to choose to open a blank VI or to open the **New** dialog box.

Click the **New** button to create a new VI. Click the arrow on the **New** button to choose to open a blank VI or to open the **New** dialog box.

Click the **Open** button to open an existing VI. Click the arrow on the **Open** button to open recent files.

Click the **Configure** button to configure your data acquisition devices. Click the arrow on the **Configure** button to configure LabVIEW.

Click the **Help** button to launch *LabView Help*. Click the arrow on the **Help** button for other Help options, including the NI Example Finder.

FIGURE 1.1
The startup screen.

- Click the **Open** button to open an existing VI. Click the arrow on the **Open** button to open recent files.

- Click the **Configure** button to configure your data acquisition devices. Click the arrow on the **Configure** button to configure LabVIEW.

- Click the **Help** button to launch *LabVIEW Help*. Click the arrow on the **Help** button for other Help options, including the NI Example Finder.

Throughout this book, use the left mouse button (if you have one) unless we specifically tell you to use the right one.

Searching the LabVIEW Examples

In this exercise you will search through the list of example VIs and demonstrations that are included with the *LabVIEW Student Edition*. Open the LabVIEW application and get to the startup screen. The search begins at the LabVIEW startup screen by clicking on the **Open** and selecting the **Examples** button, as shown in Figure 1.2. The NI Examples Finder screen displays the numerous examples available with LabVIEW, as illustrated in Figure 1.3. The examples can be browsed by Task or by Directory Structure. In Figure 1.3, the examples are sorted by Task.

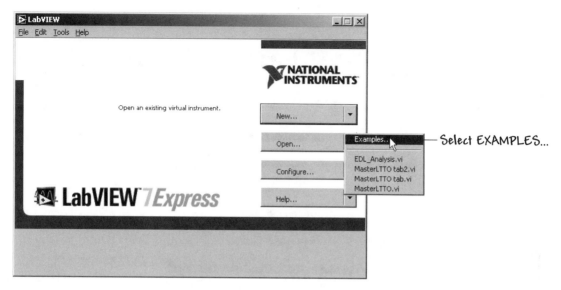

FIGURE 1.2
LabVIEW examples.

To reach the desired example—in this case, we are searching for the Temperature System Demonstration—select **Analysis**, as shown in Figure 1.3. The subdirectory will open to show three examples: Signal Generation and Processing, Temperature System Demonstration, and Vibration Analysis. Selecting the **Temperature System Demo** opens up the associated virtual instrument (VI) (more on VIs in Chapter 2). Just for fun you can start the VI running and see what happens. Start the VI by clicking on the **Run** button, as shown in Figure 1.4. Stop the VI by clicking on the **Abort Execution** button. Try it!   ◆

1.5   PANEL AND DIAGRAM WINDOWS

An untitled front panel window appears when you select **New VI** from the startup screen. The front panel window is the interface to your VI code and is one of the two LabVIEW windows that comprise a virtual instrument. The other window— the block diagram window—contains program code that exists in a graphical form (such as icons, wires, etc.).

Front panels and block diagrams consist of graphical objects that are the G programming elements. Front panels contain various types of controls and indicators (that is, inputs and outputs, respectively). Block diagrams contain terminals corresponding to front panel controls and indicators, as well as constants, functions, subVIs, structures, and wires that carry data from one object to another. Structures are program control elements (such as For Loops and While Loops).

Figure 1.5 shows a front panel and its associated block diagram. You can find the virtual instrument FirstVI.vi shown in Figure 1.5 in Chapter1 folder within

Browse examples by **Task** or **Directory Structure.**

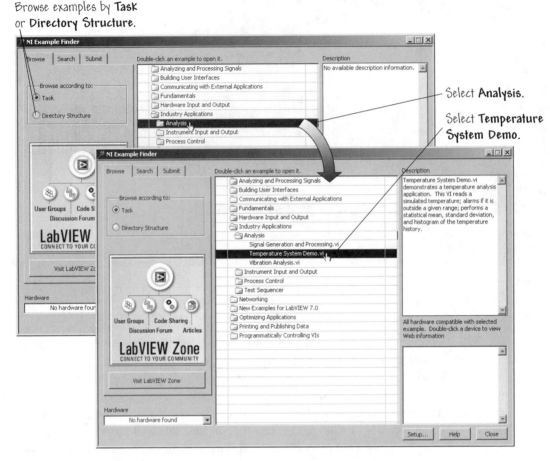

Select **Analysis.**

Select **Temperature System Demo.**

FIGURE 1.3
The **Analysis Demonstrations** screen.

the directory Learning. That VI can be located by choosing **Open VI** on the Startup screen and navigating to the Chapter1 folder in the Learning directory and then selecting FirstVI.vi. Once you have the VI front panel open, find the **Run** button on the panel toolbar and click on it. Your VI is now running. You can turn the knob and vary the different inputs and watch the output changes reflected in the graph. Give it a try! If you have difficulty getting things to work, then just press ahead with the material in the next sections and come back to this VI when you are ready.

1.5.1   Front Panel Toolbar

A toolbar of command buttons and status indicators that you use for controlling VIs is located on both the front panel and block diagram windows. The front panel toolbar and the block diagram toolbar are different, although they do each

Run button

Abort Execution button

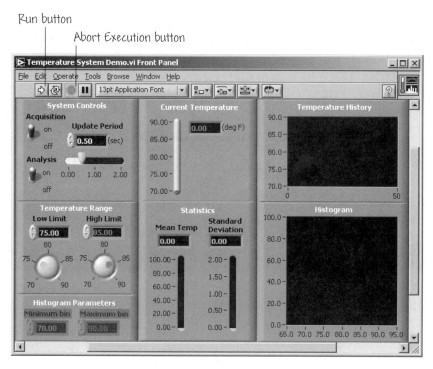

FIGURE 1.4
The temperature system demonstration front panel.

contain some of the same buttons and indicators. The toolbar that appears at the top of the front panel window is shown in Figure 1.6.

While the VI is executing, the **Abort Execution** button appears. Although clicking on the abort button terminates the execution of the VI, as a general rule you should avoid terminating the program execution this way and either let the VI execute to completion or incorporate a programmatic execution control (that is, an on-off switch or button) to terminate the VI from the front panel.

The **Broken Run** button replaces the **Run** button when the VI cannot compile and run due to coding errors. If you encounter a problem running your VI, just click on the **Broken Run** button, and a window will automatically appear on the desktop that lists all the detected program errors. And then, if you double click on one of the specific errors in the list, you will be taken automatically to the location in the block diagram (that is, to the place in the code) where the error exists. This is a great debugging feature! More discussion on the issue of debugging VIs can be found in Chapter 3.

Clicking on the **Run Continuously** button leads to a continuous execution of the VI. Clicking on this button again disables the continuous execution—the VI stops when it completes normally. The behavior of the VI and the state of the

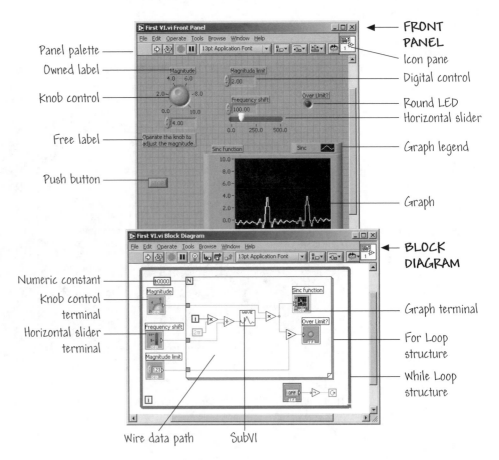

FIGURE 1.5
A front panel and the associated block diagram.

toolbar during continuous run is the same as during a single run started with the **Run** button.

The **Pause/Continue** button pauses VI execution. To continue program execution after pausing, press the button again, and the VI resumes execution.

The **Text Settings** pull-down menu, shown in Figure 1.7, sets font options—font type, size, style, and color.

The **Align Objects** sets the preferred alignment of the various objects on either the front panel or on the block diagram. After selecting the desired objects for alignment, you can set the preferred alignment for two or more objects. For example, you can align objects by their left edges or by their top edges. The various alignment options are illustrated in the **Align Objects** pull-down menu shown in Figure 1.8. Aligning objects is very useful in organizing the VI front panel (and the block diagram, for that matter). On the surface, it may appear that aligning the front panel objects, while making things "neat and pretty," does not contribute to the goal of a functioning VI. As you gain experience with

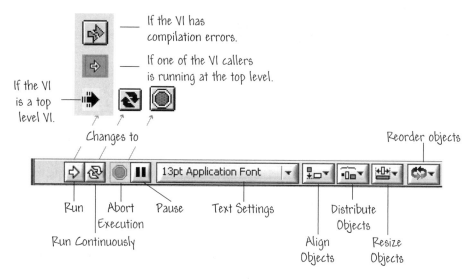

FIGURE 1.6
The front panel toolbar.

constructing VIs, you will find that they are easier to debug and verify if the interface (front panel) and the code (block diagram) are organized to allow for easy visual inspection.

The **Distribute Objects** pull-down menu, shown in Figure 1.9, sets the preferred distribution options for two or more objects. For example, you can evenly space selected objects, or you can remove all the space between the objects.

The **Resize Objects** pull-down menu, shown in Figure 1.10, is used to resize multiple front panel objects to the same size. For example, this feature allows

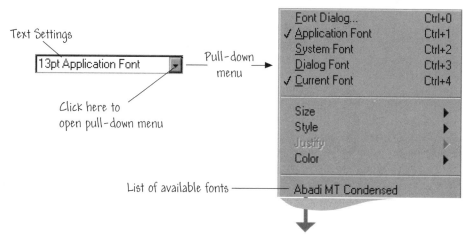

FIGURE 1.7
The **Text Settings** pull-down menu.

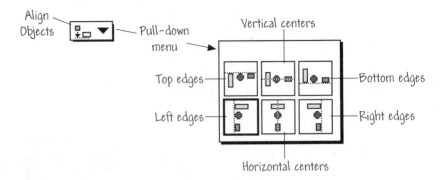

FIGURE 1.8
The **Align Objects** pull-down menu.

you to select multiple objects and resize them all to the same height as the object within the selected group with the maximum height.

1.5.2 Block Diagram Toolbar

The block diagram toolbar contains many of the same buttons as the front panel toolbar. Four additional program debugging features are available on the block diagram toolbar and are enabled via the buttons, as shown in Figure 1.11.

Clicking on the **Highlight Execution** button enables execution highlighting. As the program executes, you will be able to see the data flow through the code on the block diagram. This is extremely helpful for debugging and verifying proper execution. In the execution highlighting mode, the button changes to a brightly lit light bulb.

LabVIEW debugging capabilities allow you to single-step through the VI node to node. A **node** is an execution element, such as a For Loop or subVI. You will learn more about the different types of nodes as you proceed through the book, but for the time being you can think of a node as a section of the computer

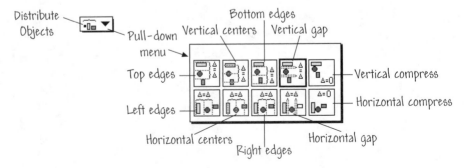

FIGURE 1.9
The **Distribute Objects** pull-down menu.

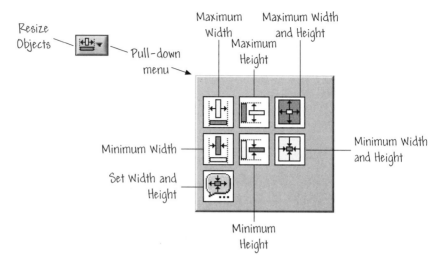

FIGURE 1.10
The **Resize Objects** pull-down menu.

code that you want to observe executing. Each node blinks to show it is ready to execute.

The **Step Over** button steps over a node. You are in effect executing the node without single stepping through the node.

The **Step Into** button allows you to step into a node. Once you have stepped into the node, you can single step through the node.

The **Step Out** button allows you to step out of a node. By stepping out of a node, you can complete the single stepping through the node and go to the next node.

The **Warning** indicator only appears when there is a potential problem with your block diagram. The appearance of the warning does not prevent you from

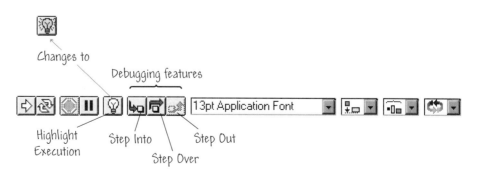

FIGURE 1.11
The block diagram toolbar.

executing the VI. You can enable the Warning indicator on the Options...≫ Debugging menu in the **Tools** pull-down menu.

1.6   SHORT CUT MENUS

LabVIEW has two types of menus—**pull-down** menus and **short cut** menus. We will focus on short cut menus in this section and on pull-down menus in the next section. Our discussions here in Chapter 1 are top-level; we reserve the detailed discussions on each menu item for later chapters as they are used.

To access a short cut menu, position the cursor on the desired object on the front panel or block diagram and click the right mouse button on a PC-compatible or hold down the <Command> key and then click the mouse button on the Mac. In most cases, a short cut menu will appear since most LabVIEW objects have short cut menus with options and commands. This process is called "popping up," and short cut menus are also know as pop-up menus. You will also find that you can pop-up on the empty front panel and block diagram space, giving you access to the **Controls** and **Functions** palettes and other important palettes. The options presented to you on short cut menus depend on the selected object—popping up on a numeric control will open a different short cut menu than popping up on a For Loop. When you construct a program in G, you will use short cut menus extensively!

Many short cut and pull-down menus contain submenus, as shown in Figures 1.12 and 1.13.

On a PC-compatible, right mouse click on the object to open a short cut menu. On a Mac, press <Command> and simultaneously click on the object.

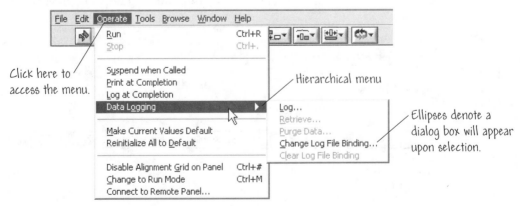

FIGURE 1.12
An example of a pull-down menu expanding into a submenu.

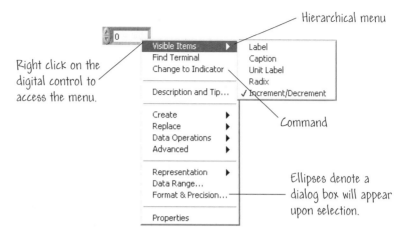

FIGURE 1.13
An example of a short cut menu.

Menu items that expand into submenus are called **hierarchical** menus and are denoted by a right arrowhead on the menu. Hierarchical menus will present you with different types of options and commands. One typical option is the so-called mutually exclusive option. This means that if you select the option (by clicking on it), a check mark will appear next, indicating the option is selected; otherwise the option is not selected.

Popping up on different areas of an object may lead to different short cut menus. If you pop up and do not see the anticipated menu selection, then pop up somewhere else on the object.

Another type of menu item opens dialog boxes containing options for you to use to modify and configure your program elements. Menu items leading to dialog boxes are denoted by ellipses (...). Menu items without right arrowheads or ellipses are generally commands that execute immediately upon selection. **Create Constant** is an example of a command that appears in many short cut menus. In some instances, commands are replaced in the menu by their inverse commands when selected. For example, after you choose **Change to Indicator**, the menu selection is replaced by **Change to Control**.

1.7   PULL-DOWN MENUS

The menu bar at the top of the LabVIEW screen, shown in Figure 1.14, contains the important pull-down menus. In this section we will introduce the pull-down menus: **File**, **Edit**, **Operate**, **Tools**, **Browse**, **Window**, and **Help**.

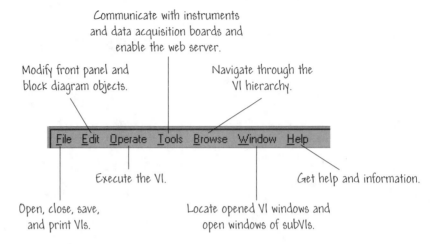

FIGURE 1.14
The menu bar.

*In **Windows**, the menus display only the most recently used items by default. Click the arrows at the bottom of a menu to display all items. You can display all menu items by default by selecting **Tools≫Options** and selecting **Miscellaneous** from the top pull-down menu.*

The level of discussion on pull-down windows given here is consistent with our previous discussions. The main goal is to introduce the menus.

Some menu items are unavailable while a VI is in run mode.

1.7.1   File Menu

The **File** pull-down menu, shown in Figure 1.15, contains commands associated with file manipulations. For example, you can create new VIs or open existing ones from the **File** menu. You use options in the **File** menu primarily to open, close, save, and print VIs. As you observe each pull-down window, notice that selected commands and options have shortcuts listed beside them. These shortcuts are keystroke sequences that can be used to choose the desired option without pulling down the menu. For example, you can open a new VI by typing and entering Ctrl+O at the keyboard, or you can access the **File** pull-down menu and select **Open**.

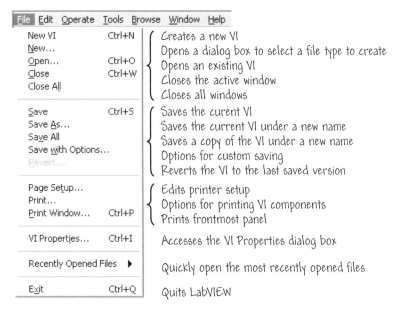

FIGURE 1.15
Pull-down menus—**File**.

1.7.2   Edit Menu

The **Edit** menu, shown in Figure 1.16, is used to modify front panel and block diagram objects of a VI. The **Undo** and **Redo** options are very useful when you are editing because it allows you to undo an action after it is performed, and once you undo an action you can redo it. By default, the maximum number of undo steps per VI is 8—you can increase or decrease this number if desired.

1.7.3   Operate Menu

The **Operate** menu, shown in Figure 1.17, can be used to run or stop your VI execution, change the default values of your VI, and switch between the run mode and edit mode.

1.7.4   Tools Menu

The **Tools** menu, shown in Figure 1.18, is used to communicate with instruments and data acquisition boards, compare VIs, build applications, enable the Web server, and access other options of LabVIEW.

An important link is to the main National Instruments website, where you can obtain general information about the company and its products. If you use LabVIEW to control external instruments, you will be interested in the **Instrument Driver Network**. . . link, which connects you to over 600 LabVIEW-ready

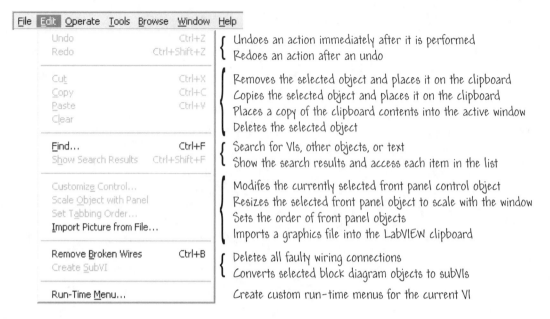

FIGURE 1.16
Pull-down menus—**Edit**.

instrument drivers. This can be found in the Tools≫Instrumentation hierarchical menu. Refer to Chapter 10 for more information on instrument drivers.

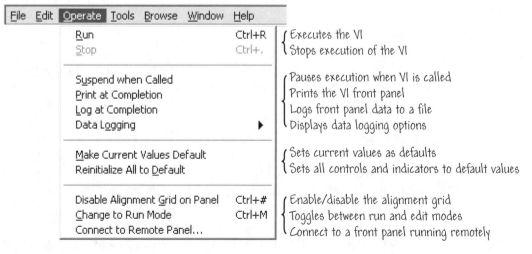

FIGURE 1.17
Pull-down menus—**Operate**.

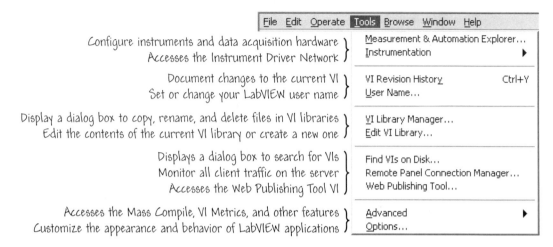

FIGURE 1.18
Pull-down menus—**Tools**.

1.7.5   Browse Menu

You use the **Browse** menu to navigate through the VI hierarchy. The **Browse** menu is depicted in Figure 1.19.

1.7.6   Windows Menu

The **Windows** menu, shown in Figure 1.20, is used for a variety of activities. You can toggle between the panel and diagram windows, and you can "tile" both windows so you can see them at the same time (one above the other or side-by-side). All the open VIs are listed in the menu (at the bottom), and you can switch between the open VIs. Also, if you want to show the palettes on the desktop, you can use the **Windows** menu to select either (or both) palettes (more on palettes in the next section).

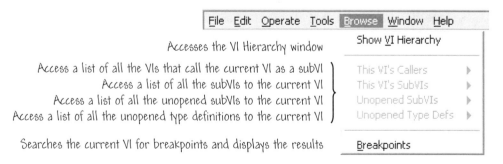

FIGURE 1.19
Pull-down menus—**Browse**.

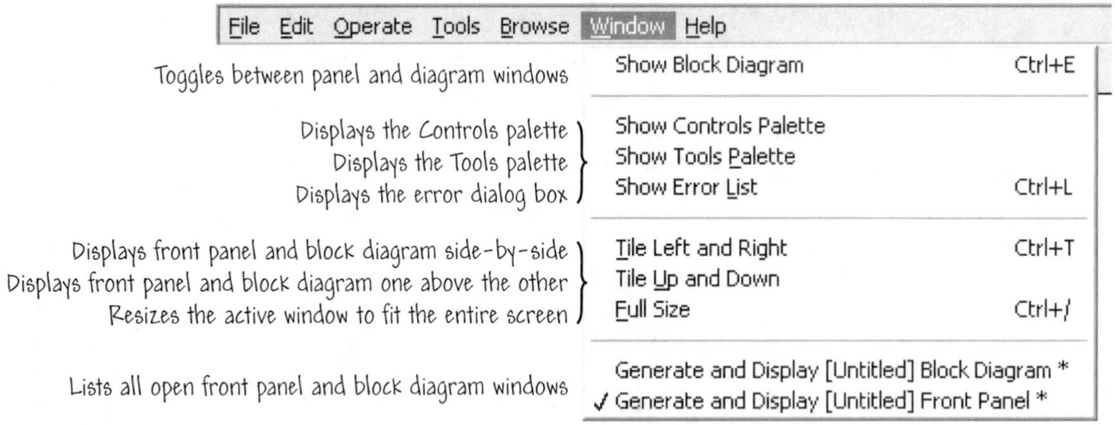

File Edit Operate Tools Browse	Window	Help	

Toggles between panel and diagram windows — Show Block Diagram — Ctrl+E

Displays the Controls palette ⎫
Displays the Tools palette ⎬ — Show Controls Palette / Show Tools Palette / Show Error List — Ctrl+L
Displays the error dialog box ⎭

Displays front panel and block diagram side-by-side ⎫
Displays front panel and block diagram one above the other ⎬ — Tile Left and Right (Ctrl+T) / Tile Up and Down / Full Size (Ctrl+/)
Resizes the active window to fit the entire screen ⎭

Lists all open front panel and block diagram windows — Generate and Display [Untitled] Block Diagram *
✓ Generate and Display [Untitled] Front Panel *

FIGURE 1.20
Pull-down menus—**Windows**.

1.7.7 Help Menu

The **Help** menu, shown in Figure 1.21, provides access to the extensive Lab-VIEW online help facilities. You can view information about panel or diagram objects, activate the online reference utilities, and view information about your LabVIEW version number and computer memory. The direct pathways to the Internet are provided by **Web Resources** and **Student Edition Web Resources**, which connect you directly to the main source of information concerning the

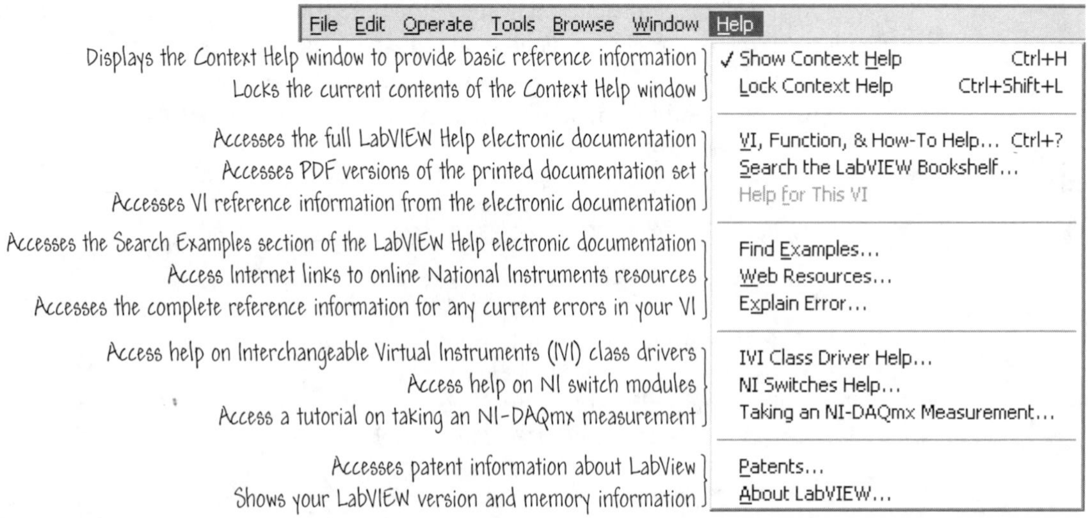

File Edit Operate Tools Browse Window	Help

Displays the Context Help window to provide basic reference information ⎫ — ✓ Show Context Help (Ctrl+H)
Locks the current contents of the Context Help window ⎭ — Lock Context Help (Ctrl+Shift+L)

Accesses the full LabVIEW Help electronic documentation ⎫ — VI, Function, & How-To Help... Ctrl+?
Accesses PDF versions of the printed documentation set ⎬ — Search the LabVIEW Bookshelf...
Accesses VI reference information from the electronic documentation ⎭ — Help for This VI

Accesses the Search Examples section of the LabVIEW Help electronic documentation ⎫ — Find Examples...
Access Internet links to online National Instruments resources ⎬ — Web Resources...
Accesses the complete reference information for any current errors in your VI ⎭ — Explain Error...

Access help on Interchangeable Virtual Instruments (IVI) class drivers ⎫ — IVI Class Driver Help...
Access help on NI switch modules ⎬ — NI Switches Help...
Access a tutorial on taking an NI-DAQmx measurement ⎭ — Taking an NI-DAQmx Measurement...

Accesses patent information about LabView ⎫ — Patents...
Shows your LabVIEW version and memory information ⎭ — About LabVIEW...

FIGURE 1.21
Pull-down menus—**Help**.

LabVIEW Student Edition available on the Web. You should make a point to surf this website—this is where the latest and greatest information, news, and updates on the both the software *Student Edition of LabVIEW* and on the book *Learning with LabVIEW* will be posted.

1.8 PALETTES

Palettes are graphical panels that contain various tools and objects used to create and operate VIs. You can move the palettes anywhere on the desktop that you want—preferably off to one side so that they do not block objects on either the front panel or block diagram. It is sometimes said that the palettes float. The three main palettes are the **Tools**, **Controls**, and **Functions** palettes.

1.8.1 Tools Palette

A **tool** is a special operating mode of the mouse cursor. You use tools to perform specific editing functions, similar to how you would use them in a standard paint program. You can create, modify, and debug VIs using the tools located in the floating **Tools** palette, shown in Figure 1.22. If the **Tools** palette is not visible, select **Show Tool Palette** from the **Windows** pull-down menu to display the palette. After you select a tool from this menu, the mouse cursor changes to the appropriate shape.

When the **Automatic Tool Selection** button (located at the top center of the **Tools** palette, as shown in Figure 1.22) is enabled, LabVIEW automatically selects the corresponding tool from the **Tools** palette as you move the cursor over objects on either the front panel or the block diagram. If necessary, you can disable the automatic tool selection feature by clicking on the **Automatic Tool Selection** button to toggle the state. The green light on the button will turn off to

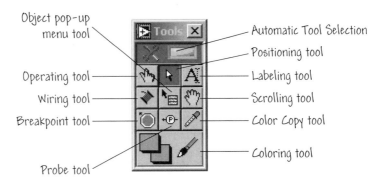

FIGURE 1.22
The **Tools** palette.

indicate that the automatic tool selection is off. You also can manually select a tool on the **Tools** palette to disable the automatic tool selection feature. This allows you to select a tool manually by clicking the tool you want on the **Tools** palette.

One way to access the online help is to place any tool found in the **Tools** palette over the object of interest in the block diagram window. If online help exists for that object, it will appear in a separate Help window. This process requires that you first select **Show Context Help** from the **Help** pull-down menu.

*On a **Windows** platform, a shortcut to accessing the **Tools** palette is to press the <shift> and the right mouse button on the panel or the diagram window. On a **Macintosh** platform, you can access the **Tools** palette by pressing <command-shift> and the mouse button on the panel or diagram window.*

1.8.2   Controls Palette

The **Controls** palette, shown in Figure 1.23, consists of top-level icons representing subpalettes, which contain a full range of available objects that you can use in creating a front panel. The **Controls** palette can be displayed in the Express view or the Advanced view. By default, the **Controls** palette starts in the Express view (as depicted in Figure 1.23). The Express palette view includes subpalettes on the top level of the **Controls** palette that contain Express VIs and other objects you need to build common measurement applications. The **All Controls** subpalette (located at the bottom right-hand side of the **Controls** palette) contains the complete set of built-in controls, VIs, and functions. The Advanced palette view includes subpalettes on the top level of the **Controls** palette that contain the complete set of built-in controls, VIs, and functions.

*The palette view can be altered in the **Tools**≫**Advanced**≫**Edit Palette Views** dialog box.*

You can access the subpalettes by clicking on the desired top-level icon. An example top-level icon is **All Controls** (see Figure 1.23). The subpalette **Numeric**, which when opened, will reveal the various numeric controls (by control we mean "input") and indicators (by indicator we mean "output") that you will utilize on your front panel as a way to move data into and out of the program code (on the block diagram). The topic of numeric controls and indicators is covered in Chapter 2. Each item on the **Controls** palette is discussed in more detail in the chapter in which it is first utilized.

If the **Controls** palette is not visible, you can open the palette by selecting **Show Controls Palette** from the **Windows** pull-down menu. You also can access the **Controls** palette by right-clicking on an open area in the front panel window.

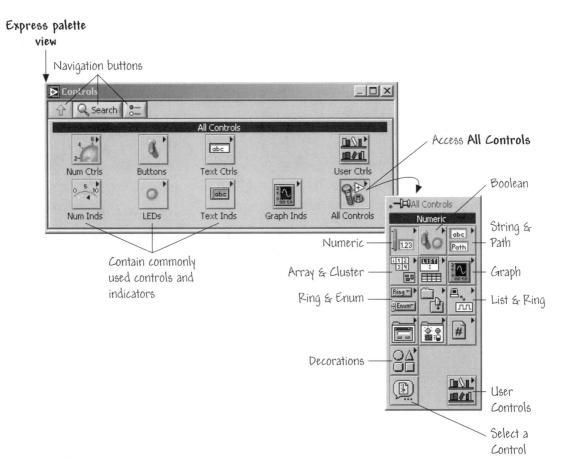

FIGURE 1.23
The **Controls** palette.

*The **Controls** palette is available only when the front panel window is active.*

1.8.3   Functions Palette

The **Functions** palettes, shown in Figure 1.24, works in the same general way as the **Controls** palette. It consists of top-level icons representing subpalettes, which contain a full range of available objects that you can use in creating the block diagram. You access the subpalettes by clicking on the desired top-level icon. Many of the G program elements are accessed through the **Functions** palette. For example, the subpalette **Structures** (see Figure 1.24) contains For Loops, While Loops, and Formula Nodes—all of which are common elements of VIs.

Express palette view

Navigation buttons

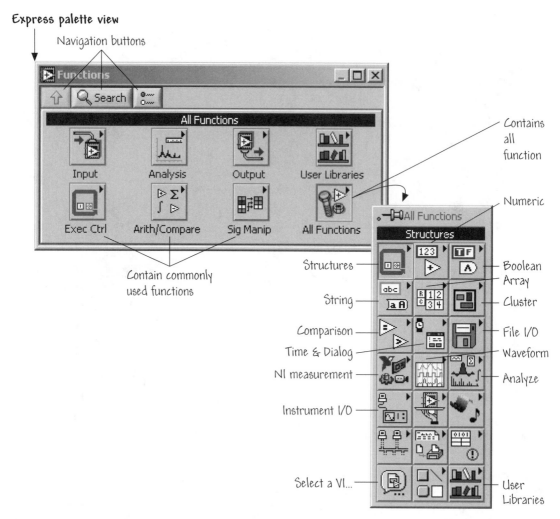

FIGURE 1.24
The **Functions** palette.

If the **Functions** palette is not visible, you can open the palette by selecting **Show Functions Palette** from the **Windows** pull-down menu. You can also access the **Functions** palette by right-clicking on an open area in the block diagram window.

*The **Functions** palette is available only when the block diagram window is active.*

Choose Blank VI
to start from
scratch

VI template
preview

Template
descriptions

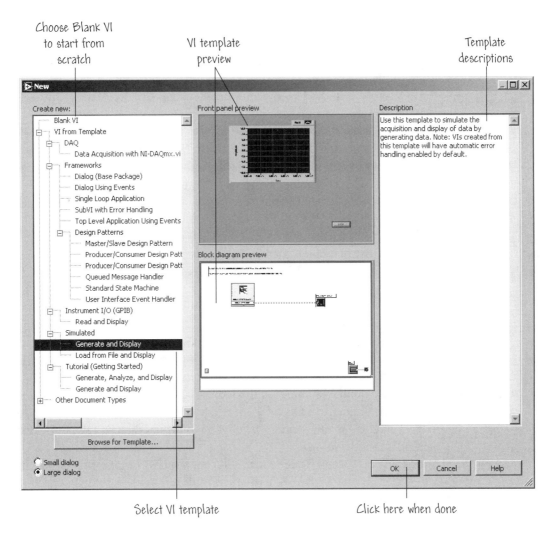

Select VI template

Click here when done

FIGURE 1.25
New dialog box.

1.9   OPENING, LOADING, AND SAVING VIS

When you click the **New** button in the LabVIEW dialog box (see Figure 1.1), the **New** dialog box appears, as shown in Figure 1.25. Selecting the Blank VI from the **Create new** list opens a blank VI front panel and block diagram.

*Remember that from the VI pull-down menu, you can select **File≫New** to display the **New** dialog box (refer to Section 1.7.1). You also can open a blank by selecting **File≫New VI**.*

The **New** dialog box can be employed to help you create LabVIEW applications by utilizing VI templates. If you prefer not to start with a blank VI to build your VI from scratch, you can start with a VI template. This may simplify your programming task. When you select a template in the **Create new** list, previews of the VI appear in the **Front panel preview** and the **Block diagram preview** sections, and a description of the template appears in the **Description** section. Figure 1.25 shows the Generate and Display VI template.

Making a selection from the **VI from Template** opens a front panel and block diagram with components you need to build different types of VIs. For example, the DAQ VI template opens a front panel and a block diagram with the components you need to measure or generate signals using the DAQ Assistant Express VI and NI-DAQmx. The Instrument I/O VI template opens a front panel and block diagram with the components you need to communicate with an external instrument attached to the computer through a port, such as a serial or GPIB-enabled device. The **Browse for Template** button displays the **Browse** dialog box so you can navigate to a VI or VI template.

You load a VI into memory by choosing the **Open** option from the **File** menu. When you choose that option, a dialog box similar to the one in Figure 1.26 appears. VI directories and VI libraries appear in the dialog box next to a representative symbol. VI libraries contain multiple VIs in a compressed format.

You can open a VI library or directory by clicking on its icon and then on **Open**, or by double clicking on the icon. The dialog box opens VI libraries as if they were directories. Once the directory or library is opened, you can locate your VI and load it into memory by clicking on it and then on **OK**, or by double-clicking on the VI icon.

If LabVIEW cannot immediately locate a subVI (think of this as a subroutine) called by the VI, it begins searching through all directories specified by the VI Search Path (**Tools≫Options≫Paths**). A status dialog box will appear (sometimes the status box disappears so fast that you cannot see it was even there!) as the VI loads. The *Searching* field in the status box lists directories or VIs as LabVIEW searches through them. The *Loading* field lists the subVIs of your VI as they are loaded into memory. If a subVI cannot be found, you can have LabVIEW ignore the subVI by clicking on **Ignore SubVI** in the status box, or you can click on **Browse** to search for the missing subVI.

You can save your VI to a regular directory or VI library by selecting **Save**, **Save As. . .**, **Save All. . .**, or **Save with Options**. . . from the **File** menu. You also can transfer VIs from one platform to another (for example, from LabVIEW for Macintosh to LabVIEW for Windows). LabVIEW automatically translates and recompiles the VIs on the new platform. Because VIs are files, you can use any file transfer method or utility to move your VIs between platforms. Porting VIs over networks using FTP protocol, Z- or X-Modem protocol, and other similar utilities eliminates the need for additional file translation software. If you port

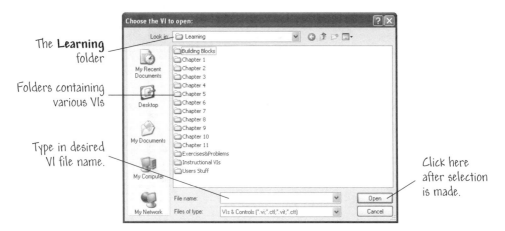

The **Learning** folder

Folders containing various VIs

Type in desired VI file name.

Click here after selection is made.

FIGURE 1.26
Locating the desired VI.

your VIs via magnetic media (such as floppy disks), you will need to use a file transfer utility (such as MacDisk or Transfer Pro).

1.10   LABVIEW HELP OPTIONS

The two common help options that you will use as you learn about LabVIEW programming are the **Context** and the **LabVIEW Help**. Both help options can be accessed in the **Help** pull-down menu.

1.10.1   Context Help Window

To display the help window, choose **Show Context Help** from the **Help** pull-down menu. If you have already placed objects on the front panel or block diagram, you can find out more about those objects by simply placing one of the tools from the **Tools** palette on block diagram and panel objects. This process causes the **Context Help Window** to appear showing the icon associated with the selected object and displaying the wires attached to each terminal. As you will discover in the next chapter, some icon terminals must be wired and others are optional. To help you locate terminals that require wiring, in the help window required terminals are labeled in bold, recommended connections in plain text, and optional connections are gray. The example in Figure 1.27 displays a help window in the so-called **Simple Context Help** mode.

On the lower left-hand side of the help window is a button to switch between the simple and detailed context help modes. The simple context emphasizes the important terminal connections—de-emphasized terminals are shown by wire stubs. The detailed help displays all terminals, as illustrated in Figure 1.28.

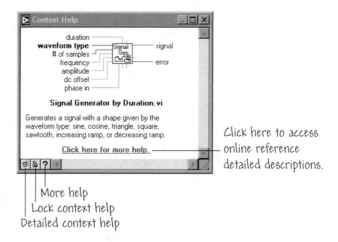

FIGURE 1.27
A simple context help window.

On the lower left-hand side of the help window is a lock icon that locks the current contents of the help window, so that moving the tool over another function or icon does not change the display. To unlock the window, click again on the lock icon at the bottom of the help window.

The **More Help** icon is the question mark located in the lower left-hand portion of the context help window. This provides a link to the description of the

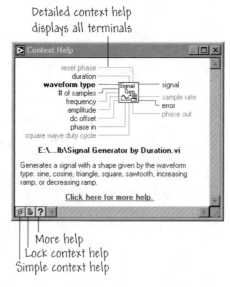

FIGURE 1.28
A detailed context help window.

object in the online reference documentation, which features detailed descriptions of most block diagram objects.

1.10.2 LabVIEW Help

The LabVIEW online reference contains detailed descriptions of most block diagram objects. This information is accessible either by clicking on the More Help icon in the Context Help window, choosing **Contents and Index** from the **Help** menu, or clicking on the sentence **Click here for more help** in the Context Help window.

BUILDING BLOCK

1.11 BUILDING BLOCKS: TRAJECTORY ANALYSIS

In each chapter of this book you will find a "Building Blocks" section. The purpose of this section is to give you the chance to apply the main principles of the chapter. In some cases the building block exercise of one chapter will continue in the next chapter; in other cases, the exercise will be new. The exercises will be short, and you will be asked to do all the work!

This first building block is an exercise in opening and running a VI. The VI that you should open is called Trajectory Analysis.vi and is included in the Building Blocks folder in the Learning directory. Find the VI and open it. The front panel is shown in Figure 1.29. Make sure the Automatic Tool Selection on the **Tools** palette is selected. Run the VI and observe the path of the projectile in the graph. Change the initial angle to 20 degrees and run the VI again. What is the maximum height achieved by the projectile? What is the distance traveled? Vary the initial velocity and run a few more numerical experiments.

1.12 RELAXED READING: REMOTE CONTROLLED EXCAVATION WITH LABVIEW

A project was recently undertaken to develop an excavator to remove chemical and explosive hazards. Safety concerns suggested that the hazardous materials be removed using a remote-controlled excavator at a remote location to minimize the exposure of chemical and explosive hazards to the equipment operator. In addition, the design of the excavator should include six independent movements, vision capabilities, operating parameters, and remote ignition and emergency stop links.

The CASE CX-160 excavator shown in Figure 1.30 operates by conventional means with two dual-axis joysticks and two single-axis foot pedals. Each of these

Click here to
run the VI.

Vary the initial
angle using the
Operating tool and
clicking on the
up/down arrows.

Select the Automatic
Tool Selection

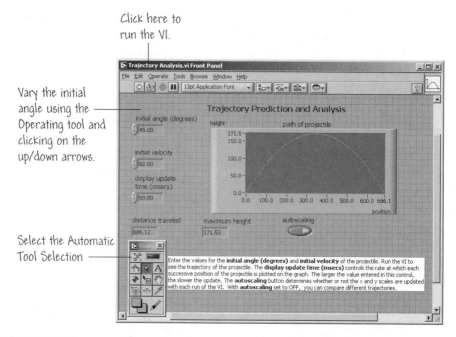

FIGURE 1.29
The Trajectory Analysis.vi front panel.

axes actuate through two stages of hydraulic spool valves. The excavator has six
independent movements: arm, boom, bucket, swing, and left and right track. To
control these movements, FieldPoint modules were placed at the operator station
and in the excavator. Programs were then loaded on the FieldPoint modules
that convert joystick inputs to excavator movements. A FieldPoint distributed

FIGURE 1.30
CASE CX-160 Excavator.

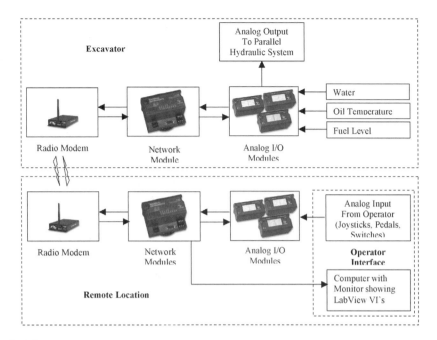

FIGURE 1.31
Wireless Distributed I/O System.

I/O system operates at 10 Hz and controls the six independent movements at the remote location. Electrical power was available in 12 and 24 volts with a maximum current of 45 amps. An independent emergency shutdown circuit was employed to fulfill the safety requirements.

To ensure stability and overall safety, the robust excavator controllers had to be remotely controlled in real time. The FieldPoint modules, in conjunction with LabVIEW and LabVIEW Real-Time, provided intelligent operation, as well as the failsafe conditions that the excavator required. An electronic sensing and acquisition system was selected that consisted of various FieldPoint distributed I/O modules that link together using FieldPoint network modules. FieldPoint distributed I/O quickly facilitated client-based systems through simple integration with LabVIEW. FieldPoint modules communicated through Ethernet and serial data transfer protocols, while data transmitted between modules through wireless serial radio modems. The basic excavator operation reads analog voltages from control input devices at the remote station. The analog voltage then travels across the network for implementation on the excavator FieldPoint modules. The wireless distributed I/O system is depicted in Figure 1.31.

At the remote operator station, the operator can remotely toggle the power of vision systems and work lights. Four joystick buttons provide pan-and-tilt capabilities for the excavator cameras. To ensure the vehicle is running at capacity, the operator must monitor vehicle parameters, such as fuel level, water and oil temperature, and warning messages in real time. These parameters are normally displayed on an on-board liquid crystal display next to the joystick. This data is reported via an available RS-232 port located on-board the excavator's computer. The status of the excavator is processed and sent back across the wireless network to the remote site where proper action is taken. If a critical error occurs, the vehicle control terminates and warning messages display for the operator. The operator can monitor the excavator through any PC with either LabVIEW or a Web browser.

Five undergraduate students and one graduate student working part-time implemented this concept in five months. FieldPoint and LabVIEW Real-Time software allowed easy implementation of a remote control for the excavator. Field-Point modules help advance the development of the teleoperated applications. Systems can run independent of each other or they can collectively work together as a team. The FieldPoint design facilitated quick and easy programming and provided watchdog timers and power-on reset tools that ensure failsafe operation. Normally, it is a challenge to digitally control an application, but with FieldPoint distributed I/O, it was easy to maintain system stability, even at great distances.

For more information contact:

Chris R. Terwelp
cterwelp@vt.edu

1.13 SUMMARY

LabVIEW is a powerful and flexible development tool designed specifically for the needs of scientists and engineers. It uses the graphical programming language G to create programs called virtual instruments (VIs) in a flowchart-like form called a block diagram. The user interacts with the program through the front panel. LabVIEW has many built-in functions to facilitate the programming process. The next chapters will teach you how to make the most of LabVIEW's many features.

KEY TERMS

Block diagram: Pictorial representation of a program or algorithm. In G, the block diagram, which consists of executable icons, called nodes, and wires that carry data between the nodes, is the source code for the VI.

Context Help window: Special window that displays the names and locations of the terminals for a function or subVI, the description of controls and indicators, the values of universal constants, and descriptions and data types of control attributes. The window also accesses the **LabVIEW Help**.

Controls palette: Palette containing front panel controls and indicators.

Front panel: The interactive interface of a VI. Modeled from the front panel of physical instruments, it is composed of switches, slides, meters, graphs, charts, gauges, LEDs, and other controls and indicators.

Functions palette: Palette containing block diagram structures, constants, and VIs.

Hierarchical menus: Menu items that expand into submenus.

LabVIEW: **Lab**oratory **V**irtual **I**nstrument **E**ngineering **W**orkbench. It is a powerful and flexible instrumentation and analysis software development application.

Nodes: Execution elements of a block diagram consisting of functions, structures, and subVIs.

Palette: Menu of pictures that represent possible options.

Pop up: To call up a special menu by clicking an object with the right mouse button (on **Windows** platforms) or with the command key and the mouse button (on **Macintosh** platforms).

Pull-down menu: Menus accessed from a menu bar. Pull-down menu options are usually general in nature.

Short cut menu: Menus accessed by popping up, usually on an object. Menu options pertain to that object specifically.

Tool: A special operating mode of the mouse cursor.

Tools palette: Palette containing tools you can use to edit and debug front panel and block diagram objects.

Toolbar: Bar containing command buttons to run and debug VIs.

Virtual instrument (VI): Program in LabVIEW; so-called because it models the appearance and function of a physical instrument.

EXERCISES

E1.1   In this exercise we want to open and run an existing VI. In LabVIEW, open
VibrationAnalysis.vi. This VI is located in Examples\Apps\demos.llb. The
front panel should look like the one shown in Figure E1.1.

(a) Run the VI by clicking on the **Run** button.

(b) Vary the Acquisition Rate on the vertical pointer slide control.

(c) Vary the desired velocity on the Set Velocity [km/hr] dial and verify that the
actual velocity, as indicated on the Actual Velocity [km/hr] gauge, matches
the desired velocity.

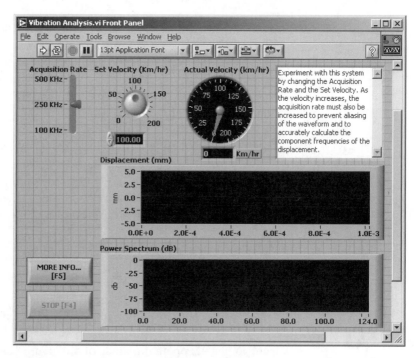

FIGURE E1.1
The Vibration Analysis.vi front panel.

E1.2   Referring to VibrationAnalysis.vi from E1.1, we can inspect the block diagram
and watch it execute using **Highlight Execution**. Under the **Window** pull-down
menu, select **Show Diagram**. The panel should switch to the block diagram
shown in Figure E1.2.

(a) Click on the **Highlight Execution** button.

(b) Run the VI by clicking on the **Run** button.

(c) Watch as the data flows through the code.

Select Execution Highlighting

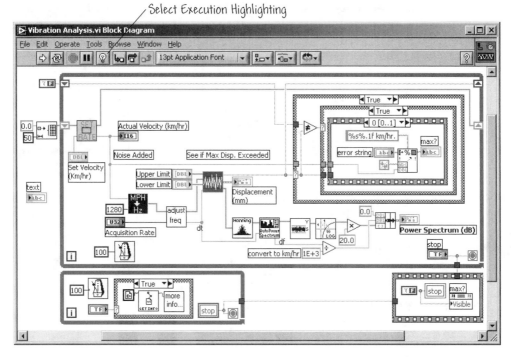

FIGURE E1.2
The Vibration Analysis.vi block diagram.

E1.3 In this exercise, we want to open and run an existing VI. In LabVIEW, go to **Help≫Find Examples** and click the **Search** tab and type "filter." Select "filter" to display the example VIs that include filter in the title. Find the VI titled "Express Filter.VI" and open it. The VI front panel is shown in Figure E1.3.

    (a) Run the VI by clicking on the **Run** button

    (b) Vary the "Simulated frequency" and watch the values change

    (c) Vary "Simulated amplitude" and "Simulated noise amplitude" and verify that the value on the indicator matches the graph

E1.4 Referring to "Express Filter.VI" from E1.3, we can inspect the block diagram and watch it execute using **Highlight Execution**. Under the **Window** pull-down menu, select **Show Diagram**. The panel should switch to the block diagram shown in Figure E1.4.

    (a) Click on the **Highlight Execution** button

    (b) Run the VI by clicking on the **Run** button

    (c) Watch as the data flows through the code and the Express VIs

    (d) Stop the VI and return to the block diagram

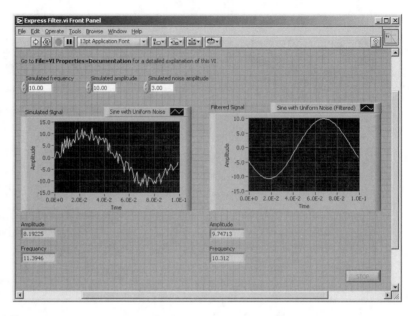

FIGURE E1.3
The Express Filter.vi front panel.

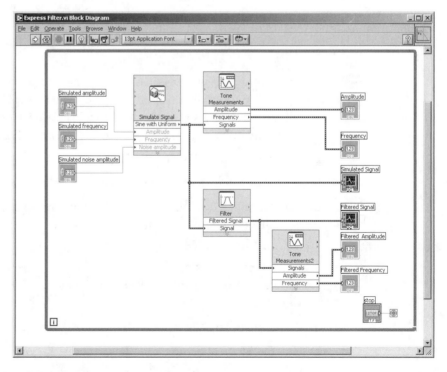

FIGURE E1.4
The Express Filter.vi block diagram.

(e) Double click on the "Simulate Signal — Express VI". Change the signal type from Sine to Square. Click **OK** and return to the front panel. Run the VI again. Notice it is now plotting a square wave instead of a sine wave.

E1.5 Open up a new blank VI. Navigate to the **Help** pull-down menu and select **Find Examples** and view the examples available in other categories. Look at some of the examples in **Browse≫New Examples for LabVIEW 7.0≫Express VIs**. The NI Example Finder is shown in Figure E1.5. Run several other VIs.

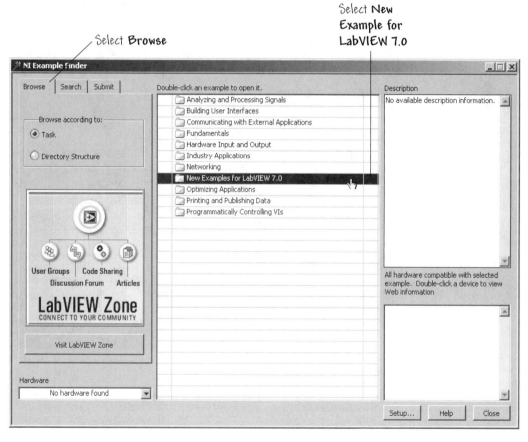

FIGURE E1.5
Select **New Example for LabVIEW 7.0**.

E1.6 On the LabVIEW Start-up Screen, select **New**. On this screen, we have the option to open a blank VI or a VI from a template. With a template, you will not have to start building your application from scratch. Browse through the available templates and then open **VI from Template≫Tutorial (Getting Started)≫Generate, Analyze, and Display**. Look at the front panel and block diagram of this VI and then run the VI.

PROBLEMS

P1.1 Complete the crossword puzzle.

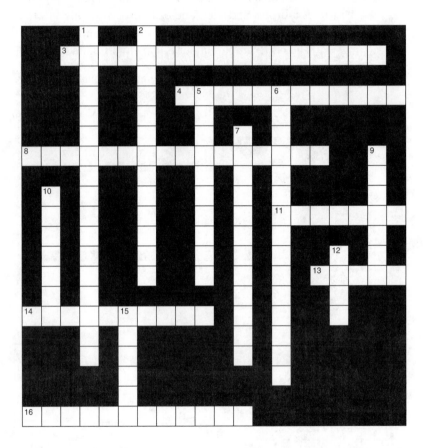

Across
3. Program in LabVIEW.
4. Menus accessed by popping up on an object.
5. Palette containing block diagram structures, constants, communication features, and VIs.
11. A powerful and flexible instrumentation and analysis software development application.
13. Execution elements of a block diagram consisting of functions, structures, and subVIs.
14. The interactive interface of a VI.
16. Bar containing command buttons to run and debug VIs.

Down
1. Menu items that expand into submenus.
2. Menus accessed from a menu bar.
5. Special window that displays helpful information.
6. Palette containing front panel controls and indicators.
7. Pictorial representation of a program or algorithm.
9. Menu of pictures that represent possible options.
10. Palette containing tools for editing and debugging.
12. A special operating mode of the mouse cursor.
15. To call up a special menu by clicking on an object.

P1.2 In the problem we want to open an existing VI from the **Learning** directory. You can open the VI by either selecting **Open VI** from the Startup screen, or if you are already in LabVIEW, you can use the **File** pull-down menu (see Figure 1.15) and select **Open.**... In both cases, you must navigate through your local file structure to find the desired VI. Find, open, and run **Running Dog.vi** located in **Learning\Instructional VIs\CompSci.llb**.

This VI is only available on the Windows platform. If you are on a Macintosh platform, locate, open, and run **Control Mixer Process.vi** *located in the library* **Examples\Apps\demos.llb**.

P1.3 You can construct games using LabVIEW! In this problem, you will download a LabVIEW game of your choice from the "LabVIEW Zone." Go to http://www.ni.com/labviewzone. Once you're in the LabVIEW Zone, click on the "Fun Stuff" icon and see the list of games. Download a game of your choice. Play the game. Document the following: Which game did you download? What was the purpose of the game? Check out the block diagram of the game to see the code.

CHAPTER 2

Virtual Instruments

Virtual instruments (VI) are the building blocks of LabVIEW programming. We will see in this chapter that VIs have three main components: the front panel, the block diagram, and the icon and connector pair. We will revisit the front panel and block diagram concepts first introduced in Chapter 1. An introduction to wiring the elements together on the block diagram is presented, although many of the debugging issues associated with wires are left to the next chapter. The important notion of data flow programming is also discussed in this chapter. Finally—you will have the opportunity to build your first VI!

GOALS

1. Gain experience by running more worked examples.
2. Understand the three basic components of a virtual instrument.
3. Begin the study of programming in G.
4. Understand the notion of data flow programming.
5. Build your first virtual instrument.

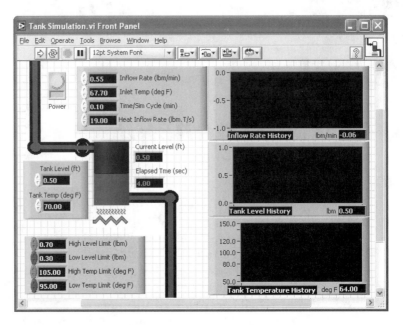

FIGURE 2.1
A virtual instrument front panel.

2.1   WHAT ARE VIRTUAL INSTRUMENTS?

LabVIEW programs are called virtual instruments (VIs) because they have the look and feel of physical systems or instruments. The illustration in Figure 2.1 shows an example of a front panel. A VI and its components are analogous to main programs and subroutines from text programming languages like C and Fortran. VIs have both an interactive user interface—known as the front panel—and the source code—represented in graphical form on the block diagram. LabVIEW provides mechanisms that allow data to pass easily between the front panel and the block diagram.

The block diagram is a pictorial representation of the program code. The block diagram associated with the front panel in Figure 2.1 is shown in Figure 2.2. The block diagram consists of executable icons (called nodes) connected (or **wired**) together. We will discuss wiring later in this chapter. The important concept to remember is that, in the G programming language, the block diagram is the source code.

The art of successful programming in G is an exercise in **modular programming**. After dividing a given task into a series of simpler subtasks (in G these subtasks are called subVIs and are analogous to subroutines), you then construct a virtual instrument to accomplish each subtask. Chapter 4 focuses on building subVIs. The resulting subtasks (remember, these are called subVIs) are then assembled on a top-level block diagram to form the complete program.

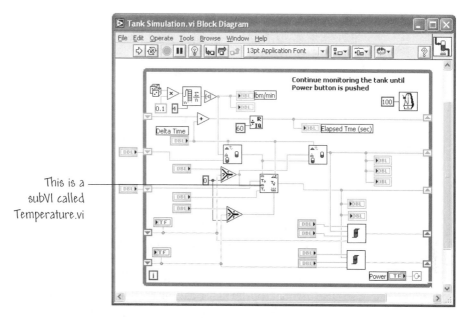

This is a
subVI called
Temperature.vi

FIGURE 2.2
The virtual instrument block diagram associated with the front panel in Figure 2.1.

Modularity means that you can execute each subVI independently, thus making debugging and verification easier. Furthermore, if your subVIs are general purpose programs, you can use them in other programs.

VIs (and subVIs) have three main parts: the front panel, the block diagram, and the icon/connector. The front panel is the interactive user interface of a VI—a window through which the user interacts with the code. When you run a VI, you must have the front panel open so you can pass inputs to the executing program and receive outputs (such as data for graphical display). The front panel is indispensable for viewing the program outputs. It is possible, as we will discuss in Chapter 9, to write data out to a file for subsequent analysis, but generally you will use the front panel to view the program outputs. The front panel contains knobs, push buttons, graphs, and many other controls (the term *controls* is interchangeable with *inputs*) and indicators (the term *indicators* is interchangeable with *outputs*).

The block diagram is the source code for the VI. The source code is "written" in the G programming language. We use the term *written* loosely, since in fact the code is made up of graphical icons, wires, and such, rather than traditional "lines of code." The block diagram is actually the executable code. The **icons** of a block diagram represent lower-level VIs, built-in functions, and program control structures. These icons are wired together to allow the data flow. As you will learn later in this chapter, the execution of a G program is governed by the

(a) icon   (b) connector

FIGURE 2.3
The icon and connector of the Temperature.vi subVI shown in Figure 2.2.

data flow and not by a linear execution of lines of code. This concept is known as **data flow programming**.

The **icons** and **connectors** specify the pathways for data to flow into and out of VIs. The icon is the graphical representation of the VI in the block diagram and the connector defines the inputs and outputs. All VIs have an icon and a connector. As previously mentioned, VIs are hierarchical and modular. You can use them as top-level (or calling) programs or as subprograms (or subVIs) within other programs. The icon and connector are shown in Figure 2.3 for the subVI Temperature.vi. This subVI can be found in the center of the Tank Simulation.vi diagram in Figure 2.2.

The subVI Temperature.vi has six inputs and one output. For proper operation of the subVI, the data flow must pass the (1) mass and (2) temperature of the initial fluid, the (3) flow rate and (4) temperature of the inlet fluid, the (5) heat flow rate, and the (6) elapsed time to the subVI. Once all the necessary input data to the subVI is available, the new temperature of the fluid is computed within the subVI and the result is output—the data "flows" out.

2.2 SEVERAL WORKED EXAMPLES

Before you construct your own VI, we will open several existing LabVIEW programs and run them to see how LabVIEW works. The first VI example—Temperature System Demo.vi—can be found in the suite of examples provided as part of LabVIEW. The second VI example—ODE Example.vi—illustrates how LabVIEW can be used to simulate linear systems. In this example, the motion of a mass-spring-cart system is simulated, and you can observe the effects of changing any of the system parameters of the resulting motion of the cart.

Temperature System Demo

In this example, you will open and run the virtual instrument called Temperature System Demo.vi. At the LabVIEW startup window, select **Open VI**. Double click on examples folder located in the LabVIEW 7 directory to open it

Set update period here.

Run button

Acquisition
switch

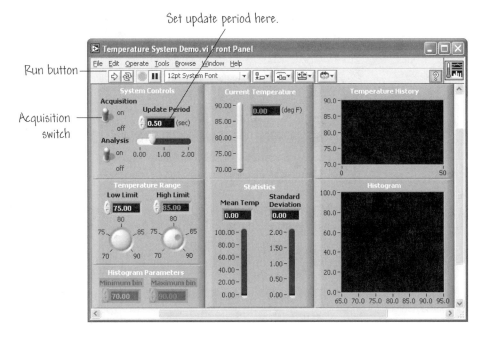

FIGURE 2.4
Temperature system demonstration front panel.

and locate the **apps** folder within. Then double click on **apps** and after it opens up, double click on the library **tempsys.llb**. A file dialog box will appear listing the available VIs and subVIs:

- Temperature System Demo.vi

- Array to Bar Graph.vi

- histogram+.vi

- Temperature Status.vi

- Update Statistics.vi

Double click on Temperature System Demo.vi to open it. The front panel window appears and should resemble the one shown in Figure 2.4. The front panel contains numeric controls, Boolean switches, slide controls, knob controls, charts, graphs, and a thermometer indicator.

Run the VI by clicking on the **Run** button. The front panel toolbar changes as the VI switches from edit mode to run mode. For example, once the VI begins executing, the stop button will change appearance on the front panel toolbar (it changes from a shaded symbol to a red stop sign). Also, the run button changes appearance to indicate that the VI is running.

This temperature system VI simulates a temperature monitoring application. The VI takes (simulated) temperature readings and displays them in both the thermometer indicator and on the chart. The simulated temperature readings are actually obtained from a pre-stored array of measurements, but with LabVIEW it is would be easy to modify the VI to acquire and process real temperature data.

The Update Period slide controls the speed at which the VI acquires the new temperature readings. The VI plots high and low temperature limits on the chart (high limit is in red, and low limit is in blue). These limits can be varied utilizing the Temperature Range knobs (located middle, left-hand side on the front panel). If the current temperature reading is out of the set range, Over Temp or Under Temp indicators will appear and light up next to the thermometer. You can also turn the data analysis on and off using the switch in the System Controls area of the front panel (upper left-hand side). The analysis section shows you a running calculation of the mean and standard deviation of the temperature values and a histogram of the temperature values. Once the VI begins to run, it will continue until you set the Acquisition switch to off.

While the VI is running, use the **Operating** tool to change the values of the high and low limits. Highlight the current high or low value of the limits, either by clicking twice on the value you want to change, or by clicking and dragging across the value with the tool. Type in the new value and click on the **Enter** button, located next to the **Run** button on the front panel toolbar.

Another input that you can modify is the Update Period, which is controlled by the slide—locate the slide control on the front panel in Figure 2.4. Place the **Operating** tool on the slider, click and drag it to a new location, and then run the VI. What changes do you observe? You can also operate slide controls by clicking on a point on the slide to snap the slider to that location, by clicking on the slider and moving it to the desired location, or by clicking in the slide's digital display and entering a number. When you are finished experimenting with the VI, terminate the execution by setting the Acquisition switch to off.

*LabVIEW does not accept values in digital displays until you press the **Enter** button or click the mouse in an open area of the window.*

Switch to the block diagram of Temperature System Demo.vi by choosing **Show Diagram** from the **Window** pull-down menu. The block diagram shown in Figure 2.5 is the underlying code for the VI. At this point in the learning process, you may not understand all of the block diagram elements depicted in Figure 2.5—but you will eventually!

As previously discussed, most VIs are hierarchical and modular. After creating a VI, you can (with a little work configuring the icon and connector) use the VI as a subVI (similar to a subroutine) in the block diagram of another VI. By

Temperature Status.vi

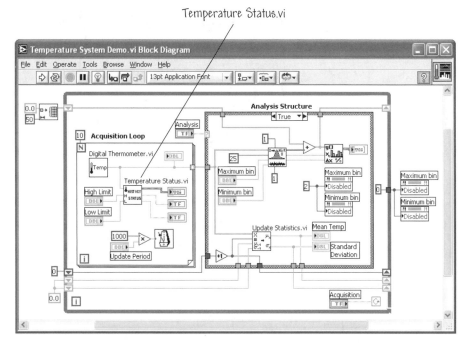

FIGURE 2.5
Temperature system demonstration block diagram—the code.

creating subVIs, you can construct modular block diagrams which make your VIs easier to debug. The Temperature System Demo VI uses several subVIs. Open the Temperature Status subVI by double clicking on the appropriate subVI icon (see Figure 2.5). The icon for the subVI is labeled History Status. The front panel shown in Figure 2.6 should appear.

The icon and connector provide the graphical representation and parameter definitions required to use a VI as a subVI in the block diagrams of other VIs. The icon and connector are located in the upper right corner of the VI front panel (see Figure 2.6). The icon is a graphical representation of the VI when used as a component in a G program, that is, when used in the block diagram of other VIs. An icon can be a pictorial representation or a textual description of the VI, or a combination of both. The icon for the subVI Temperature Status includes both text (History Status) and a graphical representation of a thermometer.

Every VI (and subVI) has a connector. The connector is a set of terminals that correspond to its controls and indicators. When you show the connector for the first time, LabVIEW will suggest a connector pattern that has one terminal for each control or indicator on the front panel—you can choose a different pattern. In Chapter 3, you will learn how to associate front panel controls and indicators with connector terminals. The connector terminals determine where you must wire the inputs and outputs on the icon. These terminals are analogous to parameters of a

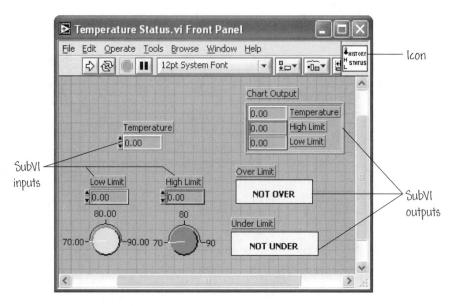

FIGURE 2.6
Temperature Status subVI.

subroutine. You might wonder where the icon is located relative to the connector. It is at the same location—the icon sits on top of the connector pattern. The icon and connector of the Temperature Status subVI are shown in Figure 2.7.

Every VI has a default icon, which is displayed in the icon panel in the upper right corner of the front panel and block diagram windows in Figure 2.8. You will learn how to edit the VI icon in the next chapter. You may want to personalize VI

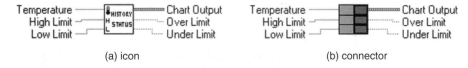

FIGURE 2.7
The icon and connector of Temperature Status subVI.

FIGURE 2.8
The default icon.

icons so that they transmit information by visual inspection about the contents of the underlying VI.

When you are finished experimenting with the Temperature System Demo.vi, close the VI and subVI by selecting **Close** from the **File** pull-down menu on each open front panel. Remember—do not save any changes!

*Selecting **Close** from the **File** pull-down menu of a block diagram closes the block diagram window only. Selecting **Close** on a front panel window closes both the front panel and the block diagram.*   ◆

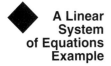

A Linear System of Equations Example

In this example we use LabVIEW to solve a set of linear, constant coefficient, ordinary differential equations. Many physical systems can be modeled mathematically as a set of linear, constant coefficient differential equations of the form

$$\dot{\mathbf{x}} = \mathbf{A}\mathbf{x}$$

with initial conditions $\mathbf{x}(0) = \mathbf{x}_o$, where the matrix $\mathbf{A}$ is a constant matrix.

Suppose we want to model the motion of a mass-spring-damper system, as shown in Figure 2.9. Let m represent the mass of the cart, k represent the spring constant, and b the damping coefficient. The position of the cart is denoted by y, and the velocity is the time derivative of the position, that is, the velocity is $\dot{y}$. Equating the sum of forces to the mass times acceleration (using Newton's Second Law) we obtain the equation of motion:

$$m\ddot{y} + b\dot{y} + ky = 0,$$

with the initial conditions $y(0) = y_o$ and $\dot{y}(0) = \dot{y}_o$. The motion of the cart is described by the solution of the second-order linear differential equation above. For this simple system we can obtain the solution analytically. For more complex systems it is usually necessary to obtain the solution numerically using

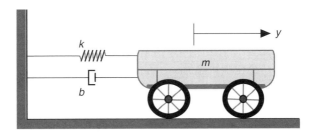

FIGURE 2.9
A simple mass-spring-damper system.

the computer. In this exercise we seek to obtain the solution numerically using LabVIEW.

It is sometimes convenient for obtaining the numerical solution to rewrite the second-order linear differential equation as two first-order differential equations. We first define the state vector of the system as

$$\mathbf{x} = \begin{pmatrix} y \\ \dot{y} \end{pmatrix},$$

where the components of the state vector are given by

$$x_1 = y \quad \text{and} \quad x_2 = \dot{y}.$$

Using the definitions of the state vector and the equation of motion, we obtain

$$\dot{x}_1 = x_2$$
$$\dot{x}_2 = -\frac{k}{m}x_1 - \frac{b}{m}x_2.$$

Writing in matrix notation yields

$$\dot{\mathbf{x}} = \mathbf{A}\mathbf{x},$$

where

$$\mathbf{A} = \begin{bmatrix} 0 & 1 \\ -k/m & -b/m \end{bmatrix}.$$

In this example, let

$$\frac{k}{m} = 2 \quad \text{and} \quad \frac{b}{m} = 4.$$

Choose the initial conditions as

$$x_1 = 10 \quad \text{and} \quad x_2 = 0.$$

Now we are ready to compute the solution of the system of ordinary differential equations numerically with LabVIEW.

- Select **Open VI** from the startup screen or by using the **File** menu to open the VI.

- Open the ODE Example.vi located in the folder Chapter 2 in the directory Learning.

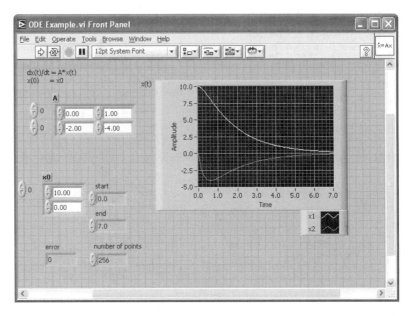

FIGURE 2.10
Linear system of equations front panel.

The front panel depicted in Figure 2.10 will appear. Verify that the initial conditions $\mathbf{x}_o$ are set correctly to $x_1 = 10$ and $x_2 = 0$. Run the VI by clicking on the **Run** button. About how long does it take for the cart position to come to rest? Change the initial position to $x_1 = 20$. Now how long does it take for the cart to come to rest? What is the maximum value of the cart velocity?

Open and examine the block diagram by choosing **Show Diagram** from the **Window** menu. The code is shown in Figure 2.11. This is a relatively complex VI—most of the ones that you develop in this book will be simpler! But if you need to solve a set of linear ordinary differential equations, you can start with this VI and modify it as necessary. The idea of starting with a VI that solves a related problem is a good approach in the early stages of learning LabVIEW. Access the **Help** pull-down menu and select **Show Context Help**. Move the cursor over various objects on the block diagram and read what the online help has to say. For example, if you are familiar with eigenvalues and eigenvectors, you should recognize the online help discussion related to EigenValues and Vectors.vi.

When you are finished experimenting, close the VI by selecting **Close** from the **File** menu.   ◆

2.3   THE FRONT PANEL

The front panel of a VI is a combination of controls and indicators. Controls simulate the types of input devices you might find on a conventional instrument,

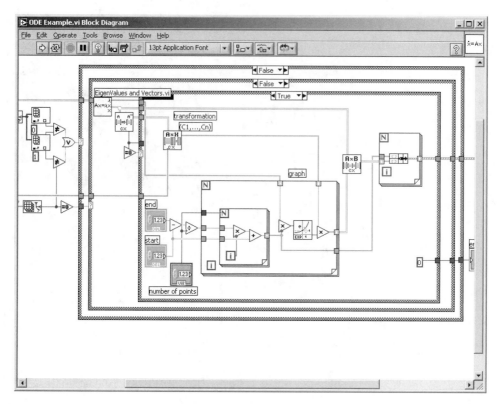

FIGURE 2.11
Linear system of equations block diagram.

such as knobs and switches, and provide a mechanism to move input from the front panel to the underlying block diagram. On the other hand, indicators provide a mechanism to display data originating in the block diagram back on the front panel. Indicators include various kinds of graphs and charts (more on this topic in Chapter 7), as well as numeric, Boolean, and string indicators. Thus, when we use the term *controls* we mean "inputs," and when we say *indicators* we mean "outputs."

You place controls and indicators on the front panel by selecting and "dropping" them from the **Controls** palette. Once you select a control (or indicator) from the palette and release the mouse button, the cursor will change to a "hand" icon, which you then use to carry the object to the desired location on the front panel and "drop" it by clicking on the mouse button again. Once an object is on the front panel, you can easily adjust its size, shape, and position (see Chapter 3). If the **Controls** palette is not visible, you can either right click on an open area of the front panel window, or select **Show Controls Palette** from the **Window** pull-down menu.

*If you right click on an open area of the front panel window, you easily access the **All Controls** palette. Similarly, you access the **Functions** palette by popping up on an open area of the block diagram.*

2.3.1   Numeric Controls and Indicators

You access the numeric controls and indicators from the **Numeric** subpalette located on the **All Controls** palette, or the **Numeric Controls** or **Numeric**

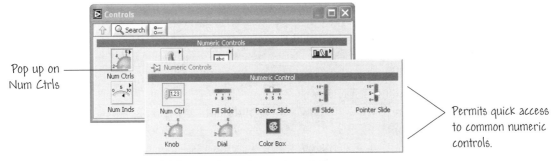

Pop up on Num Ctrls

Permits quick access to common numeric controls.

(a) Commonly used numeric controls.

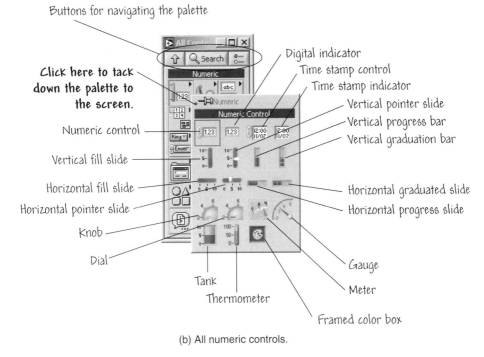

Buttons for navigating the palette

Click here to tack down the palette to the screen.

Numeric control

Vertical fill slide

Horizontal fill slide

Horizontal pointer slide

Knob

Dial

Tank

Thermometer

Digital indicator

Time stamp control

Time stamp indicator

Vertical pointer slide

Vertical progress bar

Vertical graduation bar

Horizontal graduated slide

Horizontal progress slide

Gauge

Meter

Framed color box

(b) All numeric controls.

FIGURE 2.12
Numeric controls and indicators.

Indicators on the **Controls** palette, as shown in Figure 2.12. As the figure shows, there are quite a large number of available numeric controls and indicators. The two most commonly used numeric objects are the numeric control and the numeric indicator. When you construct your first VI later in this chapter, you will get the chance to practice dropping numeric controls and indicators on the front panel. Once a numeric control is on the front panel, you click on the increment buttons (that is, the up and down arrows on the left hand-side of the control) with the **Operating** tool to enter or change the displayed numerical values. Alternatively, you can double click on the current value of the numeric control with the **Auto** tool, which will highlight the value, and you can then enter a different value.

*You can tack down the **Numeric** palette (and most other palettes) to the screen so they are visible at all times by clicking on the thumbtack on the top left corner of the palette.*

You can use the navigation buttons (see Figure 2.12 on the **Controls** and **Functions**) palettes to navigate and search for controls, VIs, and functions. When you left click a subpalette icon, the entire palette changes to the subpalette you selected. If you right click a subpalette icon, the subpalette appears, but the **Controls** (or **Functions**) palette remains in view. An example of a subpalette icon on the **Controls** palette is the **Numerics** subpalette. The **Controls** and **Functions** palettes contain three navigation buttons (as illustrated in Figure 2.13):

- **Up**—Takes you up one level in the palette hierarchy.

- **Search**—Changes the palette to search mode. In search mode, you can perform text-based searches to locate controls, VIs, or functions in the palettes.

- **Options**—Opens the **Function Browser Options** dialog box, from which you can configure the appearance of the palettes.

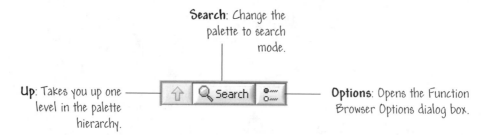

FIGURE 2.13
Navigating the **Functions** and **Controls** palettes.

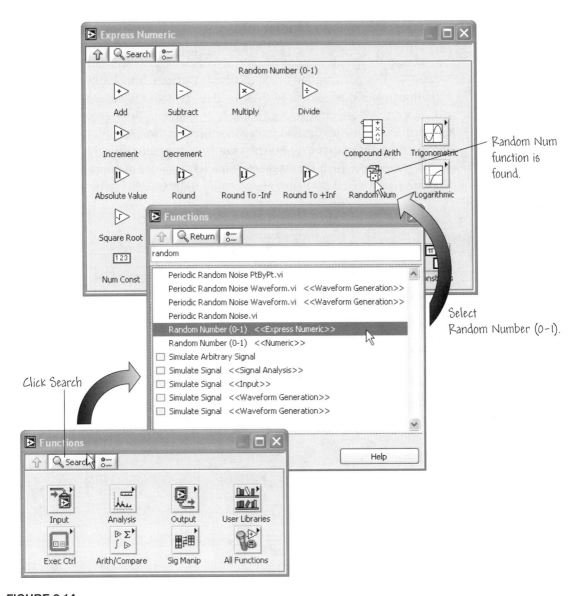

FIGURE 2.14
Searching the **Functions** palette for a Random Number VI.

Suppose you need to search for a function that provides a random number between 0 and 1. On the **Functions** palette, select the **Search** button and in the dialog box that appears, enter the word "random," as illustrated in Figure 2.14. In the results list of the search, select Random Number (0-1). Then the Express Numeric palette will appear and the Random Num function is indeed located, as shown in Figure 2.14.

2.3.2   Boolean Controls and Indicators

You access the Boolean controls and indicators from the **Boolean** subpalette located on the **All Controls** palette or the **Buttons and LED** palette on the **Functions** palette, as shown in Figure 2.15. As with the numeric controls and indicators, there are quite a large number of available Boolean controls and indicators. Boolean controls and indicators simulate switches, buttons, and LEDs and are used for entering and displaying Boolean (True-False) values. For example, you might use a Boolean LED in a temperature monitoring system as a warning signal that a measured temperature has exceeded some predetermined

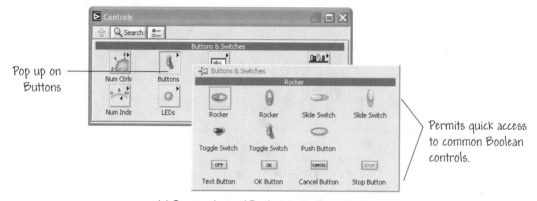

(a) Commonly used Boolean controls.

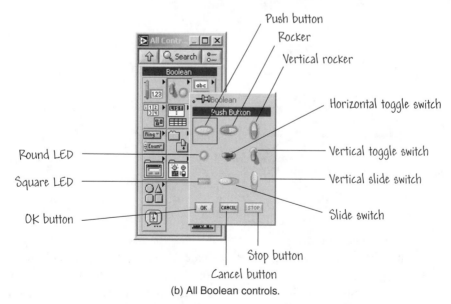

(b) All Boolean controls.

FIGURE 2.15
Boolean controls and indicators.

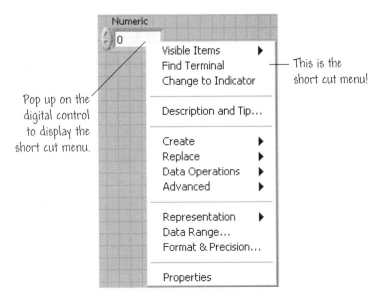

Pop up on the digital control to display the short cut menu.

This is the short cut menu!

FIGURE 2.16
Configuring a digital control using the short cut menu.

safety limit. The measured temperature is too high, so an LED indicator on the front panel turns from green to red! Before continuing, take a few moments to familiarize yourself with the types of available digital and Boolean controls and indicators shown in Figures 2.12 and 2.15, so that when you begin to construct your own VIs, you'll have a feel for what's available on the palettes.

2.3.3   Configuring Controls and Indicators

Popping up on a digital control displays the short cut menu, as shown in Figure 2.16. You can change the defaults for controls and indicators using options from the short cut menus. For example, under the submenu **Representation** you will find that you can choose from 12 representations of the digital control or indicator, including 32-bit single precision, 64-bit double precision, signed integer 8-bit, and more. The representation indicates how much memory is used to represent a number. This choice is important if you are displaying large amounts of data and you want to conserve computer memory. Another useful item on the short cut menu is the capability to switch from control and indicator using **Change to Indicator** and vice-versa. We will discuss each item in the short cut menu on an as-needed basis.

2.4   THE BLOCK DIAGRAM

The graphical objects comprising the block diagram together make up what is usually call the source code. The block diagram (visually resembling a computer

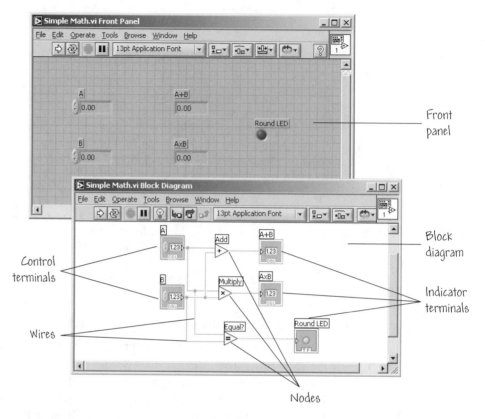

FIGURE 2.17
A typical VI illustrating nodes, terminals, and wires.

program flowchart) corresponds to the lines of text found in text-based programming languages. In fact, the block diagram is the actual executable code. The block diagram is built by wiring together objects that perform specific functions.

The components of a block diagram (a VI is depicted in Figure 2.17) belong to one of three classes of objects:

- **Nodes**: Program execution elements.
- **Terminals**: Ports through which data passes between the block diagram and the front panel and between nodes of the block diagram.
- **Wires**: Data paths between terminals.

2.4.1 VIs and Express VIs

An important new element of LabVIEW is the so-called Express VI. These VIs are provided to allow for quick construction of VIs designed to accomplish common measurement tasks, such as data acquisition. Express VIs are nodes (see Section 2.4.2 for more information on nodes) that require minimal wiring because they are configured with dialog boxes. The positive implications of minimizing the required wiring will become more evident as we proceed through the forthcoming chapters.

LabVIEW views a VI placed on the block diagram to be a subVI (more on subVIs in Chapter 4). A subVI is itself a VI that can be used as an element on a block diagram. In general, a block diagram can have VIs and Express VIs as elements. One difference in their use is that when you double-click a subVI, its front panel and block diagram appear. When you double-click an Express VI a dialog box appears in which you can configure the VI to meet your needs. The idea is that the Express VI allows you to quickly configure a VI by interacting with a dialog box rather than reconfiguring the code in a subVI block diagram. When an Express VI is available for the task you need, it will be better, in general, to consider the use of the Express VI.

VIs and Express VIs are distinguishable on the block diagram through the use of colored icons. By default, icons for Express VIs appear on the block diagram as expandable nodes (see Section 2.4.2) with icons surrounded by a blue field, while icons for VIs have white backgrounds. The other common element on the block diagram is the function, and you can easily identify functions since their icons have pale yellow backgrounds.

2.4.2 Nodes

Nodes are analogous to statements, functions, and subroutines in text-based programming languages. There are three node types—**functions**, **subVI nodes**, and **structures**. Functions are the built-in nodes for performing elementary operations such as adding numbers, file I/O, or string formatting. Functions are the fundamental operating element of a block diagram. The Add and Multiply functions in Figure 2.17 represent one type of node. SubVI nodes are VIs that you design and later call from the diagram of another VI. You can also create subVIs from Express VIs. Structures—such as For Loops and While Loops—control the program flow.

Express VIs and subVIs can be displayed as either icons or as expandable nodes. By default, most subVIs appear as icons that are not expandable. On the other hand, most Express VIs appear as expandable icons. Figure 2.18 depicts the various possibilities for displaying subVIs and Express VIs.

Expandable nodes appear as icons surrounded by a colored field. SubVIs appear with a yellow field, and Express VIs appear with a blue field. By default, subVIs appear as icons on the block diagram, and Express VIs appear as expandable nodes. To display a subVI or Express VI as an expandable node, right-click the subVI or Express VI and select **View As Icon** from the shortcut menu.

You can resize the expandable node to make wiring even easier. When you place the Positioning tool over an expandable node, resizing handles will appear at the top and bottom of the node. Placing the cursor over a resizing handle will

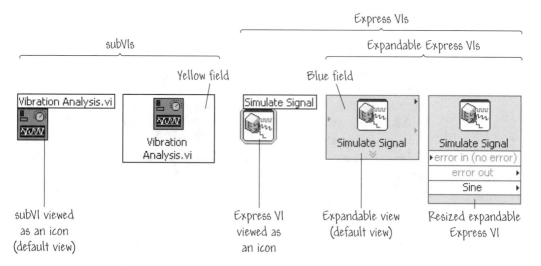

FIGURE 2.18
Expandable Icons versus Icons

transform the cursor to the resizing cursor, which can be used to drag the border of the node down to display input and output terminals. In Figure 2.18, the Simulate Signal Express VI is resized to display the inputs "error in (no error)" and outputs "error out and Sine." Resizing the expandable node takes a larger amount of space on the block diagram which can then clutter a complex program. Use icons if you want to conserve space on the block diagram and use expandable nodes to make wiring easier and to aid in documenting block diagrams.

2.4.3   Terminals

Terminals are analogous to parameters and constants in text-based programming languages. There are different types of terminals—control and indicator terminals, node terminals, constants, and specialized terminals that you will find on various structures. In plain words, a terminal is any point to which you can attach a wire to pass data.

For example, in the case of control and indicator terminals, numeric data entered into a numeric control passes to the block diagram via the control terminals when the VI executes. When the VI finishes executing, the numeric output data passes from the block diagram to the front panel through the indicator terminals. Data flows in only one direction—from a "source" terminal to one or more "destination" terminals. In particular, controls are source terminals and indicators are destination terminals. The data flow direction is from the control terminal to the indicator terminal, and not vice-versa. Clearly, controls and indicators are not interchangeable.

Control and indicator terminals belong to front panel controls and indicators and are automatically created or deleted when you create or delete the corresponding front panel control or indicator. The block diagram of the VI in Figure 2.17 shows terminals belonging to four front panel controls and indicators. The Add and Multiply functions shown in the figure also have node terminals through which the input and output data flow into and out of the functions.

You can configure front panel controls or indicators to appear as icon terminals or data type terminals on the block diagram. The terminals shown in Figure 2.18 are all icon terminals. The data type terminals were the standard representation of terminals through LabVIEW 6.1, but with LabVIEW 7.0, by default, front panel objects appear as icon terminals. To display a terminal as a data type on the block diagram, right-click the terminal and select **View As Icon** from the shortcut menu to remove the checkmark.

Control terminals have thick *borders and indicator terminal borders have* thin *borders. It is important to distinguish between thick and thin borders since they are not functionally equivalent. Additionally, small black arrows point out of (to the right) controls and into (from the right) indicators to depict data flow.*

Data types indicate what objects, inputs, and outputs you can wire together. You cannot wire together objects with incompatible data types. For example, a switch has a green border so you can wire a switch to any input with a green label on an Express VI. Similarly, a knob has an orange border so you can wire a knob to any input with an orange label. However, you cannot wire an orange knob to an input with a green label.

Wires are the same color as the terminal.

Associated with the Express VI is the **dynamic data type**. The dynamic data type stores information generated or acquired by an Express VI and appears on the block diagram as a dark blue terminal. You can wire the dynamic data type to any indicator or input that accepts numeric, waveform, or Boolean data. The objective is to wire the dynamic data type to an indicator (graph, chart, or numeric) that can best present the data.

Most subVIs and functions do not accept the dynamic data type. To use a subVI or function to process the dynamic data type data, you must convert the dynamic data type using the Convert from Dynamic Data Express VI to change the dynamic data type to numeric, Boolean, waveform, or array data types. As illustrated in Figure 2.19, the Convert to Dynamic Data Express VI and the Convert from Dynamic Data Express VI are found on the **Signal Manipulation**

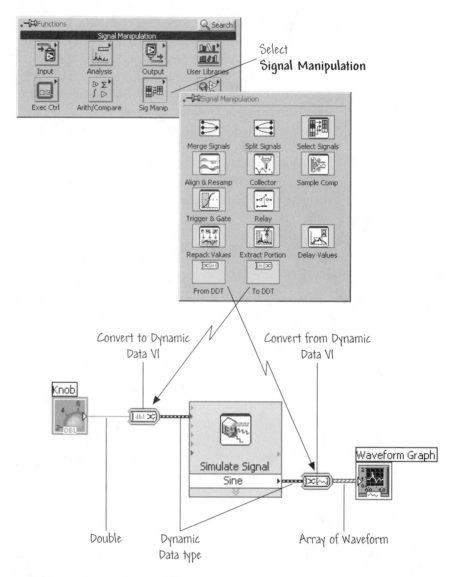

FIGURE 2.19
The Dynamic data type.

palette. When you place the Convert from Dynamic Data Express VI on the block diagram, a dialog box appears and displays options that let you specify how you want to format the data that the Convert from Dynamic Data Express VI returns. In Figure 2.19 the Convert from Dynamic Data Express VI is configured to transfer double numeric data from a knob to the dynamic data type used by the Simulate Signal Express VI.

TABLE 2.1   Common wire types.

	Scalar	1D array	2D array	Color
Numeric	————	————	═══════	Orange (floating point) &Blue (integer)
Boolean	··············	˅˅˅˅˅˅˅˅	˅˅˅˅˅˅˅˅˅	Green
String	∿∿∿∿∿∿	∷∷∷∷∷∷∷	∷∷∷∷∷∷∷∷	Pink

When you wire a dynamic data type to an array indicator, LabVIEW automatically places the Convert from Dynamic Data Express VI on the block diagram. Double-click the Convert from Dynamic Data Express VI to open the dialog box to control how the data appears in the array.

Transforming data for use in Express VIs occurs in a similar fashion. You use the Convert to Dynamic Data Express VI to convert numeric, Boolean, waveform, and array data types to the dynamic data type for use with Express VIs. When you place the Convert to Dynamic Data Express VI on the block diagram, a dialog box appears that allows you select the kind of data to convert to the dynamic data type.

2.4.4   Wiring

Wires are data paths between terminals and are analogous to variables in conventional languages. How then can we represent different data types on the block diagram? Since the block diagram consists of graphical objects, it seems appropriate to utilize different wire patterns (shape, line style, color, etc.) to represent different data types. In fact, each wire possesses a unique pattern depending on the type of data (numeric, Boolean, string, etc) that flows through the wire. Each data type appears in a different color for emphasis. To determine the data types on a given wire, match up the colors and styles with the wire types are shown in Table 2.1.

The hot spot of the **Wiring** tool is the tip of the unwound wiring segment. To wire from one terminal to another, click the hot spot of the **Wiring** tool on the first terminal (you can start wiring at either terminal!), move the tool to the second terminal, and click on the second terminal. No need to hold down the mouse button while moving the **Wiring** tool from one terminal to the other. The

FIGURE 2.20
The hot spot on the **Wiring** tool.

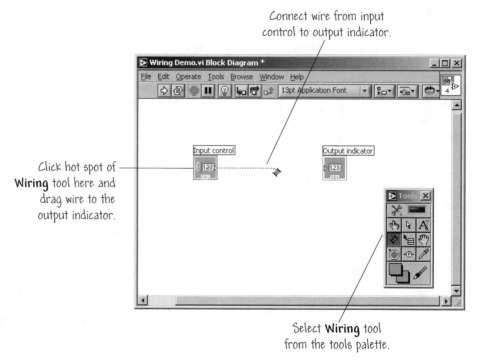

Connect wire from input control to output indicator.

Click hot spot of **Wiring** tool here and drag wire to the output indicator.

Select **Wiring** tool from the tools palette.

FIGURE 2.21
Wiring terminals.

wiring process is illustrated in Figure 2.21. When you wire two terminals, notice that moving the **Wiring** tool over one of the terminals causes that terminal to blink. This is an indication that clicking the mouse button will make the wire connection.

The VI shown in Figure 2.21 is easy to construct. It consists of one digital control and one digital indicator wired together. Open a new VI and try to build the VI! A working version can be found in the **Chapter 2** folder in the **Learning** directory—it is called Wiring Demo. The function of the VI is to set a value for **Input control** on the front panel and to display the same input at the digital indicator **Output indicator**.

To delete a wire as you are wiring:

- **Windows**—Click the right mouse button or click on the origination terminal.

- **Macintosh**—Hold down <option> and click or click on the origination terminal.

When wiring two terminals together, you may want to bend the wire to avoid running the wire under other objects. This is accomplished during the wiring process by clicking the mouse button to tack the wire down at the desired location

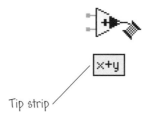

Tip strip

FIGURE 2.22
Tip strip.

of the bend, and moving the mouse in a perpendicular direction to continue the wiring to the terminal. Another way to change the direction of a wire while wiring is to press the space bar while moving the **Wiring** tool.

Also new with LabVIEW 7 is automatic wiring. Instead of tacking down the wire, simply connect the two terminals and LabVIEW will choose the best path. If you have an existing wire that you would like to fix, right-click on the wire and choose **Clean Up Wire** on the **Edit** pull-down menu.

Windows: All wiring is performed using the left *mouse button.*

Tip strips make it easier to identify function and node terminals for wiring. When you move the **Wiring** tool over a terminal, a tip strip pops up, as illustrated in Figure 2.22. Tip strips are small text banners that display the name of each terminal. When you place the **Wiring** tool over a node, each input and output will show as a wire stub—a dot at the end of the wire stub indicates an input. Tip strips should help you wire the terminals correctly.

It is possible to have objects wired automatically. A feature of LabVIEW is the capability to automatically wire objects when you first drop them on the diagram. After you select a node from the **Functions** palette, move that node close to another node to which you want to wire the first node. Terminals containing similar datatypes with similar names will automatically connect. You can disable the automatic wiring feature by pressing the space bar. You can adjust the auto wiring settings from the Tools≫Options≫Block Diagram window.

Since it is important to correctly wire the terminals on functions, LabVIEW provides an easy way to show the icon connector to make the wiring task easier. This is accomplished by popping up on the function and choosing Visible Items≫Terminals from the short cut menu, as illustrated in Figure 2.23. To return to the icon, pop up on the function and deselect Visible Items≫Terminals.

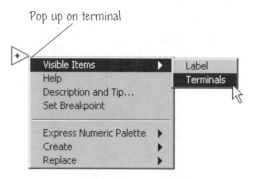

FIGURE 2.23
Showing terminals.

2.5 BUILDING YOUR FIRST VI

In this section you will create your first virtual instrument to perform the following functions:

- Add two input numbers and display the result.

- Multiply the same two input numbers and display the result.

- Compare the two input numbers and turn on an LED if the numbers are equal.

Begin by considering the front panel shown in Figure 2.24. It has two digital control inputs for the numbers A and B, two digital indicator outputs to display the results $A + B$ and $A \times B$, respectively, and a round LED that will turn on when the input numbers A and B are identical.

As with the development of most sophisticated computer programs, constructing VIs is an art, and you will develop your own style as you gain experience with programming in G. With that in mind, you should consider the following steps as only one possible path to building a working VI that carries out the desired calculations and displays the results.

1. Open a new front panel by choosing **New** from the **File** menu.

2. **Create the numeric digital controls and indicators**. The two front panel controls are used to enter the numbers, and the two indicators are used to display the results of the addition and multiplication of the input numbers.

   (a) Select **Numeric Control** from the **Numeric Controls** subpalette of the **Controls** palette. If the **Controls** palette is not visible, pop up in an open area of the front panel to gain access to the palette.

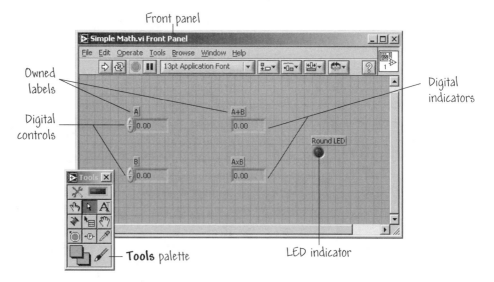

FIGURE 2.24
The front panel for your first VI.

(b) Drop the control on the front panel, as illustrated in Figure 2.25. Drag the control to the desired location and then click the mouse button to complete the drop.

(c) Type the letter *A* inside the label box (which appears above the control) and press the **Enter** button on the front panel toolbar. If you do not type the control label before starting other programming actions (such as dropping the other control on the front panel), the label box will remain labeled with the default label Numeric. If the control or indicator does not have a label, you can pop up on the control and select **Label** from the **Show** menu. The label box appears, and you can then edit the text using the Labeling tool (see Figure 2.26).

(d) Repeat the above process to create the second numeric control and the two numeric indicators. You can arrange the controls and indicators in any manner that you choose—although a neat and orderly arrangement is preferable. Add the labels to each control and indicator using Figure 2.24 as a guide.

3. **Create the Boolean LED**. This indicator will turn on if the two input numbers are identical, or remain off if they do not match.

(a) Select **Round LED** from the **LED** subpalette of the **Controls** palette. Place the indicator on the front panel, drag it to the desired location, and then click the mouse button to complete the process.

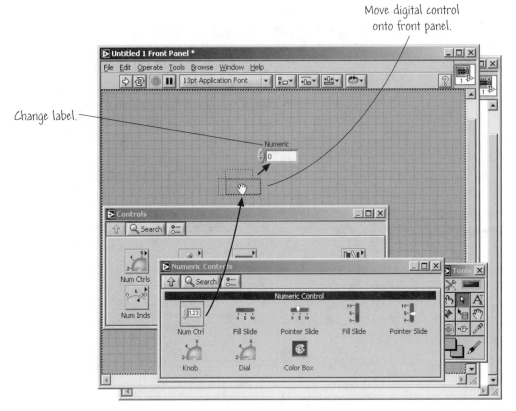

FIGURE 2.25
Placing the controls and indicators on the front panel.

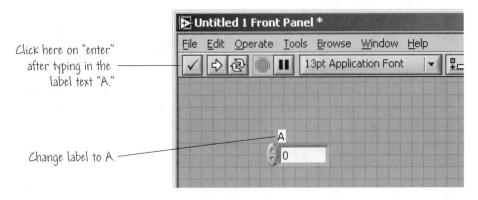

FIGURE 2.26
Labeling the digital control and indicators on the front panel.

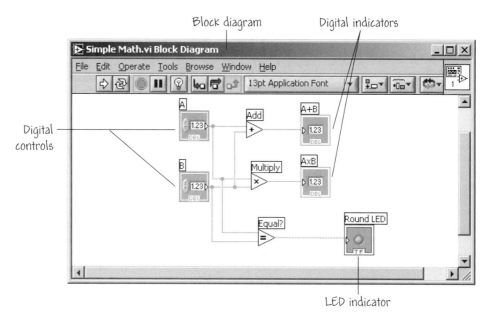

FIGURE 2.27
The block diagram window for your first VI.

(b) Type **Round LED** inside the label box and click anywhere outside the label when finished, or click on the **Enter** button.

Each time you create a new control or indicator, LabVIEW automatically creates the corresponding terminal in the block diagram. When viewed as icons, the terminals are graphical representations of the controls or indicators.

The Block Diagram

1. Switch your center of activity to the block diagram by selecting **Show Diagram** from the **Window** pull-down menu. The completed block diagram is shown in Figure 2.27. It may be helpful to display the front panel and block diagram simultaneously using either the **Tile Left and Right** or the **Tile Up and Down** options found in the **Window** pull-down menu. For this example, the up and down option works better in the sense that the all the block diagram and front panel objects can be displayed on the screen without having to use the scrollbars.

2. Now we want to place the addition and multiplication functions on the block diagram. Select the Add function from the **Arithmetic & Comparison≫Numeric** subpalette of the **Functions** palette. If the **Functions** palette is not visible, pop up on an open area of the block diagram to gain access to the palette. Drop the Add function on the block diagram in approximately the

Function label

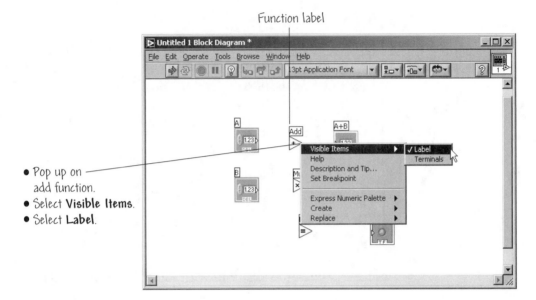

● Pop up on
  add function.
● Select **Visible Items**.
● Select **Label**.

FIGURE 2.28
Pop up to select the Visible Items≫Label option.

same position as shown in Figure 2.27. The label for the Add function can be displayed using the short cut menu and selecting **Visible Items≫Label**. This is illustrated in Figure 2.28. Following the same procedure, place the Multiply function on the block diagram and display the label.

3. Select the Equal? function from the **Arithmetic & Comparison≫Numeric** subpalette of the **Functions** palette and place it on the block diagram, as shown in Figure 2.29. The Equal? function compares two numbers and returns TRUE if they are equal or FALSE if they are not. To get more information on this function, you can activate the online help by choosing **Show Context Help** from the **Help** menu. Then placing the cursor over the Equal? function (or any of the other functions on the block diagram) leads to the display of the online help information.

4. Using the **Auto-tool** found on the **Tools** palette, wire the terminals as shown in Figure 2.27. As seen in Figure 2.30, to wire from one terminal to another, click the **Auto-tool** on the first terminal, move the tool to the second terminal, and click on the second terminal. Remember that it does not matter on which terminal you initiate the wiring. To aid in wiring, pop up on the three functions and choose **Visible Items≫Terminals**. Having the terminals shown explicitly helps to wire more quickly and accurately. Once the wiring

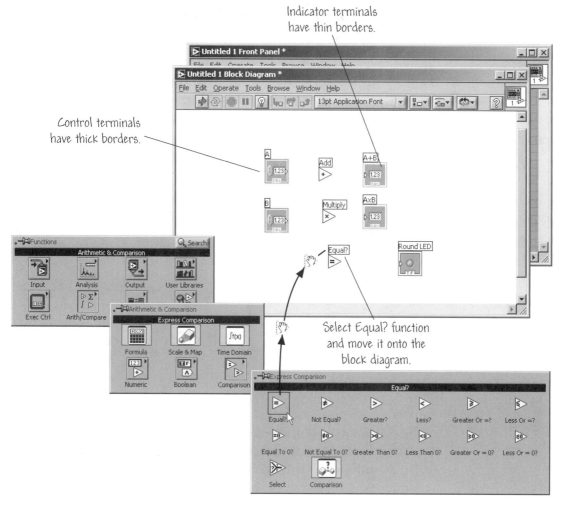

FIGURE 2.29
Adding the Equal? function to the VI.

is finished for a given function, it is best to return to the icon by popping up on the function and choosing **Visible Items** and deselecting **Terminals**.

5. Switch back to the front panel window by clicking anywhere on it or by choosing **Show Panel** from the **Window** menu.

6. Save the VI as Simple Math.vi. Select **Save** from the **File** menu and make sure to save the VI in the Users Stuff folder within the Learning directory.

In case you cannot get your VI to run properly, a working version of the VI (called Simple Math.vi) is located in the Chapter 2 folder within the Learning directory.

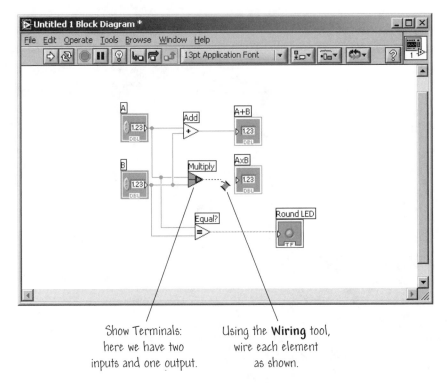

FIGURE 2.30
Wiring from one terminal to another.

7. **Enter input data**. Enter numbers in the numeric controls utilizing the **Auto-tool** by double-clicking in the numeric control box and typing in a number. The default values for *A* and *B* are 0 and 0, respectively. You can run the VI using these default values as a first try! When you use the default values, the LED should light up since $A = B$.

8. Run the VI by clicking on the **Run** button.

9. Experiment with different input numbers—make *A* and *B* identical, and verify that the LED does indeed turn on.

10. When you are finished experimenting, close the VI by selecting **Close** from the **File** menu.

2.6   DATA FLOW PROGRAMMING

The principle that governs VI execution is known as data flow. Unlike most sequential programming languages, the executable elements of a VI execute only when they have received all required input data—in other words, data flows out of the executable element only after the code is finished executing. The notion

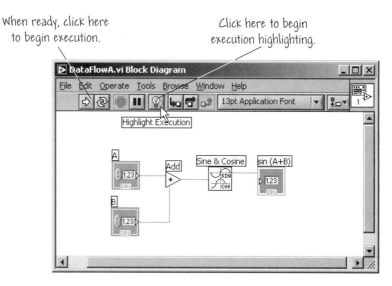

When ready, click here
to begin execution.

Click here to begin
execution highlighting.

FIGURE 2.31
Block diagram that adds two numbers and then computes the sine of the result.

of data flow contrasts with the control flow method of executing a conventional program, in which instructions execute sequentially in the order specified by the programmer. Another way to say the same thing is that the flow of traditional sequential code is instruction driven, while the data flow of a VI is data driven.

Consider the VI block diagram, shown in Figure 2.31, that adds two numbers and then computes the sine of the result. In this case, the block diagram executes from left to right, not because the objects are placed in that order, but because one of the inputs of the Sine & Cosine function is not valid until the Add function has added the numbers together and passed the data to the Sine & Cosine function. Remember that a node (in this case, the Sine function) executes only when data is available at all of its input terminals, and it supplies data to its output terminals only when it finishes execution. Open DataFlowA.vi located in the Chapter 2 folder in the Learning directory, press on execution highlighting and then run.

Consider the example in Figure 2.32. Which code segment would execute first—the one on the left or the one on the right? You cannot determine the answer just by looking at the codes. The one on the left does not necessarily execute first. In a situation where one code segment must execute before another, and there is no type of dependency between the functions, you must use a Sequence structure to force the order of execution. To observe the data flow on the code in Figure 2.32, open the DataFlowB.vi located in the folder Chapter 2 in the Learning directory. Before running the VI, click on the **Highlight Execution** button and watch the flow of the execution.

When ready, click here   Click here to begin
to begin execution.   execution highlighting.

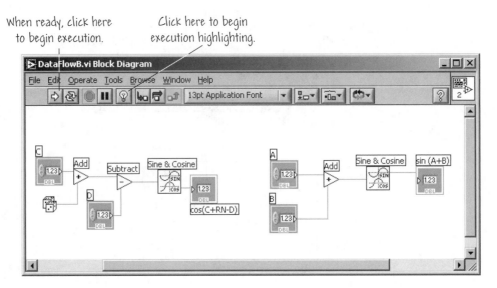

FIGURE 2.32
Which code executes first?

2.7   BUILDING A VI USING EXPRESS VIS

In this section, you will create your first VI using Express VIs. The objective is to construct a VI that generates a sawtooth signal at an amplitude that we prescribe on the front panel and displays the sawtooth signal graphically on the front panel. LabVIEW provides a VI template containing information that will help you in building this VI.

At the LabVIEW startup screen (see Figure 1.1), select the **New** button to display the **New** dialog box. Select **VI from Template≫Simulated≫Generate and Display** in the **Create new** list (see Section 1.9 and Figure 1.25). This template VI generates and displays a signal.

1. Click the **OK** button to open the template. You also can double-click the name of the template VI in the **Create new** list to open the template.

2. Examine the front panel of the VI, as shown in Figure 2.33. The front panel appears with a gray background and includes a waveform graph and a stop button. The title bar of the front panel indicates that this window is the front panel for the Generate and Display [Untitled] VI.

*If the front panel is not visible, you can display the front panel by either selecting **Window≫Show Front Panel**. Press the <Ctrl-E> keys to switch from the front panel to the block diagram or from the block diagram to the front panel.*

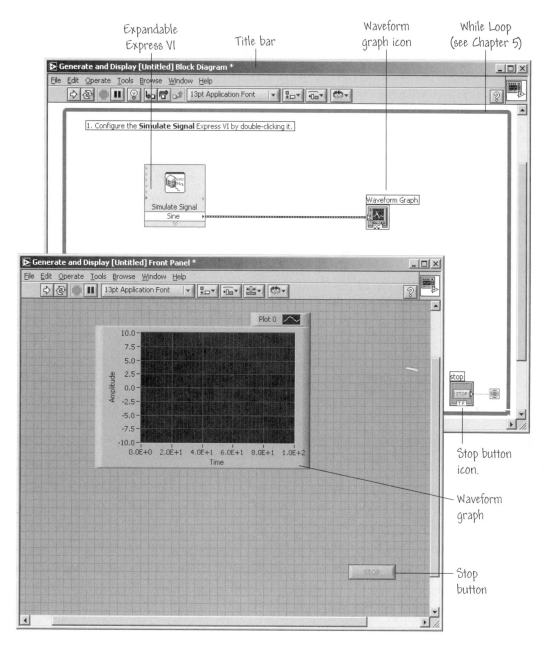

FIGURE 2.33
Building a VI using the Simulate Signal Express VI.

3. Examine the block diagram of the VI shown in Figure 2.33. The block diagram appears with a white background and includes the **Simulate Signal** Express VI, a waveform graph icon, a stop button icon, and a While Loop (more on loops in Chapter 5). The title bar of the block diagram indicates that this window is the block diagram for the Generate and Display [Untitled] VI.

*If the block diagram is not visible, you can display the block diagram by selecting **Window≫Show Block Diagram**.*

4. On the front panel toolbar, click the **Run** button and verify that a sine wave appears on the graph.

5. Stop the VI by clicking the **STOP** button located at the bottom right hand-side of the front panel.

*Although the **Abort Execution** button may seem, at first glance, to operate like a stop button, the **Abort Execution** button does not always properly close the VI. It is recommended to always stop your VIs using the **STOP** button on the front panel. Use the **Abort Execution** button only when errors prevent you from terminating the application using the **STOP** button.*

We can now add other elements to the VI template to begin the process of constructing the VI. First, we will add a **Control** to the front panel to use to vary the sawtooth signal amplitude. You can think of controls on the front panel as input devices on a physical instrument that are used to vary key instrument parameters. In our VI, the control will supply the sawtooth amplitude data to the block diagram.

6. On the front panel, select **Window≫Show Controls Palette** to display the **Controls** palette.

7. Move the cursor over the icons on the **Controls** palette to locate the **Numeric Controls** palette (should be located on the left-most area of the first row as shown in Figure 1.23). Notice that when you move the cursor over icons on the **Controls** palette, the name of that subpalette appears in the gray space above all the icons on the palette. When you idle the cursor over any icon on any palette, the full name of the subpalette, control, or indicator appears.

8. Click the **Numeric Controls** icon to access the **Numeric Controls** palette.

9. Since many physical instruments have knobs to vary the operational parameters of the instrument, we will select the **Knob** control on the **Numeric Controls** palette and place it on the front panel to the left of the waveform graph, as shown in Figure 2.34. This knob will provide the control over the amplitude of the sawtooth signal once it is properly wired.

10. Select **File≫Save As** and save this VI as Acquiring a Signal.vi in the Users Stuff folder in the Learning directory. Notice that the VI name appears in the closed brackets in the title bar (see Figure 2.34).

Save the VIs you edit or create in this book in the Learning directory in the Users Stuff folder.

Name of the VI is Acquiring a Signal.vi

Knob

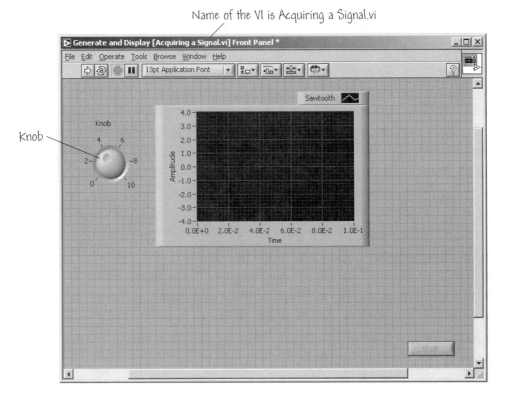

FIGURE 2.34
Adding a **Knob** to the Generate and Display template.

Examine the block diagram of the VI as it currently is configured. Notice that the block diagram has a blue icon labeled Simulate Signal. This icon represents the Simulate Signal Express VI, which simulates a sine wave by default. To meet our objectives, we must reconfigure the Express VI to simulate a sawtooth signal.

11. Display the block diagram by selecting **Window≫Show Block Diagram**. The Simulate Signal Express VI depicted in Figure 2.35 simulates a signal based on the configuration that you specify. We must interact with the Express VI dialog box to output a sawtooth signal, since the default is a sine wave signal.

12. Right-click the Simulate Signal Express VI and select **Properties** from the shortcut menu to display the **Configure Simulate Signal** dialog box, as shown in Figure 2.35.

13. Select **Sawtooth** from the **Signal type** pull-down menu. Notice that the waveform on the graph in the **Result Preview** section changes to a sawtooth wave.

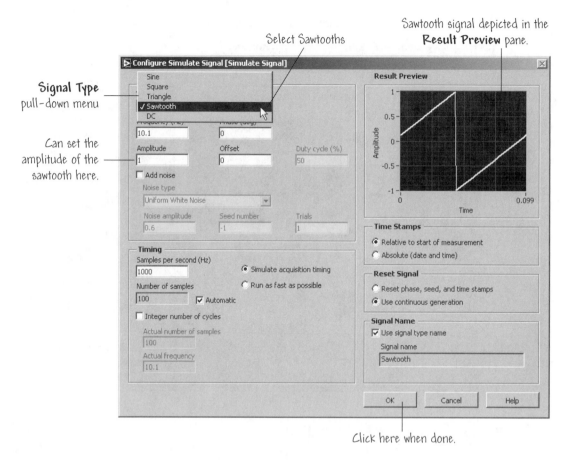

FIGURE 2.35
The **Configure Simulate Signal** dialog box.

14. Click the **OK** button to apply the current configuration and close the **Configure Simulate Signal** dialog box.

15. Now we want to expand the Simulate Signal Express VI to show the inputs and outputs to make the wiring easier. Move the cursor over the down arrows at the bottom of the Simulate Signal Express VI as illustrated in Figure 2.36.

16. When a double-headed arrow appears, click and drag the border of the Express VI until the **Amplitude** input appears. Because the **Amplitude** input appears on the block diagram, you can configure the amplitude of the sawtooth wave on the block diagram.

In Figure 2.37, observe how the signal parameter **Amplitude** is an option in the **Configure Simulate Signal** dialog box. When inputs, such as **Amplitude**, appear on the block diagram and in the configuration dialog box, you can configure the inputs in either location.

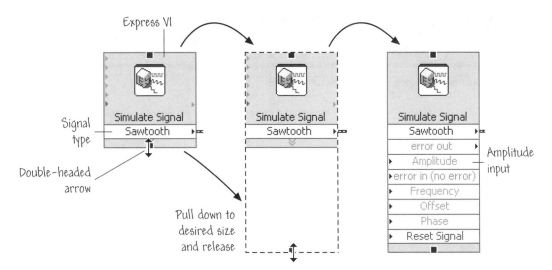

FIGURE 2.36
Input and Outputs of the **Simulate Signal** Express VI.

Now we can complete the VI by finishing up the wiring. To use the **Knob** control to change the amplitude of the signal, the **Knob** must be wired to the **Amplitude** input on the Simulate Signal Express VI.

17. Move the cursor over the **Knob** terminal until the **Positioning** tool appears (see Figure 1.22 for review of tools). The **Positioning** tool, represented by an arrow, is used to select, position, and resize objects.

18. Click the **Knob** terminal to select it, then drag the terminal to the left of the Simulate Signal Express VI. Make sure the **Knob** terminal is inside the While Loop. Deselect the **Knob** terminal by clicking a blank space on the block diagram.

The cursor does not switch to another tool while an object is selected.

19. Move the cursor over the arrow on the right hand-side of the **Knob** terminal. This will result in the cursor becoming the **Wiring** tool. Now we can use the **Wiring** tool to wire the **Knob** to the Express VI.

20. When the **Wiring** tool appears, click the arrow and then click the **Amplitude** input of the Simulate Signal Express VI to wire the two objects together. When the wire appears and connects the two objects, then data can flow along the wire from the **Knob** to the Express VI. The final block diagram is shown in Figure 2.34.

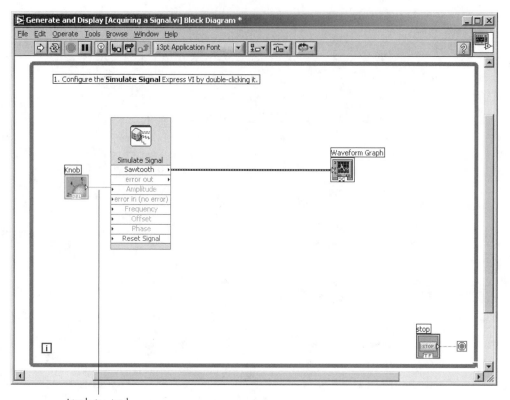

Knob is wired
to the Express VI at
the Amplitude input.

FIGURE 2.37
The block diagram of the completed VI to generate and display a signal.

21. Select **File≫Save** to save this VI.

    Now that the VI is ready for execution, we can see if we have successfully achieved the goal of generating a signal and displaying it graphically on the front panel.

22. Return to the front panel by selecting **Window≫Show Front Panel** and click the **Run** button.

23. Move the cursor over the knob control. Notice how the cursor becomes the **Operating** tool. The **Operating** tool is used to change the value of the sawtooth amplitude.

24. Using the **Operating** tool, turn the knob to adjust the amplitude of the sawtooth wave. Notice how the amplitude of the sawtooth wave changes as you turn the knob. Also notice that the y-axis on the graph autoscales to account for the change in amplitude. To indicate that the VI is running, the

Run button changes to a darkened arrow, shown at left. You cannot edit the front panel or block diagram while the VI runs.

25. Click the **STOP** button to stop the VI when you are finished experimenting.

Your VI development is now finished. The utility of using the provided VI templates is evident in this example. We were able to start from a point much closer to the desired final VI using the Generate and Display VI than starting with a blank VI. Also, the utility of the Express VIs is demonstrated with the Simulate Signal VI. This VI was easily configured to simulate a sawtooth signal and provided a quick and easy solution to our problem. Clearly, it would have taken more effort to construct the VI from scratch.

BUILDING BLOCK

2.8   BUILDING BLOCKS: DISPLACEMENT, VELOCITY, AND ACCELERATION

Like the first building block in Chapter 1, this is an exercise in opening and running an existing VI. The VI that you should open is called Displacement Velocity & Acc.vi and is included in the Building Blocks folder in the Learning directory. Find the VI and open it. The front panel is shown in Figure 2.38. This VI takes four input values, one of which is a string input: equation for displacement. We will discuss strings in Chapter 9 in detail. The default input string is $\sin(t)\exp(t/20)$. Run the VI and observe the output graphs. You should see an increasing-magnitude sinusoidal output for the displacement, velocity, and acceleration. Using the **Operating** tool, run several individual experiments by changing the input equation to the following:

- $\sin(t)\exp(-t/20)$
- $\sin(2t)/t$

To change the input equation, highlight the existing equation with the **Operating** tool, type in the desired equation, and then click on the **Enter** button on the front panel toolbar.

Switch to the block diagram window by selecting **Show Block Diagram** from the **Window** pull-down menu. The block diagram is depicted in Figure 2.38.

Locate the terminals associated with the four controls: number of points, time start, time end, and equation for displacement. Notice that the terminals all have thick borders and that the blue number of points terminal contains **I32** to indicate it is a long integer. The two terminals time start and time end are orange and contain the label DBL indicating double-precision floating-point numbers. What color is the border of the string control equation for displacement? It should be pink and contain the label abc indicating a string.

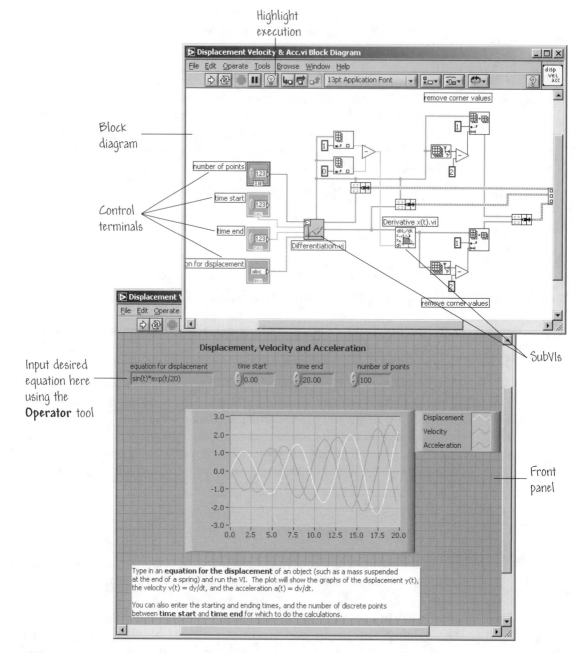

FIGURE 2.38
The Displacement Velocity & Acc.VI.

The block diagram contains two subVIs: Differentiation.vi and Derivative x(t).vi. Find them on the block diagram. If you want to see what the subVI looks like internally, simply double-click on its icon using the **Operating** tool. In case you accidentally change something in the subVI after opening it up, make sure to not save the changes when prompted upon closing the subVI!

Other interesting objects to note are the wires. Notice that some wires are solid, thick orange, some are thick pink and others are thin blue. Refer to Table 2.1 to determine the data type carried by each wire shown in Figure 2.38.

As a final experiment with this VI, on the block diagram select **Highlight Execution** and run the VI. Observe the data flow through the VI. Does the data flow strictly from left to right? You should see that the data flows at various points in the block diagram—and not left to right.

2.9   RELAXED READING: LABVIEW AUTOMATES BRAIN WAVE EXPERIMENTS

Brain potentials are divided into spontaneous and event-related. The spontaneous result from the regular brain activity (EEG potentials). The event-related potentials (ERP) result from external brain excitation (events) and are divided into evoked and anticipatory. Evoked potentials appear *after* excitation as a reflex of the brain. Anticipatory potentials appear *before* the corresponding event and represent an expectation of an excitation.

The most prominent example of the expectation-related potential is the contingent negative variation (CNV) potential extracted from a CNV experiment. In the CNV experiment, two brain stimuli, denoted by S_1 and S_2, are applied to the subject with a constant interval between stimuli. S_1 is a warning stimulus, and S_2 is an imperative stimulus to which the subject must react. The procedure is repeated tens of times, during which an ERP is produced in the EEG trace between the stimuli and shapes itself toward a specific CNV wave. The ERP after 10 to 20 trials can clearly show both the evoked (short) potential due to S_1 as well as the anticipatory (late, expectancy) potential together with the preparatory potential prior and due to S_2.

The dynamic CNV experiment is an extension of the CNV experiment involving switching S_2 on and off. The switching occurs automatically after fulfilling certain conditions in the experiment's environment, thus forcing a cyclic process of building and degrading of the CNV wave. The subject is not informed about the nature of both stimuli, so the expectation of appearance (absence) of S_2

during the experiment completely corresponds to the learning process. The CNV wave can be qualified by one of its parameters, such as amplitude or slope. After the experiment, a statistical curve of the qualifying parameter is drawn across the trials. This statistical curve is denoted as the electroexpectogram (EXG) and directly presents the subject's cognitive capabilities.

The AEP Research Tool was written in LabVIEW to perform the data acquisition, signal processing, analysis of the data traces, and reporting. The hardware used in the experiment consisted of a bandpass amplifier for mV ranges, an AT-MIO-E board, a sound card for applying the stimuli, and a push button switch with a TTL interface. The system acquires brain wave activity data on two differential analog channels.

The stimulus S_1 is an audio 1 kHz warning beep and S_2 is a longer 2 kHz imperative beep. It is essential that the subject be aware neither of the nature nor of the number of the stimuli. The data acquisition lasts for 7s. The stimulus S_1 is issued at 1s into the acquisition phase; S_2 is issued at 3s, if applied by the algorithm. The number of trials in the experiment is set to maximum 100 successful. The gap between two consecutive trials varies from 12 to 15s. During the experiment, the subject learns about the number, nature, and order of the stimuli, thus demonstrating the process of learning by shaping the ERP wave toward the expected CNV. The subject has to react upon hearing S_2 by pressing the button and immediately stopping it.

The main front panel of the AEP Research Tool is shown in Figure 2.39. The front panel displays the acquired EEG signal, the extracted CNV potential and

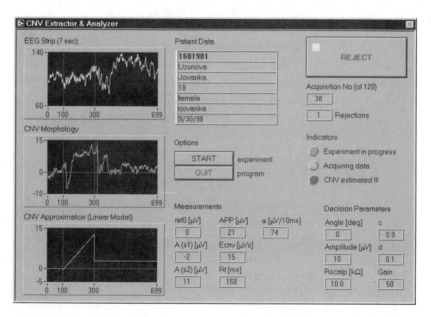

FIGURE 2.39
Main DAQ Front Panel.

its linearized model, as well as the required measurements and calculated values. The audio stimulation is performed through a sound card. The reaction time is measured by an onboard counter, started by a digital output from the card issuing pulse at the same moment with the start of S_2 and stopped by the user pressing the button or the time-out pulse, applied again by the same digital line.

The EXG curve represents a cognitive wave obtained from the human brain showing the oscillatory change of the expectancy status in the human brain and is a manifestation of the expectation process and the learning process taking place in the human brain. The AEP Research Tool successfully performs the expected task and produces an excellent nine-page report containing all required statistics. Experiments during the test phase indicated that different categories of subjects (healthy adults, groups of people with distinctive neurological disturbances, and little children) may produce quite different electroexpectograms, but very similar within the groups, because some are not able to raise their CNV waves above the thresholds. The design of this experiment illustrates the possibilities of LabVIEW-based systems in the medical field.

For more information, please contact:

Roman Golubovski
roman.golubovski@iname.com

2.10   SUMMARY

Virtual instruments (VIs) are the building blocks of LabVIEW. The graphical programming language is known as the G programming language. VIs have three main components: the front panel, the block diagram, and the icon and connector pair. VIs follow a data flow programming convention in which each executable node of the program executes only after all the necessary inputs have been received. Correspondingly, output from each executable node is produced only when the node has finished executing.

KEY TERMS

Boolean controls and indicators: Front panel objects used to manipulate and display or input and output Boolean (True or False) data.

Connector: Part of the VI or function node that contains its input and output terminals, through which data passes to and from the node.

Connector pane: Region in the upper right corner of the front panel that displays the VI terminal pattern. It underlies the icon pane.

Data flow programming: Programming system consisting of executable nodes in which nodes execute only when they have received all required input data and produce output automatically when they have executed.

G programming language: Graphical programming language used in LabVIEW.

Icon: Graphical representation of a node on a block diagram.

Icon pane: Region in the upper right corner of the front panel and block diagram that displays the VI icon.

Input terminals: Terminals that emit data. Sometimes called source terminals.

Modular programming: Programming that uses interchangeable computer routines.

Numeric controls and indicators: Front panel objects used to manipulate and display numeric data.

Output terminals: Terminals that absorb data. Sometimes called destination terminals.

String controls and indicators: Front panel objects used to manipulate and display or input and output text.

Terminals: Objects or regions on a node through which data passes.

Tip strips: Small yellow text banners that identify the terminal name and make it easier to identify function and node terminals for wiring.

> **Wire**: Data path between nodes.
>
> **Wiring tool**: Tool used to define data paths between source and sink terminals.

EXERCISES

E2.1   In this first exercise you get to play a drawing game. Open Spirograph.vi located in Chapter 2 of the Learning directory. The front panel should look like the one shown in Figure E2.1. Did you ever use a spirograph drawing kit as a kid? This VI is the computer version of that children's game. To play the drawing game, run the VI and experiment by varying the controls. After clicking on the **Run** button to start the VI execution, click on the Begin drawing push button on the front panel. Once the drawing starts, the Stop drawing label appears in the push button. The VI execution is halted by clicking on the Stop drawing button.

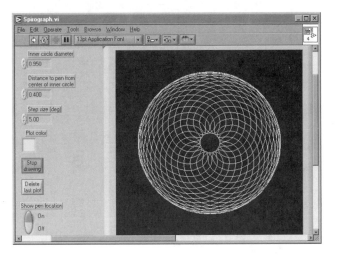

FIGURE E2.1
The Spirograph.vi front panel.

E2.2   Construct a VI that uses a round push button control to turn on a square light indicator whenever the push button is depressed. The front panel is very simple and should look something like the one shown in Figure E2.2.

E2.3   Open SimPhone.vi, located in Examples\Sound\sndExample adv.llb. The front panel and block diagram are shown in Figure E2.3. Using the **Context Help** (Ctrl-H), determine the inputs and outputs of the Snd Write Waveform subVI. Sketch the subVI icon and connector showing the inputs and outputs.

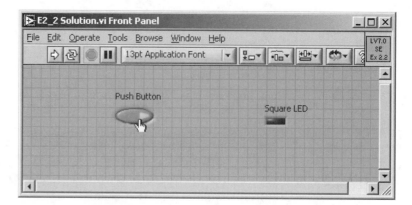

FIGURE E2.2
A front panel using a round push button control and a square light indicator.

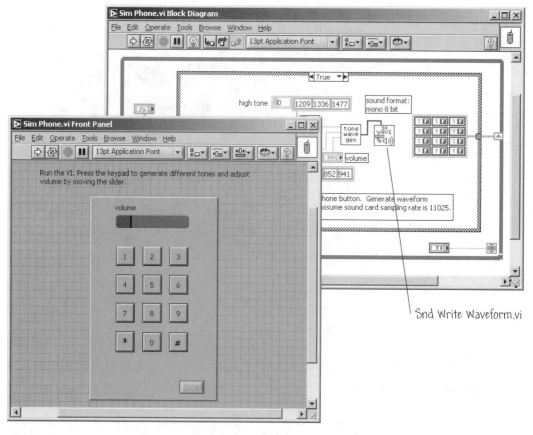

FIGURE E2.3
The SimPhone VI.

PROBLEMS

P2.1   Complete the crossword puzzle.

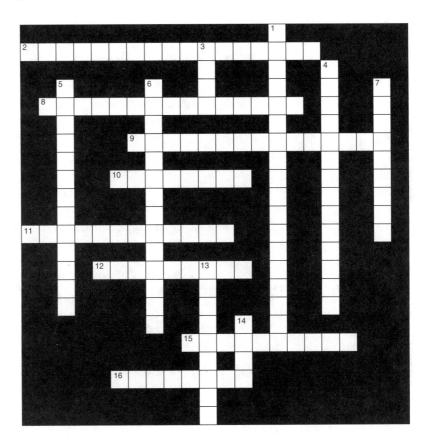

Across
2. Terminals that absorb data.
8. Front panel objects used to input Boolean data.
9. Terminals that emit data.
10. Programming system in which nodes execute only when they have received all required input data.
11. Graphical programming language used in LabVIEW.
12. Part of the VI or function node that contains terminals through which data passes to and from the node.
15. Tool used to define data paths between source and sink terminals.
16. Region in the upper right corner of the front panel and block diagram that displays the VI icon.

Down
1. Front panel objects used to manipulate and display numeric data.
3. Graphical representation of a node on a block diagram.
4. Front panel objects used to manipulate and display or input and output text.
5. Region in the upper right corner of the front panel that displays the VI terminal pattern.
6. Programming that uses interchangeable computer routines.
7. Small yellow text banners that identify the terminal name and make it easier for wiring.
13. Objects or regions on a node through which data passes.
14. Data path between nodes.

P2.2   Construct a VI that performs the following tasks:

- Takes two floating-point numbers as inputs on the front panel: X and Y.
- Subtracts Y from X and displays the result on the front panel.
- Divides X by Y and displays the result on the front panel.
- If the input $Y = 0$, a front panel LED lights up to indicate division by zero.

Name the VI Subtract and Divide.vi and save it in the Users Stuff folder in the Learning directory.

P2.3   Construct a VI that uses a vertical slide control for input and a meter indicator for output display. A front panel and block diagram that can be used as a guide are shown in Figure P2.3. Referring to the block diagram, you see a pair of dice, which is the icon for a random number function. You will find the random number function on the Functions≫Numeric palette. When running the VI, any input you provide via the vertical slide will be reflected on the meter indicator. The random number function adds "noise" to the input so that the meter output will not be exactly the same as the input. Run the VI in **Run Continuously** mode and vary the slide input.

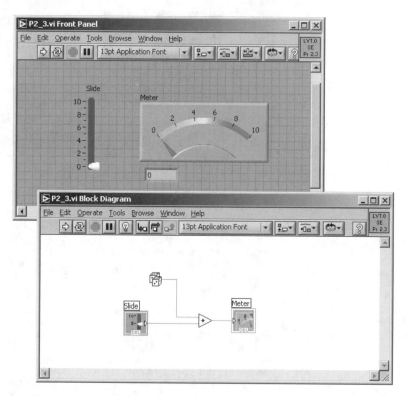

FIGURE P2.3
Using a vertical slide control and a meter indicator.

P2.4   Create a VI that has a numeric control to input a number x, uses the add and multiply functions to calculate $3x^2 + 2x + 5.0$ and display the output using a numeric indicator.

P2.5   Using a VI template and modifying Express VIs, create a program that generates a Triangle Wave with a frequency of 125 Hz and added noise.

C H A P T E R 3

Editing and Debugging Virtual Instruments

Like text-based computer programs, virtual instruments are dynamic. VIs change as their applications evolve (usually increasing in complexity). For instance, a VI that initially only performs addition may at a later time be updated to add a multiplication capability. You need debugging and editing tools to verify and test VI coding changes. Since programming in G is graphical in nature, editing and debugging are also graphical, with options available in pull-down and short cut menus and various palettes. In this chapter, we discuss how to create, select, delete, move, and arrange objects on the front panel and block diagrams. The important topic of selecting and deleting wires and locating and eliminating bad wires is also presented. Debugging subjects covered include execution highlighting (you can watch the code run!), single-stepping through code, and inserting probes to view data as the VI executes.

GOALS

1. Learn to access and practice with VI editing tools

2. Learn to access and practice with VI debugging tools

3.1   EDITING TECHNIQUES

3.1.1   Creating Controls and Indicators on the Block Diagram

As discussed in previous chapters, when building a VI you can create controls and indicators on the front panel and know that their terminals will automatically appear on the block diagram. Switching to the block diagram, you can begin wiring the terminals to functions (such as addition or multiplication functions), subVIs, or other objects. In this section we present an alternative method to create and wire controls and indicators in *one* step on the block diagram. In the following discussions, you should open a new VI and follow along by repeating the steps as presented.

1. Open a new VI and switch to the block diagram.

2. Place the square root function located in the **Numeric** subpalette on the **Arithemetic & Comparison** palette on the block diagram, as shown in Figure 3.1.

3. Now we want to add (and wire) a control terminal to the square root function. As shown in Figure 3.2, you pop up on the left side of the square root function and select **Control** from the **Create** menu. The result is that a control terminal is created and automatically wired to the Square Root function. Switch to the front panel and notice that a numeric control has appeared!

*To change a control to an indicator (or vice versa), pop up on the terminal (on the block diagram) or on the object (on the front panel) and select **Change to Indicator** (or **Change to Control**).*

4. In fact, you can wire indicators and constants by popping up on the node terminal and choosing the desired selection. For example, popping up on the right side of the square root function and selecting **Indicator** from the **Create** menu creates and automatically wires a front panel indicator, as illustrated in Figure 3.3. In many situations, you may wish to create and automatically wire a constant to a terminal, and you accomplish this by popping up and choosing **Constant** from the **Create** menu.

*To change a control or indicator (for example, from a knob to a dial), pop up on the object on the front panel and select **Replace**. The **Controls** palette will appear, and you can navigate to the desired new object.*

5. The resulting control and indicator for the square root function are shown in Figure 3.4.

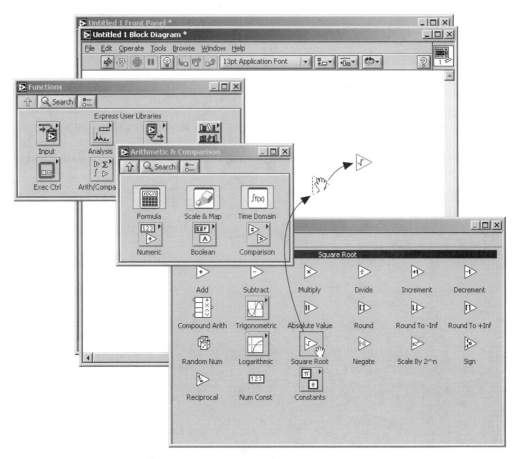

FIGURE 3.1
Add the square root function to the block diagram.

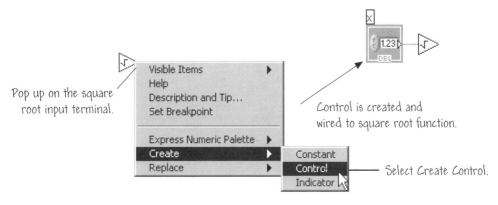

FIGURE 3.2
Front panel control created for the Square Root function.

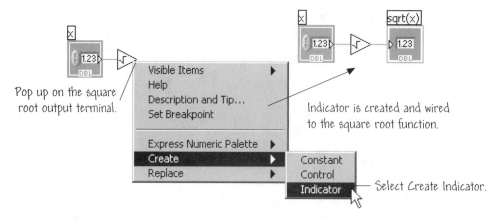

FIGURE 3.3
Front panel indicator created for the Square Root function.

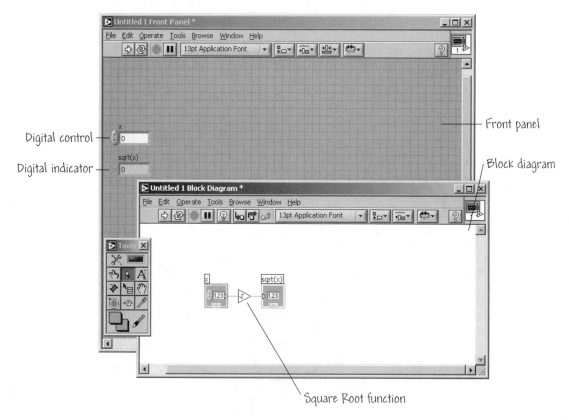

FIGURE 3.4
Front panel and block diagram for the Square Root function.

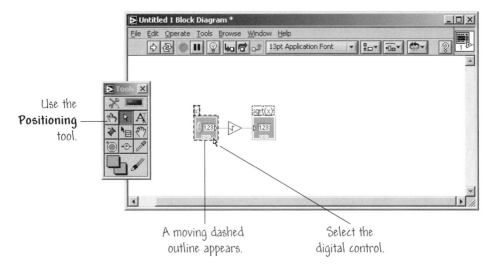

FIGURE 3.5
Selecting an object using the **Positioning** tool.

3.1.2   Selecting Objects

> The **Positioning** tool selects objects in the front panel and block diagram windows. In addition to selecting objects, you use the **Positioning** tool to move and resize objects (more on these topics in the next several sections). To select an object, click the left mouse button while the **Positioning** tool is over the object. When the object is selected, a surrounding dashed outline appears, as shown in Figure 3.5. To select more than one object, shift-click (that is, hold down the <shift> key and simultaneously click) on each additional object you want to select. You also can select multiple objects by clicking in a nearby open area and dragging the cursor until all the desired objects lie within the selection rectangle that appears (see Figure 3.6).

Sometimes after selecting several objects, you may want to deselect just one of the objects (while leaving the others selected). This is accomplished by shift-clicking on the object you want to deselect—the other objects will remain selected, as desired.

3.1.3   Moving Objects

You can move an object by clicking on it with the **Positioning** tool and dragging it to a desired location. The objects of Figure 3.6 are selected and moved as shown in Figure 3.7. Selected objects can also be moved using the up/down and right/left arrow keys. Pressing the arrow key once moves the object one pixel; holding down the arrow key repeats the action. In this manner, you can move and locate your objects very precisely.

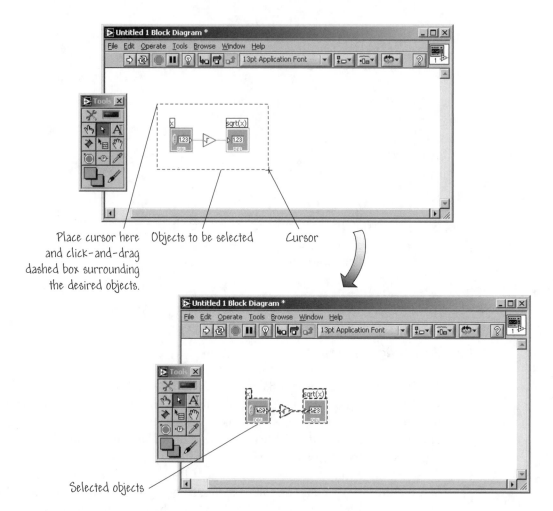

FIGURE 3.6
Selecting a group of objects.

 The direction of movement of an object can be restricted to being either horizontal or vertical by holding down the <shift> key when you move the object. The direction you initially move decides whether the object is limited to horizontal or vertical motion.

If you change your mind about moving an object while you are in the midst of dragging it to another location, continue to drag until the cursor is outside all open windows and the dashed line surrounding the selected object disappears—then release the mouse button. This will cancel the move operation and the object will not move. Alternatively, if the object is dragged and dropped to an undesirable

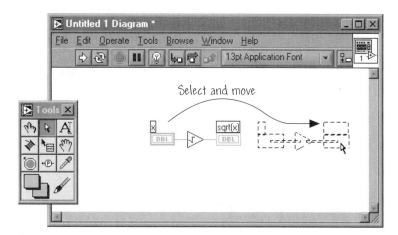

FIGURE 3.7
Selecting and moving a group of objects.

location, you can select **Undo Move** from the **Edit** menu to undo the move operation.

3.1.4   Deleting and Duplicating Objects

You can delete objects by selecting the object(s) and choosing **Clear** from the **Edit** menu or pressing the <backspace> key in Windows or the <delete> key on the Macintosh. Most objects can be deleted; however you cannot delete certain components of a control or indicator, such as the label or digital display. You must hide these components by popping up and deselecting the Visible Items≫ Label or Visible Items≫Digital Display from the short cut menu.

Most objects can be duplicated, and there are three ways to duplicate an object—by copying and pasting, by cloning, and by dragging and dropping. In all three cases, a complete new copy of the object is created, including, for example, the terminal belonging to a front panel control and the control itself.

 You can copy text and pictures from other applications and paste them into LabVIEW.

To clone an object, click the **Positioning** tool over the object while pressing <ctrl> for Windows (or <option> on the Mac) and drag the object to its new location. After you drag the selection to a new location and release the mouse button, a copy of the object appears in the new location, and the original object remains in the original location. When you clone or copy objects, the copies are labeled by the same name with the word *copy* appended (or *copy 1*, *copy 2*, and so forth for copies of copies).

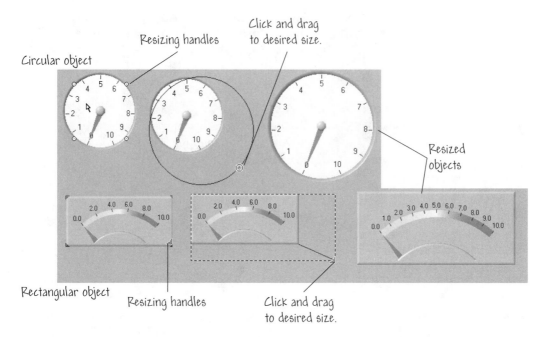

FIGURE 3.8
Resizing rectangular and circular objects.

You also can duplicate objects using Edit≫Copy and then Edit≫Paste from the **Edit** menu. First, place the **Positioning** tool over the desired object and choose Edit≫Copy. Then click at the location where you want the duplicate object to appear and choose Edit≫Paste from the **Edit** menu. You can use this process to copy and paste objects within a VI or even copy and paste objects between VIs.

The third way to duplicate objects is to use the drag-and-drop capability. In this way, you can copy objects, pictures, or text between VIs and from other applications. To drag and drop an object, select the objects, pictures, or text file with the **Positioning** tool and drag it to the front panel or block diagram of the target VI. You can also drag VIs from the file system (in Windows and on the Mac) to the active block diagram to create subVIs (we will discuss subVIs in Chapter 4).

3.1.5   Resizing Objects

You can easily resize most objects. **Resizing handles** appear when you move the **Positioning** tool over a resizable object, as illustrated in Figure 3.8. On rectangular objects the resizing handles appear at the corners of the object; resizing circles appear on circular objects. Passing the **Positioning** tool over a resizing handle transforms the tool into a resizing cursor. To enlarge or reduce the size of an object, place the **Positioning** tool over the resizing handle and click and drag the resizing cursor until the object is the desired size. When you release the mouse button, the object reappears at its new size. The resizing process is illustrated in Figure 3.8.

To cancel a resizing operation, continue dragging the frame corner outside the active window until the dotted frame disappears—then release the mouse button and the object will maintain its original size. Alternately, if the object has already been resized and you want to undo the resizing, you can use the **Undo Resize** command found in the **Edit** pull-down menu.

Some objects only change size horizontally or vertically, or keep the same proportions when you resize them (e.g., a knob). In these cases, the resizing cursor appears the same, but the dotted resize outline moves in only one direction. To restrict the resizing of any object in the vertical or horizontal direction (or to maintain the current proportions of the object) hold down the <shift> key as you click and drag the object.

3.1.6   Labeling Objects

Labels are blocks of text that annotate components of front panels and block diagrams. There are two kinds of labels—free labels and owned labels. Owned labels belong to and move with a particular object and describe that object only. You can hide these labels but you cannot copy or delete them independently of their owners. Free labels are not attached to any object, and you can create, move, or dispose of them independently. Use them to annotate your front panels and block diagrams. Free labels are one way to provide accessible documentation for your VIs.

You use the **Labeling** tool to create free labels or to edit either type of label. To create a free label, choose the **Labeling** tool from the **Tools** palette, and then click anywhere in an open area and type the desired text in the bordered box that appears. An example of creating a free label is shown in Figure 3.9.

When finished entering the text, click on the **Enter** button on the toolbar, which appears on the toolbar to remind you to end your text entry. You can also end your text entry by pressing the <Enter> key on the numeric keypad (if you have a numeric keypad). Remember that if you do not type any text in the label, the label disappears as soon as you click somewhere else.

When you add a control or an indicator to the front panel, an owned label automatically appears. The label box is ready for you to enter the desired text. If you do not enter text immediately, the default label remains. To create an owned label for an existing object, pop up on the object and select Visible Items≫Label from the short cut menu (see Figure 1.13). You can then enter your text in the bordered box that appears. If you do not enter the text immediately, the label disappears.

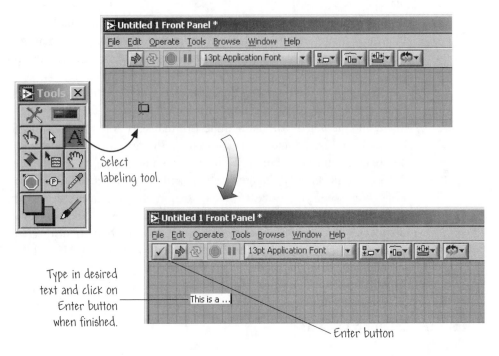

FIGURE 3.9
Creating a free label.

If you place a label or any other object over (or partially covering) a control or indicator, it slows down screen updates and could make the control or indicator flicker. To avoid this problem, do not overlap a front panel object with a label or other objects.

You can copy the text of a label by double-clicking on the text with the **Labeling** tool or dragging the **Labeling** tool across the text to highlight the desired text. When the desired text is selected, choose Edit≫Copy to copy the text onto the clipboard. You can then highlight the text of a second label and use Edit≫Paste to replace the highlighted text in the second label with the text from the clipboard. To create a new label with the text from the clipboard, click on the screen with the **Labeling** tool where you want the new label positioned and then select Edit≫Paste.

The resizing technique described in the previous section also works for labels. You can resize labels as you do other objects by using the resizing cursor. Labels normally autosize; that is, the label box automatically resizes to contain the text you enter. If for some reason you don't want the text in labels to automatically resize to the entered text, pop up on the label and select **Size to Text** to toggle autosizing off.

The text in a label remains on one line unless you enter a carriage return to resize the label box. By default, the <enter> (or <return>) key is set to add a new line. This can be changed in Tools≫Options≫Front Panel so that the text input is terminated with the <enter> (or <return>) key.

3.1.7   Changing Font, Style, and Size of Text

Using the **Text Settings** in the toolbar, you can change the font, style, size, and alignment of any text displayed in a label or on the display of controls and indicators. Certain controls and indicators display text in multiple locations—for example, on graphs (a type of indicator), the graph axes scale markers are made up of many numbers, one for each axes tick. You have the flexibility to modify each text display independently.

Text settings were discussed in Chapter 1 and is shown in Figure 1.7. Notice the word **Application** showing in the **Text Settings** pull-down menu, which also contains the **System**, **Dialog**, and **Current** options. The last option in the ring— the **Current Font**—refers to the last font style selected. The predefined fonts are used for specific portions of the various interfaces:

- The **Application** font is the default font. It is used for the **Controls** and **Functions** palettes.
- The **System** font is the font used for menus.
- The **Dialog** font is the font used for text in dialog boxes.

These fonts are predefined so that they map "best" when porting your VIs to other platforms.

Text Settings has size, style, justify, and color options. Selections made from any of these submenus (that is, size, style, etc.) apply to all selected objects. For example, if you select a new font while you have a knob selected, the labels, scales, and digital displays all change to the new font. Figure 3.10 illustrates the situation of changing the style of all the text associated with a knob from plain text to bold text.

If you select any objects or text and make a selection from the **Text Settings** pull-down menu, the changes apply to everything selected. The process of selecting just the text of an owned label of a knob is illustrated in Figure 3.11. Once the desired text is selected (remember to use the **Labeling** tool to select the desired text), you can make any changes you wish by selecting the proper pull-down submenu from the **Text Settings**. If no text is selected, the font changes apply to the default font, so that labels created from that point on will reflect the new default font, while not affecting the font of existing labels.

When working with objects that have multiple pieces of text (e.g. slides and knobs), remember that text selections affect the objects or text currently selected. For example, if you select the entire knob while selecting bold text, the scale,

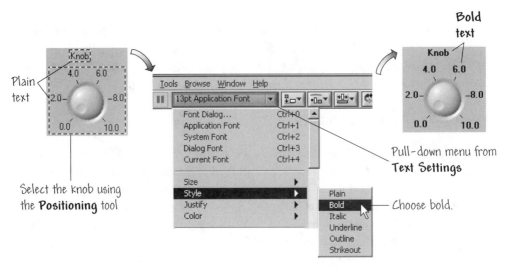

FIGURE 3.10
Changing the font style on a knob control.

digital display, and label all change to a bold font, as shown in Figure 3.10. As shown in Figure 3.12a, when you select the knob label, followed by selecting bold text from the **Style** submenu of the **Text Settings** pull-down menu, only the knob label changes to bold. Similarly, when you select text from a scale marker while choosing bold text, all the markers change to bold, as shown in Figure 3.12b.

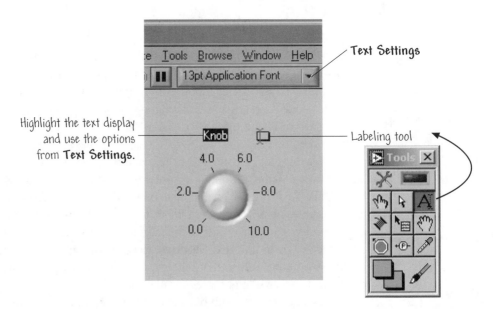

FIGURE 3.11
Selecting text for modification of style, font, size, and color.

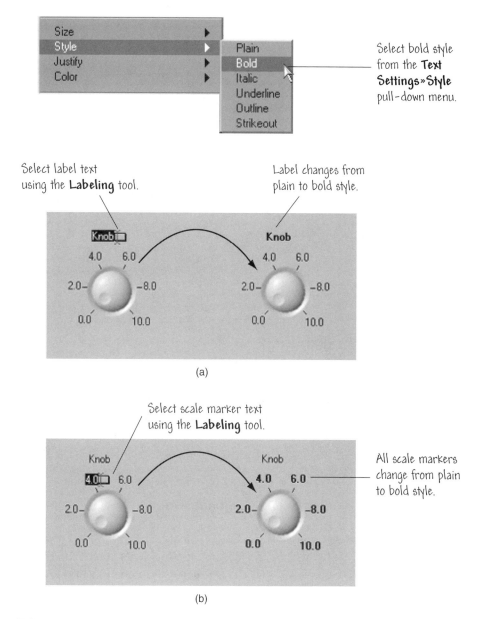

FIGURE 3.12
Changing text attributes on a knob.

If you select **Font Dialog**. . . in the **Text Settings** while a front panel is active, the dialog box shown in Figure 3.13 appears. If a block diagram is active instead, the **Panel Default** option at the bottom of the dialog box is checked. With either the **Panel Default** or **Diagram Default** checkbox selected, the other selections made in this dialog box will be used with new labels on the front panel or block diagram. In other words, if you click the **Panel Default** and **Diagram Default**

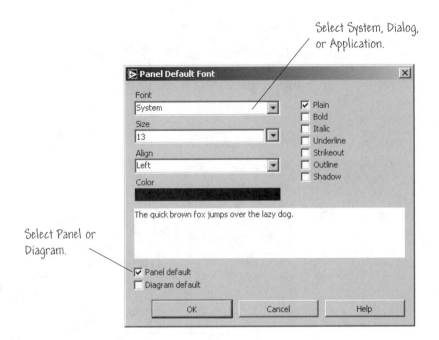

FIGURE 3.13
Using the **Font Dialog** box to change the font, size, alignment, color and style.

checkboxes, the selected font becomes the current font for the front panel, the block diagram, or both. The current font is used on new labels. The checkboxes allow you to set different fonts for the front panel and block diagram. For example, you could have a small font on the block diagram and a large one on the front panel.

3.1.8   Selecting and Deleting Wires

A single horizontal or vertical piece of wire is known as a **wire segment**. The point where three or four wire segments join is called a **junction**. A **wire branch** contains all the wire segments from one junction to another, from a terminal to the next junction, or from one terminal to another if there are no junctions in between. You select a wire segment by clicking on it with the **Positioning** tool. Clicking twice selects a branch, and clicking three times selects the entire wire. See Figure 3.14 for an example of selecting a branch, segment, or an entire wire.

3.1.9   Wire Stretching and Broken Wires

Wired objects can be moved individually or in groups by dragging the selected objects to a new location with the **Positioning** tool. The wires connecting the objects will stretch automatically. If you want to move objects from one diagram to another, the connecting wires will not move with the selected objects, unless

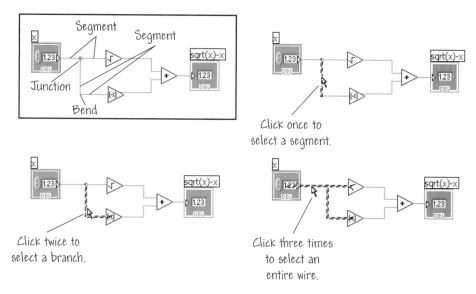

FIGURE 3.14
Selecting a segment, branch, or an entire wire.

you select the wires as well. After selecting the desired objects and the connecting wires, you can cut-and-paste the objects as a unit to another VI.

An example of stretching a wire is shown in Figure 3.15. In the illustration, the object (in this case, a numeric indicator) is selected with the **Positioning** tool and stretched to the desired new location.

Wire stretching occasionally leaves behind loose ends. Loose ends are wire branches that do not connect to a terminal. Your VI will not execute until you

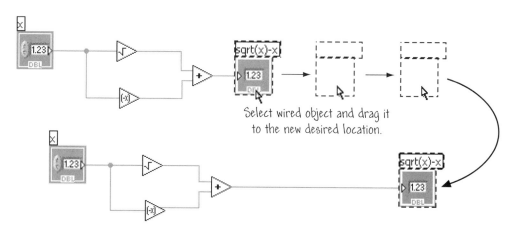

FIGURE 3.15
Moving wired objects.

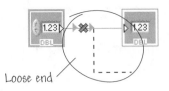

Loose end

FIGURE 3.16
Loose ends.

remove all loose ends. An example of a loose end is shown in Figure 3.16. Loose ends can be easily removed using the Edit≫Remove Broken Wires command.

When you make a wiring mistake, a broken wire (indicated by a dashed line with a red X) appears. Figure 3.17 shows a dashed line that represents a broken wire. It is inevitable that broken wires will occur in the course of programming in G. One common mistake is to attempt to connect two indicator terminals together or to connect a control terminal to an indicator terminal when the data types do not match (for example, connecting a Boolean to a numeric). If you have a broken wire, you can remove it by selecting it with the **Positioning** tool and eliminating it by pressing the <delete> key. If you want to remove all broken wires on a block diagram at one time, choose **Remove Broken Wires** from the **Edit** menu.

There are many different conditions leading to the occurrence of broken wires. Some examples are:

- **Wire type, dimension, unit, or element conflicts**—A wire type conflict occurs when you wire two objects of different data types together, such as a numeric and a Boolean, as shown in Figure 3.18a.

- **Multiple wire sources**—You can wire a control to multiple output destinations (or indicators), but you cannot wire multiple data sources to a single destination. In the example shown in Figure 3.18b, we have attempted to wire two data sources (that is, the random number and the constant ln 2) to one indicator. This produces a broken wire and must be fixed by disconnecting the random number (represented by the dice) or disconnecting the ln 2 constant. Another example of a common multiple sources error occurs during

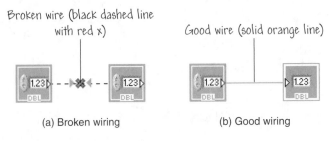

Broken wire (black dashed line with red x)

Good wire (solid orange line)

(a) Broken wiring   (b) Good wiring

FIGURE 3.17
Locating broken wires.

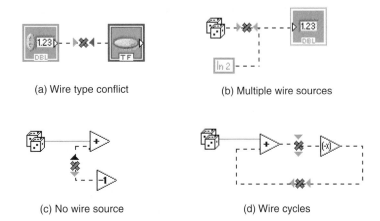

(a) Wire type conflict   (b) Multiple wire sources

(c) No wire source   (d) Wire cycles

FIGURE 3.18
Typical wiring errors leading to broken wire indications.

front panel construction when you inadvertently place a numeric control on the front panel when you meant to place a numeric indicator. If you do this and then try to wire an output value to the terminal of the front panel control, you will get a multiple sources error. To fix this error, just pop up on the terminal and select **Change To Indicator**.

- **No wire source** —An example of a wire with no source is shown in Figure 3.18c. This problem is addressed by providing a control. Another example of a situation leading to a no-source error is if you attempt to wire two front panel indicators together when one should have been a control. To fix this error, just pop up on a terminal and select **Change to Control**.

- **Wire cycles**—Wires must not form closed loops of icons or structures, as shown in Figure 3.18d. These closed loops are known as cycles. LabVIEW cannot execute cycles because each node waits on the other to supply it data before it executes (remember data flow!). Shift registers (discussed in Chapter 5) provide the proper mechanism to feed back data in a repetitive calculation.

Sometimes you have a faulty wiring connection that is not visible because the broken wire segment is very small or is hidden behind an object. If the **Broken Run** button appears in the toolbar, but you cannot see any problems in the block diagram, select Edit≫Remove Broken Wires—this will remove all broken wires in case there are hidden, broken wire segments. If the **Run** button returns, you have corrected the problem. If the wiring errors have not all been corrected after selecting Edit≫Remove Broken Wires, click on the **Broken Run** button to see a list of errors. Click on one of the errors listed in the **Error List** dialog box, and you will automatically be taken to the location of the erroneous wire in the block diagram. You can then inspect the wire, detect the error and fix the wiring problem.

LabVIEW automatically finds a wire route around existing objects on the block diagram when you are wiring components. The natural inclination of the automatic routing is to decrease the number of bends in the wire. When possible, automatically routed wires from control terminals exit the right side of the terminal, and automatically routed wires to indicator terminals enter the left side of the terminal.

If you find that you have a messy wiring situation on your block diagram you can easily fix it. You can right-click any wire and select **Clean Up Wire** from the shortcut menu to automatically route an existing wire. This may help in debugging your VI since it will be easier to see the wire routes on the block diagram.

If for some reason you want to temporarily disable automatic wire routing and route a wire manually, first use the **Wiring** tool to click a terminal and release the mouse. Then press the <A> key to temporarily disable automatic wire routing for the current wire. Click another terminal to complete the wiring. Once the wiring is complete, the automatic wire routing is resumed. You also can temporarily disable automatic routing after you click to start wiring by holding down the mouse button while you wire to another terminal and then releasing the mouse button. After you release the mouse button, automatic wire routing resumes.

3.1.10 Aligning, Distributing, and Resizing Objects

To align a group of objects, first select the desired objects. Then choose the axis along which you want to align them from **Align Objects** in the pull-down menu in the toolbar. You can align objects along the vertical axis using left, center, or right edge. You also can align objects along a horizontal axis using its top, center, or bottom edge. Open a new VI and place three objects (such as, three numeric controls) on the front panel and experiment with different aligning options. Figure 3.19 illustrates the process of aligning three objects by their left edges.

In a similar fashion, you can distribute a group of objects by selecting the objects and then choosing the axis along which you want to distribute the selected objects from the **Distribute Objects** pull-down menu in the toolbar (see Figure 1.9). In addition to distributing selected objects by making their edges or centers equidistant, four menu items at the right side of the ring let you add or delete gaps between the objects, horizontally or vertically. For example, the three objects shown in Figure 3.20 are distributed at different distances from each other. If we want them to be equally spaced by their top edges, we can use the **Distribute Objects** pull-down menu, as shown in Figure 3.20, to rearrange the elements at equal spacing.

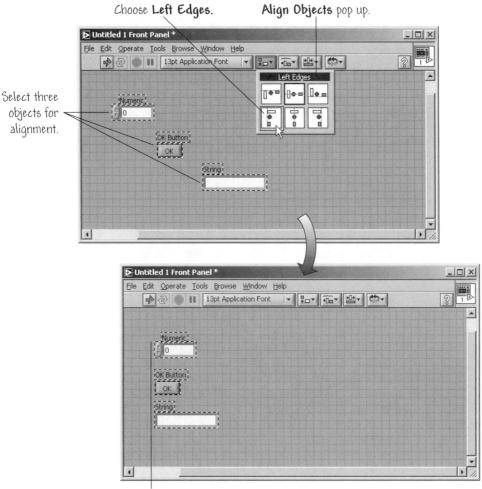

Choose **Left Edges**.

Align Objects pop up.

Select three objects for alignment.

Three objects are aligned by their left edges!

FIGURE 3.19
Aligning objects.

You can use the **Resize Objects** pull-down menu to resize multiple front panel objects to the same size. For example, in Figure 3.21, three objects of different widths are selected. Then using the **Resize Objects** pull-down menu, the **Maximum Width** icon is selected resulting in the three objects taking the width of the String, which originally had the largest width.

3.1.11 Coloring Objects

You can customize the color of many LabVIEW objects. However, the coloring of objects that convey information via their color is unalterable. For example,

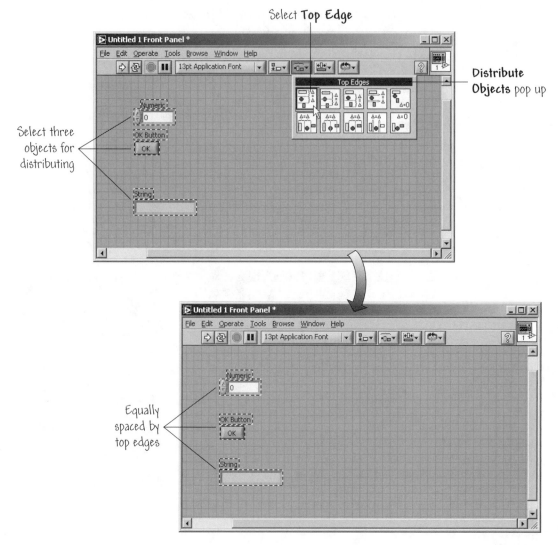

FIGURE 3.20
Distributing objects.

block diagram terminals of front panel objects and wires use color codes for the type and representation of data they carry, so you cannot change their color.

To change the color of an object (or the background of a window), pop up on the object of interest with the **Coloring** tool from the **Tools** palette, as seen in Figure 3.22. Choose the desired color from the selection palette that appears. If you keep the mouse button pressed as you move through the color selection palette, the object or background you are coloring redraws with the color the cursor currently is touching. This gives you a "real-time" preview of the object in the new color. If you release the mouse button on a color, the selected

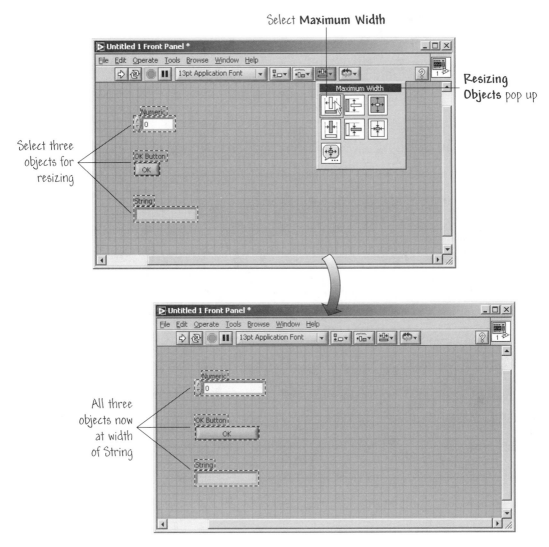

FIGURE 3.21
Resizing objects.

object retains the chosen color. To cancel the coloring operation, move the cursor out of the color selection palette before releasing the mouse button, or use the Edit≫Undo Color Change command after the undesired color change has been made.

If you select the box with a **T** in it, the object is rendered transparent. One use of the **T** (transparent) option is to hide the box around labels. Another interesting use is to create numeric controls without the standard three-dimensional border. Transparency affects the appearance but not the function of the object!

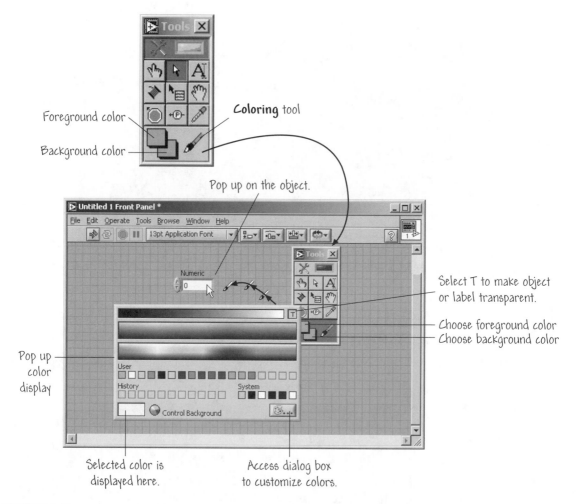

FIGURE 3.22
Customizing the color of objects.

Some objects have both a foreground and a background color that you can set separately. The foreground color of a knob, for example, is the main dial area, and the background color is the base color of the raised edge. On the **Coloring** tool, you select the foreground or background, as illustrated in Figure 3.22. Clicking the upper left box on the **Coloring** tool and then clicking on an object on the front panel will color the object's foreground. Similarly, clicking the lower right box on the **Coloring** tool and then clicking on an object on the front panel will color the object's background.

Selecting the button on the lower right-hand side in the palette (see Figure 3.23) accesses a dialog box with which you can customize the colors. Each of the three color components, red, green, and blue, describes eight bits of a 24-bit color (in the lower right-hand side of the **Color** dialog box). Therefore,

Click on the color
palette to select desired color.

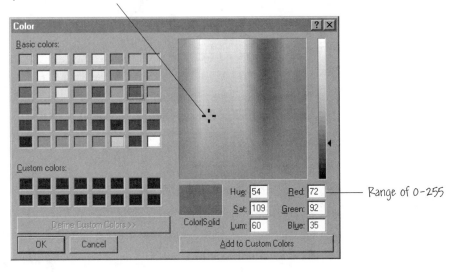

Range of 0-255

FIGURE 3.23
The **Color** dialog box.

each component has a range of 0 to 255. The last color you select from the palette becomes the current color. Clicking on an object with the **Color** tool sets that object to the current color.

Using the **Color Copy** tool, you can copy the color of one object and transfer it to a second object without using the **Color** palette. To accomplish this, click with the **Color Copy** tool on the object whose color you want to transfer to another object. Then, select the **Color** tool and click on another object to change its color.

 Practice with Editing

In this exercise you will edit and modify an existing VI to look like the panel shown in Figure 3.24. After editing the VI, you will wire the objects in the block diagram and run the program.

1. Open the VI Editing by choosing **Open** from the **File** menu (or using the **Open VI** button on the startup window) and searching in the Chapter 3 folder in the Learning directory. The front panel of the Editing VI contains a number of objects depicted in Figure 3.25. The objective of this exercise is to make the front panel of the Editing VI look like the one shown in Figure 3.24. If you have a color monitor, you can see the final VI in color by opening the VI Editing Done located in the Chapter 3 folder in the Learning directory.

2. Add an owned label to the numeric control using the **Positioning** tool by popping up on the numeric control and selecting Visible Items≫Label from

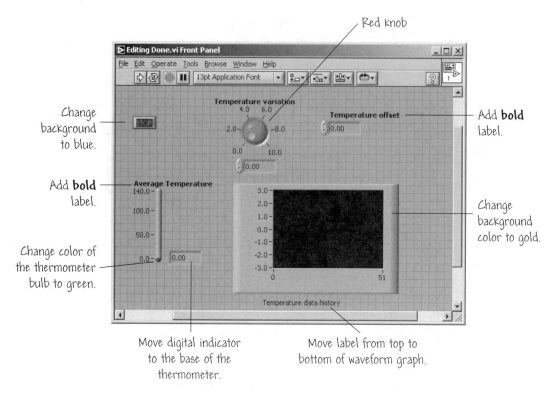

Red knob

Change background to blue.

Add **bold** label.

Add **bold** label.

Change background color to gold.

Change color of the thermometer bulb to green.

Move digital indicator to the base of the thermometer.

Move label from top to bottom of waveform graph.

FIGURE 3.24
The front panel for the Editing VI.

the pop up menu. Type in the text Temperature offset inside the bordered box and click the mouse outside the label or click the **Enter** button on the left-hand side of the toolbar.

3. Reposition the waveform graph and numeric control.

(a) Choose the **Positioning** tool from the **Tools** palette.

(b) Click on the waveform graph and drag it to the lower center of the front panel (see Figure 3.24 for the approximate location). Then click on the numeric control and drag it to the upper right of the front panel (see Figure 3.24 for approximate location).

Notice that as you move the numeric control, the owned label moves with the control. If the control is currently selected, click on a blank space on the front panel to deselect the control and then click on the label and drag it to another location. Notice that in this case the control does not follow the move. You can position an owned label anywhere relative to the control. If you move the owned label to an undesirable location, you can move it back by selecting Edit≫Undo Move to place the label back above the digital control.

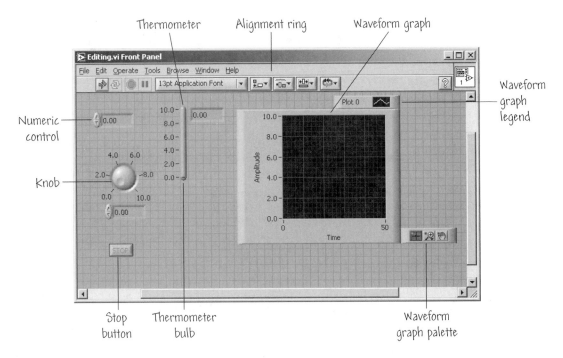

FIGURE 3.25
The Editing VI with various objects: knob, waveform chart, thermometer, and digital control.

4. Reposition the stop button, the knob, and the thermometer to the approximate locations shown in Figure 3.24.

5. Add labels to the knob and to the thermometer. To accomplish this task, pop up on the object and choose Visible Items≫Label. When the label box appears, type in the desired text. In this case, we want to label the knob Temperature variation and the thermometer Average Temperature.

6. Move the numeric indicator associated with the thermometer to a location near the bottom of the thermometer. Then reposition the thermometer label so that it is better centered above the thermometer (see Figure 3.24).

7. In the next step, we will reformat the waveform graph.

   (a) To remove the waveform graph legend, pop up on the waveform graph and select **Properties**. On the **Appearance** menu, deselect the check mark next to **Show Plot Legend**.

   (b) To remove the waveform graph palette, deselect **Show Graph** palette from the **Appearance** menu.

   (c) Remove the waveform x-axis scale by selecting the **Scales** menu. On the **Scales** menu, be sure to select the x-axis from the drop-down box. Deselect the **Show Scale** option.

(d) Add a label to the wavefrom graph—label the object **Temperature data history**. Select the owned label and move it to the bottom of the waveform graph, as illustrated in Figure 3.24. Note that this can also be done on the Properties page.

8. Align the stop button and the thermometer.

(a) Select both the stop button and the thermometer using the **Positioning** tool. Pick a point somewhere to the upper left of the stop button and drag the dashed box down until it encloses both the stop button and the thermometer. Upon release of the mouse button, both objects will be surrounded by moving dashed lines.

(b) Click on **Align Objects** and choose **Left Edges**. The stop button and the thermometer will then align to the left edges. If the objects appear not to move at all, this indicates that they were essentially aligned to the left edges already.

9. Align the stop button, the knob, and the digital control horizontally by choosing **Vertical Centers** axis from the **Align Objects** pull-down menu in the toolbar. Remember to select all three objects beforehand!

10. Space the stop button, the knob, and the digital control evenly by choosing the **Horizontal Centers** axis from the **Distribute Objects** pull-down menu in the toolbar. Again, remember to select all three objects beforehand!

11. Change the color of the stop button.

(a) Using the **Coloring** tool, pop up on the stop button to display the color palette.

(b) Choose a color from the palette. The object will assume the last color you selected. In the Editing Done VI you will see that a dark blue color was selected—you can choose a color that you like.

12. Change the color of the knob. Using the **Coloring** tool, pop up on the knob to display the color palette and then choose the desired color from the color palette. In the Editing Done VI you will find that a red/pink color was selected—you can choose a color that suits you.

13. Change the color of the waveform graph. In the Editing Done VI you will see that a gold color was selected.

14. Change the color of the thermometer bulb. Using the **Coloring** tool, pop up on the thermometer bulb (at the bottom of the thermometer) to display the color palette and then choose the desired color. In the Editing Done.vi you will find that a green color was selected.

15. Change the font style of the owned labels. Use the **Labeling** tool to highlight each of the labels and then select Style≫Bold from **Text Settings**. Do this for three of the owned labels: thermometer, knob, and digital control.

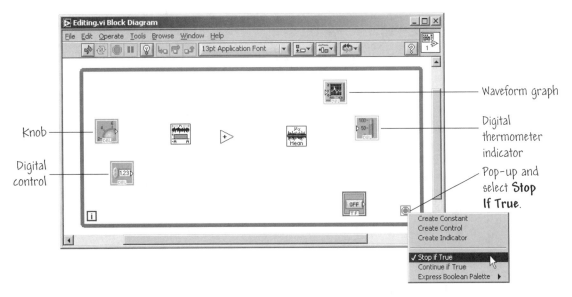

FIGURE 3.26
The block diagram for the **Editing.vi**—before wiring.

At this point, the front panel of the Editing VI should look very similar to the front panel shown in Figure 3.24.

The block diagram for the Editing.vi is shown in Figure 3.26 after the editing of the front panel is finished—but before wiring the block diagram.

We can now wire together the various objects to obtain a working VI. Notice that several additional objects are on the block diagram: a uniform white noise subVI, a mean subVI that computes the mean (or average) of a signal, a While Loop, and an Add function. In later chapters, you will learn to use While Loops and learn how to use preexisting subVIs packaged with LabVIEW. For instance, the Uniform White Noise.vi is located on the **Functions** palette under the All Functions≫Analyze≫Signal Processing≫Signal Generation subpalette.

Go ahead and wire the Editing.vi block diagram so that it looks like the block diagram shown in Figure 3.27. If you run into difficulties, you can open and examine the block diagram for Editing Done.vi (which can be found in Chapter 3 of the Learning directory).

A few wiring tips:

1. To wire the objects together with the **Wiring** tool, click and release on the source terminal and drag the **Wiring** tool to the destination terminal. When the destination terminal is blinking, click and release the left mouse button.

2. To identify terminals on the Add function, pop up on the icon and select Visible Items≫Terminal to see the connector. When the wiring is finished, pop up again on the function and choose Visible Items≫Terminal to show the icon once again.

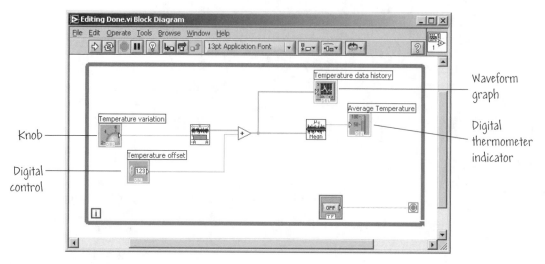

FIGURE 3.27
The block diagram for the **Editing Done.vi**—after wiring.

3. To bend the wires as you connect two objects, click with left mouse button on the bend location with the **Wiring** tool.

After you have finished wiring the objects together, switch to the front panel by selecting **Show Front Panel** from the **Window** menu. Use the **Operating** tool to change the value of the front panel controls. Run the VI by clicking on the **Run** button on the toolbar.

The Average Temperature indicator should be approximately the value that you select as the Temperature offset. The amount of variation in the temperature history as shown on the Temperature data history waveform graph, should be about the same as the setting on the Temperature variation knob.

When you are finished editing and experimenting with your VI, save it by selecting **Save** from the **File** menu. Remember to save all your work in the Users Stuff folder in Learning directory. Close the VI by selecting **Close** from the **File** menu. ◆

3.2 DEBUGGING TECHNIQUES

In this section we discuss LabVIEW's basic debugging elements, which provide an effective programming debugging environment. Most features commonly associated with good interactive debugging environments are provided—and in keeping with the spirit of graphical programming, the debugging features are accessible graphically. Execution highlighting, single stepping, breakpoints, and probes helps debug your VIs easily by tracing the flow of data through the VI. You can actually watch your program code as it executes!

Click **Broken Run** button
to see list of
program errors.

Warning
button

Bad
wire

List of
program
errors

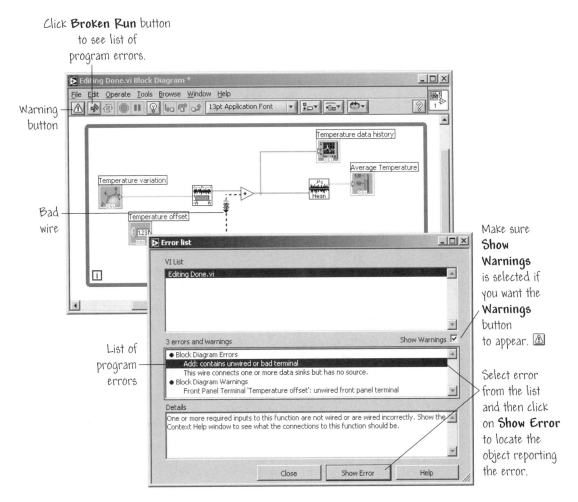

Make sure
**Show
Warnings**
is selected if
you want the
Warnings
button
to appear.

Select error
from the list
and then click
on **Show Error**
to locate the
object reporting
the error.

FIGURE 3.28
Locating program errors.

3.2.1   Finding Errors

When your VI cannot compile or run due to a programming error, a **Broken Run** button appears on the toolbar. Programming errors typically appear during VI development and editing stages and remain until you properly wire all the objects in the block diagram. You can list all your program errors by clicking on the **Broken Run** button. An information box called **Error List** appears listing all the errors. This box is shown in Figure 3.28 for the Editing Done VI with a broken wire.

Warnings make you aware of potential problems when you run a VI, but they do not inhibit program execution. If you want to be notified of any warnings,

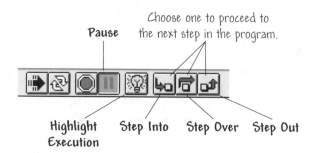

FIGURE 3.29
The **Highlight Execution** and step buttons located on the toolbar.

click the **Show Warning** checkbox in the **Error List** dialog. A warning button then appears on the toolbar whenever a warning condition occurs.

If your program has any errors that prevent proper execution, you can search for the source of a specific error by selecting the error in the **Error List** (by clicking on it) and then clicking on **Show Error** (lower right-hand corner of the **Error List** dialog box). This process will highlight the object on the block diagram that reported the error, as illustrated in Figure 3.28. Double clicking on an error in the error list will also highlight the object reporting the error.

Some of the most common reasons for a VI being broken during editing are:

1. A function terminal requiring an input is unwired. For example, an error will be reported if you do not wire all inputs to arithmetic functions.

2. The block diagram contains a broken wire because of a mismatch of data types or a loose, unconnected end.

3. A subVI is broken.

3.2.2   Highlight Execution

You can animate the VI block diagram execution by clicking on the **Highlight Execution** button located in the block diagram toolbar, shown in Figure 3.29.

For debugging purposes, it is helpful to see an animation of the VI execution in the block diagram, as illustrated in Figure 3.30. When you click on the **Highlight Execution** button, it changes to a bright light to indicate that the data flow will be animated for your visual observation when the program executes. Click on the **Highlight Execution** button at any time to return to normal running mode.

Execution highlighting is commonly used with single-step mode (more on single stepping in the next section) to trace the data flow in a block diagram in an effort to gain an understanding of how data flows through the block diagram. Keep in mind that when you utilize the highlight execution debugging feature, it greatly reduces the performance of your VI—the execution time increases significantly! The data flow animation shows the movement of data from one

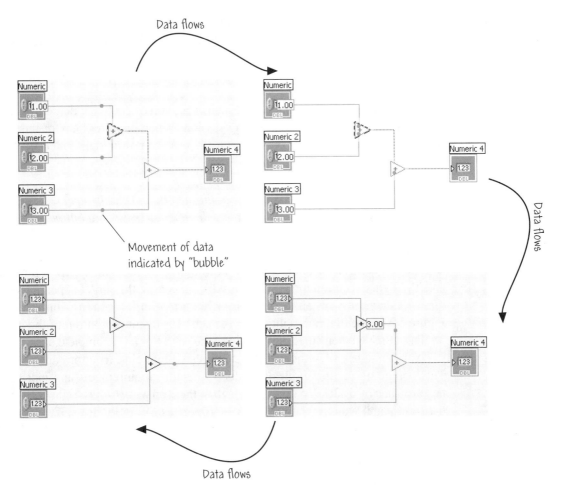

FIGURE 3.30
Using the highlight execution mode to watch the data flow through a VI.

node to another using "bubbles" to indicate data motion along the wires. This process is illustrated in Figure 3.30. Additionally, in single-step mode, the next node to be executed blinks until you click on the single-step button.

3.2.3 Single Stepping Through a VI and Its SubVIs

For debugging purposes, you may want to execute a block diagram node by node. This is known as single stepping. To run a VI in single-step mode, press any of the debugging step buttons on the toolbar to proceed to the next step. The step buttons are shown on the toolbar in Figure 3.29. The step button you press determines where the next step executes. You click on either the **Step Into** or **Step Over** button to execute the current node and proceed to the next node. If the node is a structure (such as a While Loop) or a subVI, you can select the

Step Over button to execute the node, but not single step through the node. For example, if the node is a subVI and you click on the **Step Over** button, you execute the subVI and proceed to the next node, but cannot see how the subVI node executed internally. To single step through the subVI you would select the **Step Into** button.

Click on the **Step Out** button to finish execution of the block diagram nodes or finish up the single-step debugging session. When you press any of the step buttons, the **Pause** button is pressed as well. You can return to normal execution at any time by releasing the **Pause** button.

If you place your cursor over any of the step buttons, a tip strip will appear with a description of what the next step will be if you press that button.

You might want to use highlight execution as you single-step through a VI, so that you can follow data as it flows through the nodes. In single-step mode and highlight execution mode, when a subVI executes, the subVI appears on the main VI diagram with either a green or red arrow in its icon, as illustrated in Figure 3.31. The diagram window of the subVI is displayed on top of the main VI diagram. You then single step through the subVI or let it complete executing.

You can save a VI without single stepping or highlight execution capabilities. This compiling method typically reduces memory requirements and increases performance by 1–2%. To do this, pop up in the icon pane (upper-right corner of the front panel window) and select **VI Properties**. As in Figure 3.32, from the **Execution** menu, deselect the **Allow Debugging** option to hide the **Highlight Execution** and **Single Step** buttons.

3.2.4   Breakpoints and Probes

You may want to halt execution (set **breakpoints**) at certain locations of your VI (for example, subVIs, nodes, or wires). Using the **Breakpoint** tool, click on any item in the block diagram where you want to set or clear a breakpoint. Breakpoints are depicted as red frames for nodes and red dots for wires.

You use the **Probe** tool to view data as it flows through a block diagram wire. To place a **probe** in the block diagram, click on any wire in the block diagram where you want to place the probe. Probes are depicted as numbered yellow boxes with associated pop-up windows in which the values passing through the wire at runtime are displayed. You can place probes all around the block diagram to assist in the debugging process.

Conditional probes allow you to set conditional breakpoints for common cases of each data type. For example, creating a conditional probe on a numeric allows you to set a pause condition using Equal to, Greater than, or Less than. This gives you flexibility in your debugging.

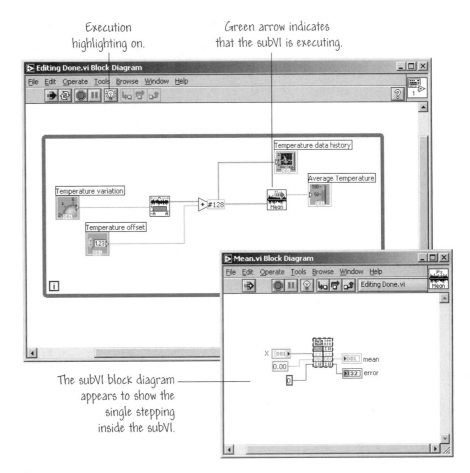

FIGURE 3.31
Single stepping into a subVI with execution highlighting selected.

Practicing with Debugging

A nonexecutable VI—named Debug.vi— has been developed for you to debug and to fix. You will get to practice using the single-step and execution highlighting modes to step through the VI, and to insert breakpoints and probes to regulate the program execution and to view the data values.

1. Open the Debug VI in Chapter 3 of the Learning directory by choosing **Open** from the **File** menu. Notice the **Broken Run** button in the toolbar indicating the VI is not executable.

2. Switch to the block diagram by choosing **Show Block Diagram** from the **Window** pull-down menu. You should see the block diagram shown in Figure 3.33.

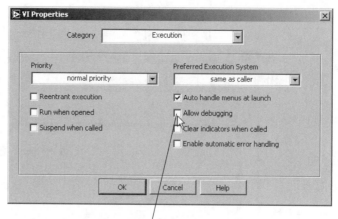

Deselect **Allow Debugging** to increase program
speed and reduce memory usage.

FIGURE 3.32
Turning off the debugging options using the **VI Setup** pop-up menu.

(a) The **Random Number (0-1)** function (represented by the two die) can be found in the Functions≫Arithmetic & Comparison≫Express Numeric subpalette and returns a random number between zero and one.

(b) The Add function (which can also be found in the Functions≫Arithmetic & Comparison≫Express Numeric subpalette) adds the random number

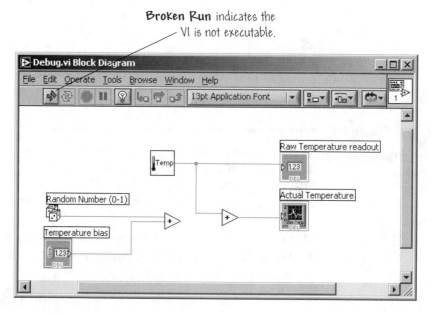

Broken Run indicates the
VI is not executable.

FIGURE 3.33
Block diagram showing the debugging exercise.

to a bias number represented by the variable Temperature bias, which accepts input by the user on the front panel.

(c) The subVI Temp (which can be found in LabVIEW7.0\activity\Digital Thermometer.vi) simulates acquiring temperature data one value at a time from a data acquisition board. It uses the Demo Voltage Read subVI to acquire the simulated data. As you single step through the code, you will see the program execution inside both subVIs.

(d) The second Add function adds the simulated temperature data point to the random number plus the bias number to create a variable named Actual Temperature.

3. Investigate the source of the programming error by clicking on the **Broken Run** button to obtain a list of the programming errors. You should find one error—**Add: contains unwired or bad terminal**.

4. Highlight the error in the **Error List** and then click on **Show Error** to locate the source of the error within the block diagram. You should find that the second Add function (the one on the right side) is shown to be the source of the programming error.

5. Fix the error by properly wiring the two Add functions together. Once this step is successfully completed, the **Broken Run** button should reappear as the **Run** button.

6. Select **Highlight Execution** and then run the VI by pressing the **Run Continuously** button. You can watch the data flow through the code! Returning to the front panel, you will see the waveform chart update as the new temperature is calculated and plotted. Notice that the simulation runs very slowly in highlight execution mode. Click the **Highlight Execution** button off to see everything move much faster!

7. Terminate the program execution by clicking on the **Abort Execution** button.

8. Enable single stepping by clicking on one of the step buttons. You can enable **Highlight Execution** if you want to see the data values as they are computed at each node.

9. Use the step buttons to single step through the program as it executes. Remember that you can let the cursor idle over the step buttons and a tip strip will appear with a description of what the next step will be if you press that button. Press the **Pause** button at any time to return to normal execution mode.

10. Enable the probe by popping up on the wire connecting the two Add functions and selecting **Probe**. Continue to single step through the VI and watch how the probe displays the data carried by the wire. Alternatively, you could have used the **Probe** tool to place the probe on the wire. Try placing a probe on the block diagram (say between the Temp subVI and the Add function) using the **Probe** tool.

11. Place a few more probes around the VI and repeat the single-stepping process to see how the probes display the data.

12. When you are finished experimenting with the probes, close all open probe windows.

13. Set a breakpoint by selecting the **Breakpoint** tool from the **Tools** palette and clicking on the wire between the two Add functions. You will notice that a red ball appears on the wire indicating that a breakpoint has been set at that location.

14. Run the VI by clicking on the **Run Continuously** button. The VI will pause each time at the breakpoint. To continue VI execution, click on the **Pause** button. It helps to use highlight execution when experimenting with the breakpoints; otherwise the program executes too fast to easily observe the data flow.

15. Terminate the program execution by pressing the **Abort Execution** button. Remove the breakpoint by clicking on the breakpoint (that is, on the red ball) with the **Breakpoint** tool.

16. Save the working VI by selecting **Save** from the **File** menu. Remember to save the working VI in the folder Users Stuff within the Learning directory. Close the VI and all open windows by selecting **Close** from the **File** menu. ◆

3.3   A FEW SHORTCUTS

Frequently used menu options have equivalent command key shortcuts. For example, to save a VI you can choose **Save** from the **File** menu, or press the control key equivalent <ctrl-S> (Windows) or <command-S> (Mac). Some of the main key equivalents are shown in Table 3.1.

TABLE 3.1   Frequently used command key shortcuts.

Windows	Macintosh	Function
<Ctrl-S>	<Command-S>	Save a VI
<Ctrl-R>	<Command-R>	Run a VI
<Ctrl-E>	<Command-E>	Toggle between the front panel and the block diagram
<Ctrl-H>	<Command-H>	Toggle the **Context Help** window on and off
<Ctrl-B>	<Command-B>	Remove all bad wires
<Ctrl-W>	<Command-W>	Close the active window
<Ctrl-F>	<Command-F>	Find objects and VIs

The shortcut access to the **Tools** palette on the front panel and block diagram is given by:

- **Windows**—Press <shift> and the right mouse button
- **Macintosh**—Press <Command-shift> and the mouse button

BUILDING BLOCK

3.4   BUILDING BLOCKS: MEASURING VOLUME

Two common measuring units for volume are liters and gallons. In this building block exercise you will construct and debug a VI that simulates reading a volume measurement in user-selectable units of liters or gallons.

The front panel and block diagram are shown in Figure 3.34. Using the figure as a guide, construct and debug a VI that simulates a volume measurement and allows the user to select the units of the measurement as either liters or gallons. The VI utilizes the subVI Temp & Vol to simulate the measurement of volume. The subVI is located in the Building Blocks folder found in the Learning directory.

Practice using the various debugging tools as you develop the VI—highlight execution, the step buttons, probes, and breakpoints. Also, notice that you will have to edit the front panel to add free labels to the horizontal switch indicating Liters on the left hand-side and Gallons on the right hand-side. Both the horizontal switch and the tank need to be labeled Volume and Tank Volume, respectively. The vertical scale of the tank should be changed to vary from 0 to 1000 (the default is 0 to 10).

On the block diagram, use the **Align Objects** pull-down menu to organize the objects in a pleasing arrangement for ease in visualization. Remember that for more complex VIs it will become increasingly difficult to debug the code if the VI objects are arranged haphazardly with crossing wires and so forth.

When you are satisfied that your VI is wired properly and ready to run, begin the execution of the VI by choosing **Run**. With the **Operating** tool, select Liters and run the VI. Then, select Gallons and run the VI again. To view the process in a continuous run mode, select **Run Continuously** and while the VI is executing, switch the output from liters to gallons, and back.

When you are done experimenting with your new VI, save it as Volume1.vi in the Users Stuff folder in the Learning directory. You will use this VI as a building block in later chapters—so make sure to save your work!

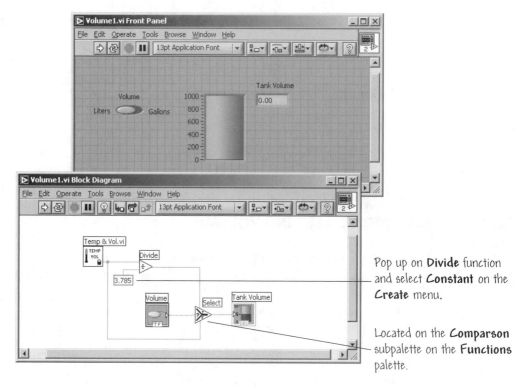

FIGURE 3.34
Front panel and block diagram for a volume measuring system.

 A working version of **Volume1.vi** *can be found in the* **Building Blocks** *folder of the* **Learning** *directory.*

3.5 RELAXED READING: LABVIEW HELPS FABRICATION PROCESS OF NEXT GENERATION MICROPROCESSORS

Moore's Law speculates that the numbers of transistors on a microchip will double every 18 months. Extreme Ultraviolet Lithography (EUVL) is leading the race to push Moore's Law to its current physical limitations. By using smaller wavelengths of light in the chip lithography process, smaller traces can be etched into a semiconductor wafer. This will allow chips to be produced with nanoscale circuitry, meaning microprocessors with a billion transistors. Each transistor on the chip can be as small as 40 atoms in width. The EUVL process, shown in Figure 3.35, uses a mirror coated with a non-reflective negative image, or mask, of a microprocessor circuit. An EUV light source, which produces light peaking at 134 Angstroms, is used to illuminate the mask. A magnified image of the

FIGURE 3.35
The EUVL laser plasma reflectometer (LPR) measures the reflectivity of the mask mirrors vs. wavelength.

circuit on the mask is projected onto the semiconductor wafer, burning away conductive material and leaving a microprocessor circuit.

The mask mirror is a critical component in the EUVL process. Lawrence Livermore National Labs (LLNL) is one of the only producers of EUVL mask mirrors. To test these mirrors, LLNL called upon EUV Technology (www.euvl.com), a small Silicon Valley-based company, to build the first reflectometer that was small enough to be placed in a clean room and able to meet LLNL's testing requirements. EUV Technologies partnered with Cal-Bay Systems to develop the reflectometer's control and data acquisition system. This small reflectometer, known as a laser plasma reflectometer (LPR), uses a laser-induced gold plasma as its EUV light source. It is one of the keys to bringing EUVL out of the lab and into the semiconductor fab for large-scale chip production. Previously, EUVL mask mirrors were tested for reflectivity using a particle accelerator ring known as a synchrotron as an EUV light source. A synchrotron costs hundreds of millions of dollars and is the size of a football field. Test time had to be scheduled months in advance, took a setup time of one hour, and had a turnaround time of eight hours. EUV Technology's LPR, shown in Figure 3.36, occupies only 1.5 square meters, easily fits into a clean room, has a test setup time of ten seconds, and can test the mirrors in five minutes. The EUV Technologies LPR is even a candidate for the 2001 R&D 100 award.

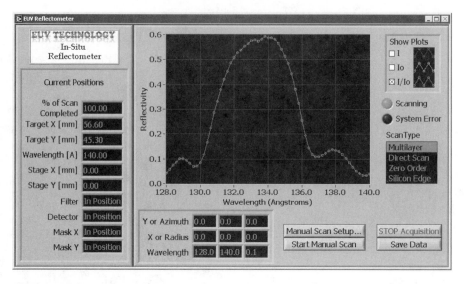

FIGURE 3.36
The EUV Technology Laser Plasma Reflectometer LabVIEW User Interface.

The research and development of any complicated first-generation system requires a flexible set of tools. The LPR's laser must be fired in synchrony with data acquisition and motion control for precise measurements to be made. Two stepper motors and five servomotors are used to position various filters, the EUV light monochromator, a gold target, and the mask mirror. Digital timing signals must be generated to trigger cascading events such as flashing the laser lamp, firing the laser, and acquiring analog data. Motorized micrometers and New Focus, Inc. Picomotors are used to align optical components with nanometric precision to focus the laser and EUV light. Analog voltages from optical power sensors must be read to acquire the reflectivity data. System interlocks had to be checked to ensure that the system was in a safe operating state. Finally, the system had to communicate with other production tools over an Ethernet network.

We chose National Instruments data acquisition and motion control hardware because of their advanced triggering features and the ability to synchronize events via the Real Time Systems Integration (RTSI) bus. We used the National Instruments PCI-6052E multifunction data acquisition card to create several precisely spaced timing signals that cause the laser to fire and analog signal acquisition to occur. We achieved the seven axes of coordinated motion with the two PCI-7344 FlexMotion control cards and we used a FlexMotion stepper and servomotor power drives to drive them. The entire control system runs on a rack-mounted industrial PC and is housed in a 19 in. industrial-rack enclosure.

We chose NI LabVIEW as the software development environment because of its ease of use, flexibility, network-ready functionalities, and tight integration of motion control and data acquisition. We used the LabVIEW data acquisition and motion control example programs to quickly prototype a solution to the complex hardware timing issues. After the prototyping software was written, we developed an operator interface, which allowed scanning optical power levels vs. any motion axis. This provided characterization information that allowed the LPR to be fine tuned. This was an invaluable feature. Having all of the motion control functions available in LabVIEW via FlexMotion allowed the code to be modular and flexible. Finally a network interface was developed to allow remote operation of the device by other tools in the fabrication system.

As the LPR's physical and functional design evolved, the LabVIEW graphical code was quickly modified to meet the new needs of the system; motion control axes were added and removed, scanning methodology changed, subsystems and interlocks evolved, and calibration and characterization methods were refined. Having all of the hardware routines developed in LabVIEW allowed all the flexibility we required. LabVIEW and National Instruments hardware proved to be an excellent platform for developing a successful system in a highly experimental R&D environment.

For further information contact:

Jim Kring
jim@jimkring.com

3.6 SUMMARY

The subjects of editing and debugging VIs were the main subjects of this chapter. Just as you would edit a C program or debug a Fortran subroutine, you must know how to edit and debug VIs. We discussed how to create, select, delete, move, and arrange objects on the front panel and block diagrams. The important topic of selecting and deleting wires and locating and eliminating bad wires was also discussed. The program debugging topics covered included execution highlighting, and how to step into, through, and out of the code and how to use probes to view data as it flow through the code.

KEY TERMS

Breakpoint: A pause in execution used for debugging. You set a breakpoint by clicking a VI, node, or wire with the **Breakpoint** tool.

Breakpoint tool: Tool used to set a breakpoint on a node or wire.

Broken VI: A VI that cannot compile and run.

Coloring tool: Tool used to set foreground and background colors.

Execution highlighting: Debugging feature that animates the VI execution to illustrate the data flow within the VI.

Label: Text object used to name or describe other objects or regions on the front panel or block diagram.

Labeling tool: Tool used to create labels and enter text.

Operating tool: Tool used to enter data into controls and operate them—resembles a pointing finger.

Positioning tool: Tool used to move, select, and resize objects.

Probe: Debugging feature for checking intermediate values in a VI during execution.

Probe tool: Tool used to create probes on wires.

Resizing handles: Angled handles on the corners of objects that indicate resizing points.

EXERCISES

E3.1 Construct a VI to accept five numeric inputs, add them up and display the result on a gauge, and light up a round light if the sum of the input numbers is less than 8.0. The light should light up in green, and the gauge dial should be yellow. The VI in Figure E3.1 can be used as a guide.

E3.2 Open Change to Indicator.vi in Learning\Exercises&Problems\Chapter 3. Click on the **Broken Run** button. Select the error, and click on the **Show Error** button. This wire connects more than one data source. The wire between the output of the Add function and the control Result is highlighted and has an "x" over it. Both the output and the control are "sources of data." You can only connect a data source to a data display. Right click on the control Result and select **Change to Indicator**. The error disappears and you can run the VI. Input $x = 2$ and $y = 3$ and run the VI. Verify that the output result is 5.

E3.3 Open the virtual instrument Multiple Controls-1 Terminal.vi. You can find this VI in Learning\Exercises&Problems\Chapter 3. Click on the **Broken Run** button, select the error **This Wire Connects to More than One Data Source**, and click on the **Show Error** button. The wire from the two controls are highlighted. If you look closely, you'll see that both wires are going to the same terminal. That means two controls are defining the value of one input value. Delete the wire from the y control. Wire the y control to the other input terminal on the Add function. Both errors disappear and you can run the VI. Input $x = 1$ and $y = 4$ and run the VI. Verify that the output $x + y = 5$.

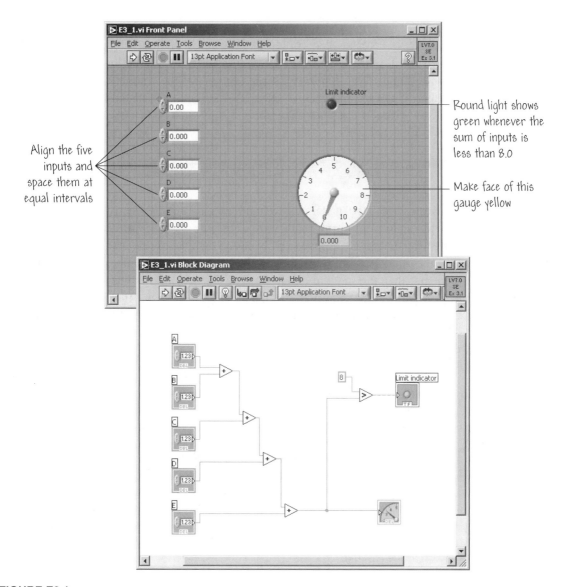

FIGURE E3.1
A VI to add five numeric inputs and light up a round LED if the sum is less than 8.0.

E3.4 Open the virtual instrument Crazy Wires.vi. You can find this VI in Learning\Exercises&Problems\Chapter 3. Go to the block diagram and see the wires running everywhere. There is an easy way to fix this! Right click on the wire running from **Knob** to the top terminal of the comparison function. Select **Clean Up Wire** and watch the wire get straightened instantly. Do the same for all the wires on the block diagram.

PROBLEMS

P3.1 Construct a VI that generates two random numbers (between 0 and 1) and displays both random numbers on meters. Label the meters Random number 1 and Random number 2, respectively. Make the face of one meter blue and the face of the other meter red. When the value of the random number on the red meter is greater than the random number on the meter with the blue face, have a square LED show green; otherwise have the LED show black. Run the VI several times and observe the results. On the block diagram select **Highlight Execution** and watch the data flow through the code.

P3.2 In this problem you will construct a stop light display. Create a dial control that goes from 0 to 2, with three LED displays: one green, one yellow, and one red. Have the VI turn the LED green when the dial is on 0, yellow when the dial is on 1, and red when the dial is on 2.

P3.3 Create a front panel that has 8 LED indicators and a vertical slider control that is an 8 bit unsigned integer. Display a digital indicator for the slider, and make sure that the LEDs are evenly spaced and aligned at the bottom. The problem is to turn the 8 LEDs into a binary (base 2) representation for the number in the slider. For example, if the slider is set to the number 10 (which in base 2 is $00001010 = 1 * (2^3) + 1 * (2^1)$), the LED's 1 and 3 should be on. To test your solution, check the number 131. LED's 0, 1 and 7 should be on since 131 is 10000011 in base 2.

P3.4 In this problem, we want to open and run an existing VI. In LabVIEW, go to Help≫Find Examples and click the **Search** tab and type "probes." Select "probes" to display the example VIs that include probes in the title. Find the VI entitled Using Supplied Probes.vi and open it.

(a) Change to the block diagram. Follow the instructions written on the diagram for each probe.

(b) What probes are available on the block diagram?

(c) On the Custom array probe, what happens if you set the number of elements equal to 250?

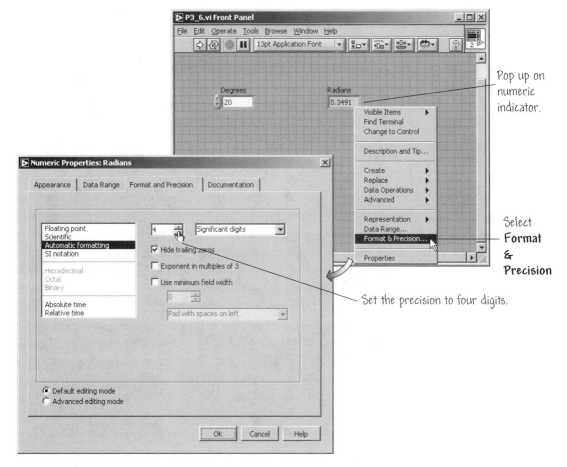

FIGURE P3.1
Converting degrees to radians with four digits of output precision.

(d) To continue after a break, press the **Pause** button on the menu bar.

(e) Continue probing each red box. What information is supplied to you in the Dynamic Data Type probe?

(f) What happens if you implement a conditional breakpoint of zero on the Error Probe?

(g) Probe the **Stop** button/Boolean indicator. Have the VI running. Then stop the VI from the front panel. What is the state of the Boolean probe?

P3.5 Complete the crossword puzzle.

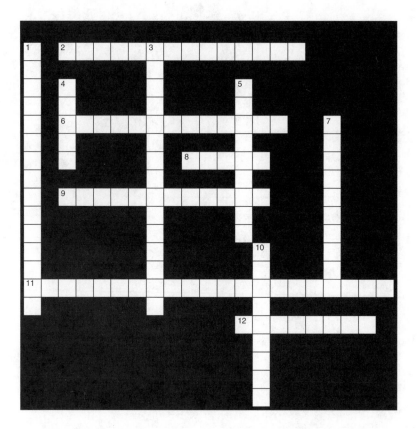

Across

2. Tool used to set a breakpoint on a node or wire
6. A tool (resembling a pointing finger) used to enter data into controls and operate.
8. Text object used to name or describe other objects or regions on the front panel or block diagram.
9. Tool used to create labels and enter text.
11. Debugging feature that animates the VI execution to illustrate the data flow within the VI.
12. A VI that cannot compile and run.

Down

1. Angles handles on the corner of objects that indicate resizing points.
3. Tool used to move, select, and resize objects.
4. Debugging feature for checking intermediate values in a VI during execution.
5. Tool used to create probes on wires.
7. A pause in execution used for debugging. You set a breakpoint by clicking a VI, node, or wire with the Breakpoint tool.
10. Tool used to set foreground and background colors.

P3.6 Develop a VI that converts an input value in degrees to radians with four digits of precision. Refer to Figure P3.1 for help on changing the output display to four digits.

SubVIs

CHAPTER 4

In this chapter we learn to build subVIs. SubVIs are VIs used by other VIs—like subroutines. One of the keys to constructing successful VIs is understanding how to build and use subVIs. The hierarchical design of LabVIEW applications (that is, of virtual instruments) depends on the use of subVIs. We will discuss two basic ways to create and use a subVI: creating subVIs from VIs and creating subVIs from selections. The **Icon Editor** is presented as a way to personalize subVI icons so the information about the function of the subVI is apparent from visual inspection. The editor has a tool set similar to that of most common paint programs. The **Hierarchy Window** will be introduced as a helpful tool for managing the hierarchical nature of your programs.

GOALS

1. Learn to build and use subVIs.

2. Understand the hierarchical nature of VIs.

3. Practice with the **Icon Editor** and with assigning terminals.

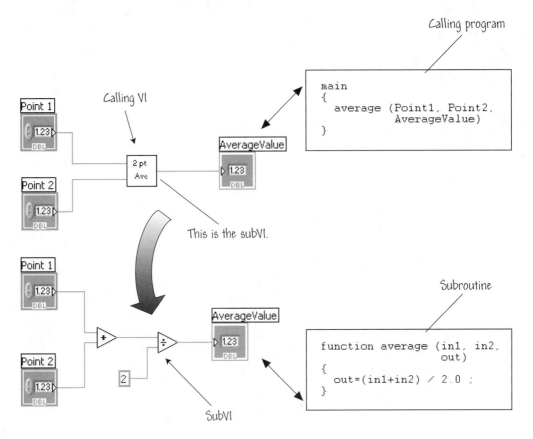

FIGURE 4.1
Analogy between subVIs and subroutines.

4.1   WHAT IS A SUBVI?

It is important to understand and appreciate the hierarchical nature of virtual instruments. **SubVIs** are critical components of a hierarchical and modular VI that is easy to debug and maintain. A subVI is a stand-alone VI that is called by other VIs—that is, a subVI is used in the block diagram of a **top-level VI**. SubVIs are analogous to subroutines in text-based programming languages like C or Fortran, and the subVI node is analogous to a subroutine call statement. The **pseudocode** and block diagram shown in Figure 4.1 demonstrate the analogy between subVIs and subroutines. There is no limit to the number of subVIs you can use in a calling VI. Using subVIs is an efficient programming technique in that it allows you to reuse the same code in different situations. The hierarchical nature of programming in G follows from the fact that you can call a subVI from within a subVI.

You can create subVIs from VIs, or create them from selections (selecting existing components of a VI and placing them in a subVI). When creating a new subVI from an existing VI, you begin by defining the inputs and outputs of the subVI, and then you "wire" the subVI connector properly. This allows calling VIs to send data to the subVI and to receive data from the subVI. On the other hand, if an existing complex block diagram has a large number of icons, you may choose to group related **functions** and icons into a lower-level VI (that is, into a subVI) to maintain the overall simplicity of the block diagram. This is what is meant by using a modular approach in your program development.

4.2   REVIEW OF THE BASICS

Before moving on to the subject of subVIs, we present a brief review of some of the basics presented in the first three chapters as part of an exercise in constructing a VI. Your VI will ultimately be used as a subVI later in the chapter.

Everyone who follows baseball knows that keeping track of statistics is one of the main features of the game for players and fans alike. And one of the most important statistics is the batting average. In this section, we will build a VI that computes a batting average and later turn the VI into a subVI. As with most things in baseball, the calculation of a batting average, while conceptually simple, has many nuances that are not accounted for in our VI. The main aspects of calculating the batting average are retained, however, and the primary goal of learning about programming is enhanced.

If you are not familiar with the game of baseball, then view the following discussion as an exercise is constructing a VI that performs elementary mathematical calculations and is suitable for use as a subVI. It is not necessary to understand the game of baseball to understand how to construct a VI to compute batting averages!

During an appearance at the plate, we suppose that one of two things happens: the batter hits the ball, or the batter does not hit the ball. We consider first the situation where the batter does not hit the ball. In this case, one of three things occurs: (1) the batter strikes out, (2) the batter receives a base on balls (i.e., a walk), or (3) the batter is hit by the pitch. Even though the batter reaches first base safely, the walk and being hit by the pitch do not count as official appearances at the plate; hence they do not factor into the batting average. On the other hand, a strikeout is an official at-bat and reduces the batting average.

If the batter hits the ball, one of four situations can arise: (1) the batter gets a clean hit, (2) the batter reaches base due a fielding error by a defensive player, (3) the batter is put out (e.g., flies out or grounds out), or (4) the batter is out on a sacrifice (that is, a base runner advances on the play, but the batter is out nonetheless). A sacrifice does not count against the batting average, and only hitting safely increases the batting average. That means that hitting the ball does not necessarily improve the batting average—an error by the defensive player on a ball hit in play will actually reduce the batting average!

Let H denote the number of clean hits, K the number of strikeouts, E the number of errors committed by the defense, and O the number of fly-outs, ground outs, and other episodes wherein the batter is thrown out. The total number of times a batter appears at the plate is $N = H + K + E + O + S + BB + HP$; however, the number of walks (BB), hit by pitches (HP), and sacrifices (S) do not contribute to the number of official appearances at the plate; hence they do not contribute to the batting average computation. Therefore, according to our simplistic evaluation of how to compute the batting average, we utilize the formula

$$Batting\ Average = 1000 \frac{H}{H + K + E + O}.$$

The factor of 1000 is incorporated into the batting average to account for the fact that, by tradition, a perfect batting average is 1000.

Your goal is to develop a VI to compute the batting average according to the formula above. The following list serves as a step-by-step guide. Feel free to deviate from the provided list and program according to your own style. As you proceed through this book you will find that the step-by-step lists will slowly disappear, and you will have to construct your VIs on your own (with help, of course!).

1. Open a new VI. Save the untitled VI as **Batting Average.vi** in the folder **Users Stuff** in the **Learning** directory. In this case you are saving the VI before any programming has actually occurred. This is a matter of personal choice, but saving a VI frequently during development may save you lots of rework if your system freezes up or crashes.

2. Add a numeric **control** to the front panel and label it **H**. This will be where the number of hits is input.

3. Make three duplicates (or clones) of the numeric control. Using the **Positioning** tool, select the numeric control while pressing the <Ctrl> (**Windows**) or <Option> (**Macintosh**) key and drag the numeric control to a new location. When you release the mouse button, a copy labeled **H 2** appears in the new location. Repeat this process until four numeric controls are visible on the front panel.

4. Relabel the four numeric controls: **H, K, E,** and **O**.

5. Place the numeric controls in a column on the left side of the front panel. Using the **Align Objects** tool, align the four numeric controls by their left edges.

6. Add a numeric **book** to the front panel and label it **Batting Average**. This will be where the batting average is displayed.

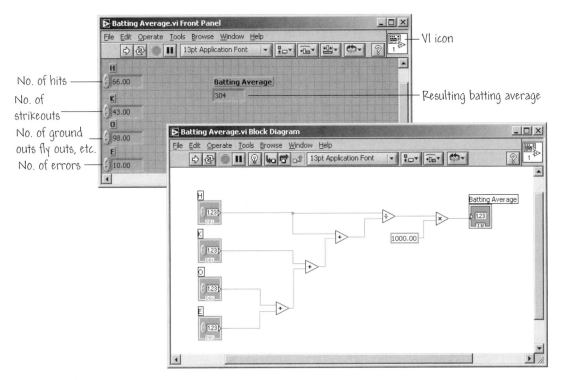

No. of hits

No. of strikeouts

No. of ground outs fly outs, etc.

No. of errors

VI icon

Resulting batting average

FIGURE 4.2
The Batting Average VI.

7. Switch to the block diagram. Arrange the control and indicator terminals so that the four control terminals appear aligned in a vertical column on the left side and the batting average indicator appears on the right side.

8. Program the batting average formula by wiring the block diagram. You can use the VI shown in Figure 4.2 as a guide.

9. The batting average should be rounded off and displayed as an integer value. To accomplish this in the VI, you need to change the format of the indicator. Pop up on the digital indicator and select **Representation**; then select **I32** as shown in Figure 4.3.

10. Once the block diagram has been wired, switch back to the front panel and input values into the various digital controls. For example, let $H = 66$, $K = 43$, $O = 98$, and $E = 10$.

11. Run the program with **Run Continuously** and select **Highlight Execution** to watch the data flow through the program. Change the various values of the input parameters to see how the batting average changes. When you are finished experimenting, you can stop the program execution.

12. Practice single stepping through the program using the step buttons on the block diagram.

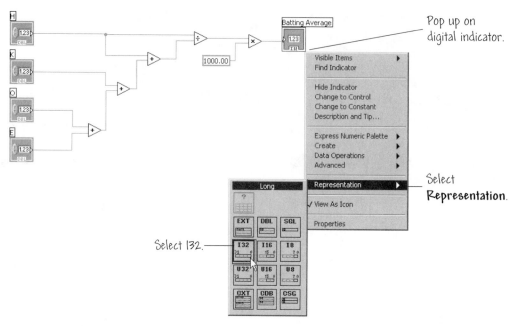

FIGURE 4.3
Changing the representation of the digital indicator to I32.

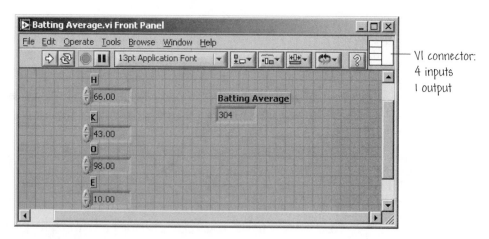

FIGURE 4.4
The Batting Average icon terminal.

13. Once the VI execution is complete, switch back to the front panel and pop up on the icon in the upper right corner. Select **Show Connector** from the menu. Observe the connector that appears—it should indicate four input terminals and one output terminal, as illustrated in Figure 4.4. We will be learning how to edit the icon and connector in the next section.

14. Save the VI by selecting **Save** from the **File** menu.

This exercise provided the opportunity to practice placing objects on the front panel, wiring them on the block diagram, using editing techniques to arrange the objects for easy visual inspection of the code, and using debugging tools to verify proper operation of the VI. The Batting Average VI can be used as a subVI (requiring some effort to wire the VI connector properly), and that is the subject of the next sections.

 A working version of this Batting Average VI can be found in the Chapter 4 *folder of the* Learning *directory. It is called* Batting Average.vi.

4.3   EDITING THE ICON AND CONNECTOR

A subVI is represented by an icon in the block diagram of a calling VI. The subVI also must have a connector with terminals properly wired to pass data to and from the top-level VI. In this section we discuss icons and connectors.

4.3.1   Icons

Every VI has a default icon displayed in the upper-right corner of the front panel and block diagram windows. The default icon is a picture of the LabVIEW logo and a number indicating how many new VIs you have opened since launching LabVIEW. You use the **Icon Editor** to customize the icon. To activate the editor, pop up on the default icon in the front panel and select **Edit Icon** as illustrated in Figure 4.5. You can also display the **Icon Editor** by double clicking on the default icon.

 You can only open the **Icon Editor** *when the VI is not in the run mode. If necessary, you can change the mode of the VI by selecting* **Change to Edit Mode** *from the* **Operate** *menu.*

The **Icon Editor** dialog box is shown in Figure 4.6. You use the tools on the palette at the left of the dialog box to create the icon design in the pixel editing area. An image of the icon (in actual size) appears in a vertical column just right of center in the dialog box and to the right of the pixel editing area. Depending on the type of monitor you are using, you can design and save a separate icon for B&W, 16-color, and 256-color mode. The editor defaults to **Black & White**, but you can click on one of the other color options to switch modes. To copy the icon from B&W to 16 colors, select the 16-color box and then click on the **Copy from: Black & White** button located at the upper right side of the **Icon Editor**

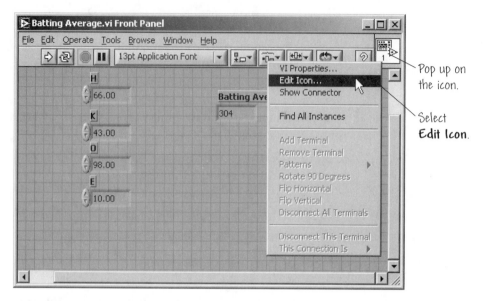

FIGURE 4.5
Activating the **Icon Editor**.

dialog box. Similarly, you can use the appropriate **Copy from** menu items to copy icons from 16 colors to 256 colors, 16 colors to B&W, and so forth.

The tools in the **Icon Editor** palette to the left of the editing area perform many functions. If you have used any other paint program, you should be familiar with the tools available in the editor. Table 4.1 describes the various tools. The options at the right of the editing screen perform the following functions:

- **Show terminal**—Click on this option to display the terminal pattern of the connector overlayed on the icon.

- **OK**—Click on this button to save your drawing as the VI icon and return to the front panel window.

- **Cancel**—Click on this button to return to the front panel window without saving any changes.

You can use the **Icon Editor** menu to cut, copy, and paste images from and to the icon. When you paste an image and a portion of the icon is selected, the image is resized to fit into the selection.

As an exercise, open Batting Average.vi and edit the icon. Remember that you developed the Batting Average.vi at the beginning of this chapter—it should be saved in User Stuff. In case you did not save it before or cannot find it now, you can open the VI located in the Chapter 4 folder in the Learning directory. An example of an edited icon is shown in Figure 4.7. Can you replicate the new icon? Check out the Batting Average Edited.vi to see the edited icon shown in Figure 4.7.

TABLE 4.1   Icon Editor tools.

	Draws and erases pixel by pixel.
	Draws straight lines. Use <shift> to restrict drawing to horizontal, vertical, and diagonal lines.
	Selects the foreground color from an element in the icon.
	Fills an outlined area with the foreground color.
	Draws a rectangle bordered with the foreground color. Double click on this tool to frame the icon in the foreground color.
	Draws a rectangle bordered with the foreground color and filled with the background color. Double click on this tool to frame the icon in the foreground color and fill it with the background color.
	Selects an area of the icon for moving, cloning, deleting, or performing other changes. Double click on this tool and press Delete on the keyboard to delete the entire icon at once.
	Enters text into the icon. Double click on this tool to select a different font. Small fonts is commonly used as the fonts in icons.
	Displays the current foreground and background colors. Click on each to get a palette from which you can choose new colors.

4.3.2 Connectors

The connector is a set of terminals that correspond to the VI controls and indicators. This is where the inputs and outputs of the VI are established so that it can be used as a subVI. A connector receives data at its input terminals and passes the data to its output terminals when the VI is finished executing. Each terminal corresponds to a particular control or indicator on the front panel. The connector terminals act like parameters in a subroutine parameter list of a function call.

If you use the front panel controls and indicators to pass data to and from subVIs, these controls or indicators need terminals on the connector pane. In this section, we will discuss how to define connections by choosing the number of terminals you want for the VI and assigning a front panel control or indicator to each of those terminals.

You can view and edit the connector pane from the front panel only.

Icon Editor
tool palette

Use these buttons to copy icon from B & W to
16 colors, 16 colors to 256 colors, and so forth.

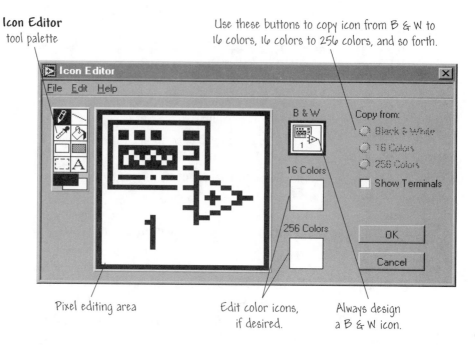

Pixel editing area

Edit color icons,
if desired.

Always design
a B & W icon.

FIGURE 4.6
The **Icon Editor**.

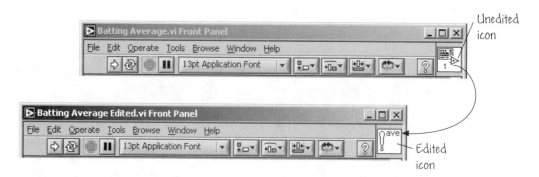

Unedited
icon

Edited
icon

FIGURE 4.7
Editing the Batting Average.vi icon.

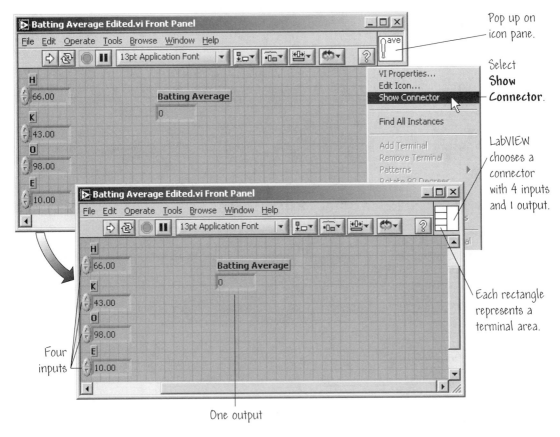

FIGURE 4.8
Defining a connector.

To define a connector, you select **Show Connector** from the icon pane pop-up menu on the front panel window as shown in Figure 4.8. The connector replaces the icon in the upper-right corner of the front panel window. By default, LabVIEW displays a terminal pattern based on the number of controls and indicators on your front panel, as shown in Figure 4.8. Control terminals are on the left side of the connector pane, and indicator terminals are on the right. If desired, you can select a different terminal pattern for your VI.

4.3.3 Selecting and Modifying Terminal Patterns

To select a different terminal pattern for your VI, pop up on the connector and choose **Patterns** from the pop-up menu. This process is illustrated in Figure 4.9. To change the pattern, click on the desired pattern on the palette. If you choose a new pattern, you will lose any assignment of controls and indicators to the terminals on the old connector pane.

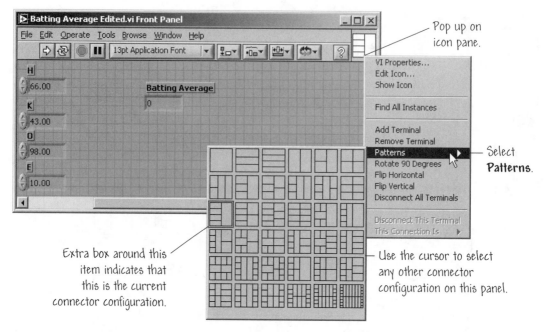

FIGURE 4.9
Changing your terminal pattern.

The maximum number of terminals available for a subVI is 28.

If you want to change the spatial arrangement of the connector terminal patterns, choose one of the following commands from the connector pane pop-up menu: **Flip Horizontal**, **Flip Vertical**, or **Rotate 90 Degrees**. If you want to add a terminal to the connector pattern, place the cursor where the terminal is to be added, pop-up on the connector pane window and select **Add Terminal**. If you want to remove an existing terminal from the pattern, pop up on the terminal and select **Remove Terminal**.

Think ahead and plan your connector patterns well. For example, select a connector pane pattern with extra terminals if you think that you will add additional inputs or outputs at a later time. With these extra terminals, you do not have to change the connector pane for your VI if you find you want to add another input or output. This flexibility enables you to make subVI changes with minimal effect on your hierarchical structure. Another useful hint is that if you create a group of subVIs that are often used together, consider assigning the subVIs a consistent connector pane with common inputs. You then can easily remember each input location without using the **Context Help** window—this will save you time. If you create a subVI that produces an output that is used as the input to another subVI, try to align the input and output connections. This technique simplifies your wiring patterns and makes debugging and program maintenance easier.

 Place inputs on the left and outputs on the right of the connector when linking controls and indicators—this prevents complex wiring patterns.

4.3.4   Assigning Terminals to Controls and Indicators

Front panel controls and indicators are assigned to the connector terminals using the **Wiring** tool. The following steps are used to associate the connector pane with the front panel controls and indicators.

1. Click on the connector terminal with the **Wiring** tool. The terminal turns black, as illustrated in Figure 4.10a.

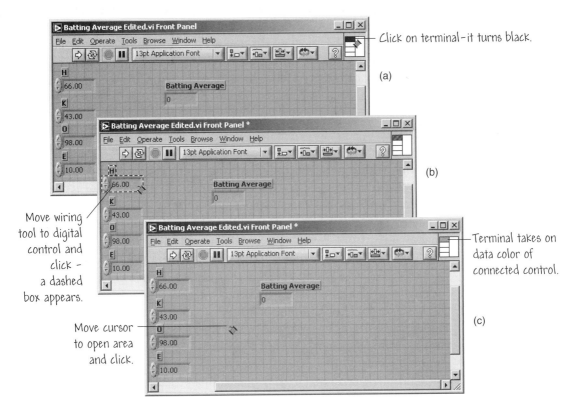

FIGURE 4.10
Assigning connector terminals to controls and indicators.

2. Click on the front panel control or indicator that you want to assign to the selected terminal. As shown in Figure 4.10b, a dotted-line **marquee** frames the selected control.

3. Position the cursor in an open area of the front panel and click. The marquee disappears, and the selected terminal takes on the data color of the connected object, indicating that the terminal is assigned. This process is illustrated in Figure 4.10c.

The connector terminal turns white to indicate that a connection was not made. If this occurs, you need to repeat steps 1 through 3 until the connector terminal takes on the proper data color.

4. Repeat steps 1–3 for each control and indicator you want to connect.

The connector terminal assignment process also works if you select the control or indicator first with the **Wiring** tool and then select the connector terminal. As already discussed, you can choose a pattern with more terminals than you need since unassigned terminals do not affect the operation of the VI. It is also true that you can have more front panel controls or indicators than terminals. Once the terminals have been connected to controls and indicators, you can disconnect them all at one time by selecting **Disconnect All Terminals** in the connector pane short cut menu. Note that although the **Wiring** tool is used to assign terminals on the connector to front panel controls and indicators, no wires are drawn.

4.4 THE HELP WINDOW

You enable the **Context Help** window by selecting Help≫Show Context Help. When you do this, you find that whenever you move an editing tool across a subVI node, the **Context Help** window displays the subVI icon with wires attached to each terminal. An example is shown in Figure 4.11 where the cursor was moved over the subVI Temp & Vol.vi, and the **Context Help** window appears and shows that the subVI has two outputs: Temp and Volume.

LabVIEW has a help feature that can keep you from forgetting to wire subVI connections—indications of required, recommended, and optional connections in the connector pane and the same indications in the **Context Help** window. For example, by classifying an input as **Required**, you can automatically detect whether you have wired the input correctly and prevent your VI from running if you have not wired correctly. To view or set connections as **Required**, **Recommended**, or **Optional**, click a terminal in the connector pane on the VI front panel and select **This Connection Is**. A checkmark indicates its status, as shown in Figure 4.12. By default, inputs and outputs of VIs you create are set to **Recommended**—if a change is desired, you must change the default to either **Required** or **Optional**.

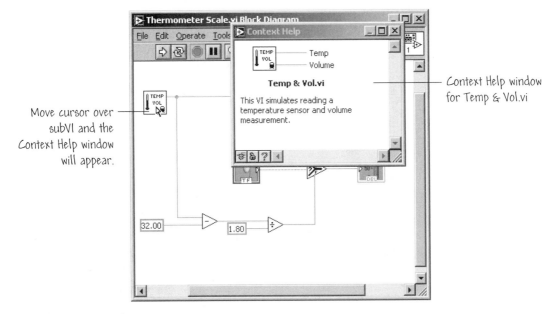

Move cursor over
subVI and the
Context Help window
will appear.

Context Help window
for Temp & Vol.vi

FIGURE 4.11
The **Context Help** window.

When you make a connection **Required**, then you cannot run the VI as a subVI unless that connection is wired correctly. In the **Context Help** window, **Required** connections appear in bold text. When you make a connection **Recommended**, then the VI can run even if the connection is not wired, but the error list window will list a warning. In the **Context Help** window, **Recommended** connections appear in plain text. A VI with **Optional** connections can run without the optional terminals being wired. In the **Context Help** window, **Optional** connections are grayed text for detailed diagram help and hidden for simple diagram help.

**Building
a SubVI**

A VI designed to compute a baseball batting average was presented in Section 4.2. In this exercise you will assign the connector terminals of that VI to the digital controls and digital indicator so that the VI can be used as a subVI by other programs. Begin by opening Batting Average Edited.vi in the Chapter 4 folder of the Learning directory. Recall that the difference between the two VIs Batting Average Edited and Batting Average is that the edited version has a custom icon rather than the default icon. A subVI with an edited icon (rather than the default icon) enables easier visual determination of the purpose of the subVI because you can inspect the icon as it exists in the code.

When you have completed assigning the connector terminals, the front panel connector pane should resemble the one shown in Figure 4.13. To display the terminal connectors, pop up on the VI icon in the front panel and select **Show Connector**. You should notice that a terminal pattern appropriate for the VI has

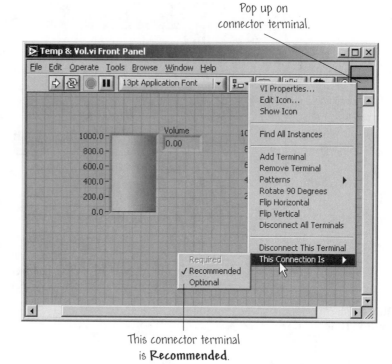

FIGURE 4.12
Showing the status of terminal connectors.

been automatically selected with four inputs on the left side of the connector pane and one output on the right. The number of terminals selected by default depends on the number of controls and indicators on the front panel.

Using the **Wiring** tool, assign the four input terminals to the four numeric controls: H, K, O, and E. Similarly, assign the output terminal to the numeric indicator **Batting Average**.

There are two levels of code documentation that you can pursue. It is important to document your code to make it accessible and understandable to other users. You can document a VI by choosing File≫VI Properties≫Documentation. A **Description** box will appear as illustrated in Figure 4.14. You type a description of the VI in the dialog box, and then whenever **Show Context Help** is selected and the cursor is placed over the VI icon, your description will appear in the **Context Help** window.

You can also document the objects on the front panel by popping up on an object and choosing **Description and Tip**... from the object short cut menu. Type the object description in the dialog box that appears as in Figure 4.15. You cannot edit the object description in the run mode—change to edit mode.

This VI is now ready
to be used as a subVI.   4 inputs (orange)   1 output (blue)

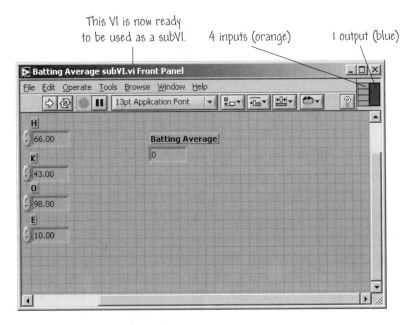

FIGURE 4.13
Assigning terminals for the Batting Average Edited.vi.

Add description
in this block area.

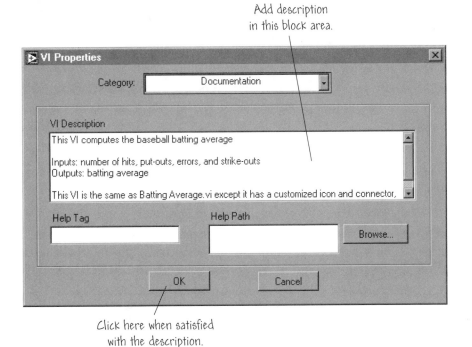

Click here when satisfied
with the description.

FIGURE 4.14
Typing the description for the subVI.

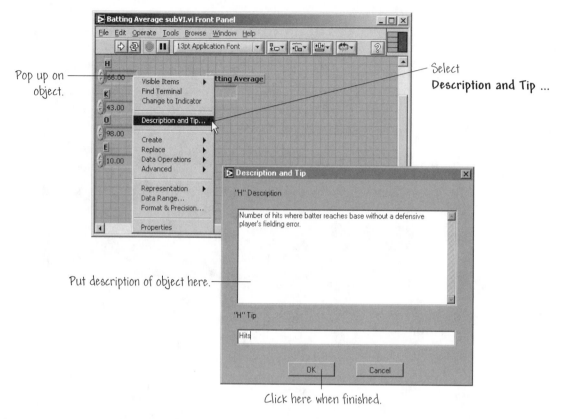

Pop up on object.

Put description of object here.

Select
Description and Tip ...

Click here when finished.

FIGURE 4.15
Documenting objects on a subVI.

Continuing with the exercise, you should add a description for each object on the front panel. Some suggested descriptions follow:

- *H*: Number of hits where batter reaches base without a defensive player's fielding error.

- *K*: Number of strikeouts.

- *O*: Number of put-outs, including ground outs, fly outs, pop ups, etc.

- *E*: Number of times the batter reaches base on a fielding error by a defensive player.

- *Batting Average*: Computed batting average.

When you have finished assigning the connector terminals to the controls and indicators, and you have documented the subVI and each of the controls and indicators, save the subVI as **Batting Average subVI.vi**. Make sure to save your work in the **Users Stuff** folder.

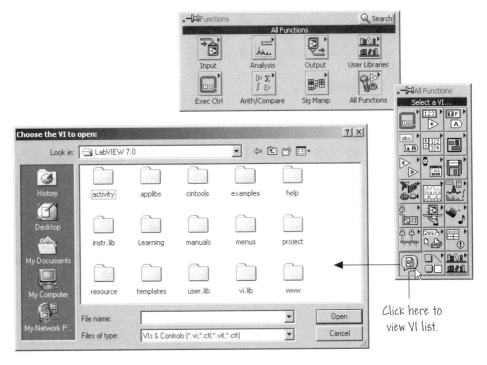

FIGURE 4.16
The **Select a VI**... palette.

   *A working version of Batting Average subVI can be found in the **Chapter 4** folder in the **Learning** directory. You may want to open the VI, read the documentation, and take a look at the icon and the connector just to verify that your construction of the same subVI was done correctly.* ◆

4.5   USING A VI AS A SUBVI

There are two basic ways to create and use a subVI: creating subVIs from VIs and creating subVIs from selections. In this section we concentrate on the first method, that is, using a VI as a subVI. Any VI that has an icon and a connector can be used as a subVI. In the block diagram, you can select VIs to use as subVIs from Functions≫All Functions≫Select a VI... palette. Choosing this option produces a file **dialog box** from which you can select any available VI in the system, as shown in Figure 4.16.

The *LabVIEW Student Edition* comes with many ready-to-use VIs. In Chapter 1 you searched around the LabVIEW examples and found many of the example and demonstration VIs. In the following example you will use one of the preexisting VIs called Generate Waveform.vi as a subVI.

**Using
a VI as a
SubVI**

1. Open a new front panel.

2. Select a vertical slide switch control from the Controls≫Buttons & Switches palette and label it Temperature Scale. Place free labels on the vertical slide switch to indicate Fahrenheit and Celsius using the **Labeling** tool, as shown in Figure 4.17, and arrange the labels as shown in the figure.

3. Select a thermometer from Controls≫Numeric Indicators and place it on the front panel. Label the thermometer Temperature.

4. Change the range of the thermometer to accommodate values ranging between 0.0 and 100.0. With the **Operating** tool, double click on the high limit and change it to 100.0 if necessary.

5. Switch to the block diagram by selecting Window≫Show Block Diagram.

6. Pop up in a free area of the block diagram and choose Functions≫All Functions≫Select a VI. . . to access the dialog box. Select Temp & Vol.vi in the Learning\Building Blocks directory. Click **Open** in the dialog box to place Temp & Vol.vi on the block diagram.

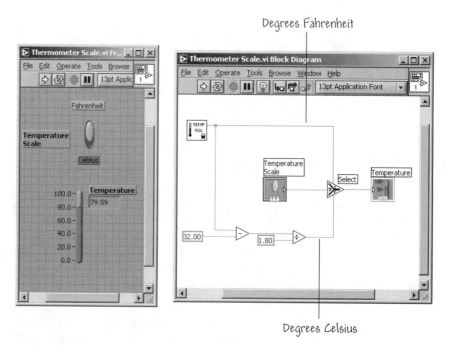

FIGURE 4.17
Calling the subVI Temp & Vol.vi.

7. Add the other objects to the block diagram using Figure 4.17 as a guide.

   ■ Place a Subtract function and a Divide function on the block diagram.
     These are located in the **Functions** palette on the Arithmetic & Com-
     parison≫Express Numeric subpalette. On the Subtract function, add
     a numeric constant equal to 32. You can add the constant by popping up
     on the Subtract function and selecting **Create Constant**. Then using the
     Labeling tool change the constant from the default 0.0 to 32.0. Similarly,
     add a numeric constant of 1.8 on the Divide function. These constants
     are used to convert from degrees Fahrenheit to Celsius according to the
     relationship

$$°C = \frac{°F - 32}{1.8}.$$

   ■ Add the Select function (located on the Arithmetic & Comparison≫Ex-
     press Comparison subpalette on the **Functions** palette). The Select
     function returns the value wired to the TRUE or FALSE input, depend-
     ing on the Boolean input value. Use **Show Context Help** for more
     information on how this function works.

8. Wire the diagram objects as shown in Figure 4.17.

9. Switch to the front panel and click the **Run Continuously** button in the
   toolbar. The thermometer shows the value in degrees Fahrenheit or degrees
   Celsius, depending on your selection.

10. Switch the scale back and forth to select either Fahrenheit or Celsius.

11. When you are finished experimenting with your VI, save it as Thermometer
    Scale.vi in the Users Stuff directory.   ◆

*A working version of Thermometer Scale.vi exists in the Chapter 4 folder
located in the Learning directory. When you open this VI, notice the use of color
on the vertical switch (orange denotes Fahrenheit and blue represents Celsius).*

In the previous exercise, the stand-alone Temp & Vol.vi was used in the role
of a subVI. Suppose that you have a subVI on your block diagram and you want
to examine its contents—to view the code. You can easily open a subVI front
panel window by double-clicking on the subVI icon. Once the front panel opens,
you can then open the subVI block diagram by selecting **Show Block Diagram**
from the **Window** menu. At that point, any changes you make to the subVI code
alter only the version in memory—until you save the subVI. Also note that, even
before saving the subVI, the changes affect all calls to the subVI and not just the
node you used to open the VI.

4.6   CREATING A SUBVI FROM A SELECTION

The second way to create a subVI is to select components of the main VI and group them into a subVI. You capture and group related parts of VIs by selecting the desired section of the VI with the **Positioning** tool and then choosing **Create SubVI** from the **Edit** pull-down menu. The selection is automatically converted into a subVI, and a default icon replaces the entire section of code. The controls and indicators for the new subVI are automatically created and wired to the existing wires. Using this method of creating subVIs allows you to modularize your block diagram, thereby creating a hierarchical structure.

Building a Sub VI Using the Selection Technique

In this exercise you will modify the Thermometer Scale VI developed in Section 4.5 to create a subVI that converts Fahrenheit temperature to Celsius temperature. The subVI selection process is illustrated in Figure 4.18.

Open Thermometer Scale VI by selecting **Open** from the **File** menu. The VI is located in the Chapter 4 folder of the Learning directory. For reference, the front panel and block diagram are shown in Figure 4.17.

To create a subVI that converts Fahrenheit to Celsius, begin by switching to the block diagram window by choosing **Show Block Diagram** from the **Window** menu. The goal is to modify the existing block diagram to call a subVI created using the **Create SubVI** option.

Select the block diagram elements that comprise the conversion from Fahrenheit to Celsius, as shown in Figure 4.18. A moving dashed line will frame the chosen portion of the block diagram. Now select **Create SubVI** in the **Edit** menu. A default subVI icon will appear in place of the selected group of objects. You can use this selection method of creating subVIs to modularize your VI.

The next step is to modify the icon of the new subVI. Open the new subVI by double-clicking on the default icon called **Untitled 1 (SubVI)**. Two front panel objects should be visible: one numeric control labeled Temp and one unlabeled numeric indicator. Pop up on the numeric indicator and select Visible Items≫Label from the short cut menu. Type Celsius Temp in the text box and then align the two objects on the front panel by their bottom edges.

Use the **Icon Editor** to create an icon similar to the one shown in Figure 4.19. Invoke the editor by popping up in the **Icon Pane** in the front panel of the subVI and selecting **Edit Icon** from the pop-up menu. Erase the default icon by double clicking on the Select tool and pressing <Delete>. Redraw the icon frame by double clicking on the rectangle tool which draws a rectangle around the icon. The easiest way to create the text is using the **Text** tool and the font Small Fonts. The arrow can be created with the **Pencil** tool.

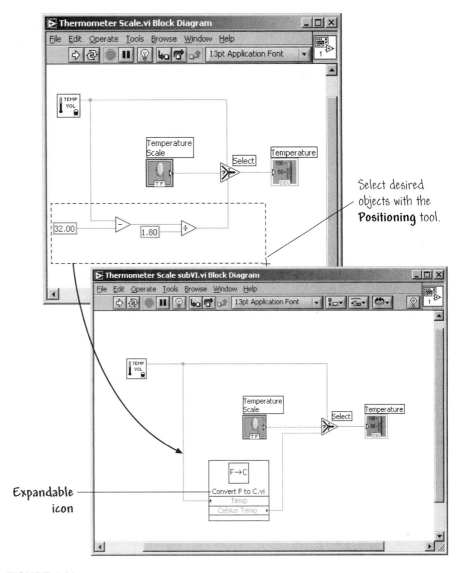

FIGURE 4.18
Selecting the code that converts from Fahrenheit to Celsius.

The connector is automatically wired when a subVI is created using the **Create SubVI** *option.*

When you are finished editing the icon, close the **Icon Editor** by clicking on **OK**. The new icon appears in the upper-right corner of the front panel, as shown in Figure 4.19. Right click on the Convert F to C.vi and click on **View as Icon** to remove the checkmark. Then use the **Positioning** tool to expand the icon to

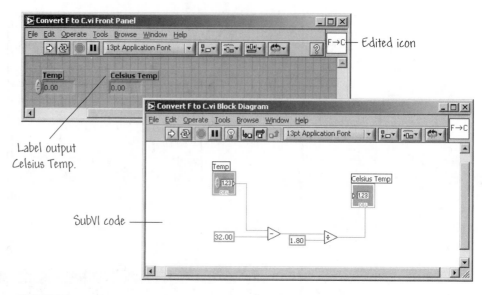

Edited icon

Label output
Celsius Temp.

SubVI code

FIGURE 4.19
The new subVI to convert degrees Fahrenheit to degrees Celsius.

show the input **Temp** and the output **Celsius Temp**, as shown in Figure 4.18. Save the subVI by choosing **Save** from the **File** menu. Name the VI Convert F to C and save it in the Users Stuff folder.

> *You cannot build a subVI from a section of code with more than 28 inputs and outputs because 28 is the maximum number of inputs and outputs allowed on a connector pane.* ◆

4.7   SAVING YOUR SUBVI

It is highly recommended that you save your subVIs to a file in a directory rather than in a library. While it is possible to save multiple VIs in a single file called a **VI library**, this is not desirable. Saving VIs as individual files is the most effective storage path because you can copy, rename, and delete files easier than when using a VI library.

VI libraries have the same load, save, and open capabilities as other directories, but they are not hierarchical. That is, you cannot create a VI library inside of another VI library, nor can you create a new directory inside a VI library. After you create a VI library, it appears in the **File** dialog box as a file with an icon that is somewhat different from a VI icon.

4.8   THE HIERARCHY WINDOW

When you create an application, you generally start at the top-level VI and define the inputs and outputs for the application. Then you construct the subVIs that you will need to perform the necessary operations on the data as it flows through the block diagram. As discussed in previous sections, if you have a very complex block diagram, you should organize related functions and nodes into subVIs for desired block diagram simplicity. Taking a modular approach to program development creates code that is easier to understand, debug, and maintain.

The **Hierarchy Window** displays a graphical representation of the hierarchical structure of all VIs in memory and shows the dependencies of top-level VIs and subVIs. There are several ways to access the **Hierarchy Window**:

- You can select Browse≫Show VI Hierarchy to open the **Hierarchy Window** with the VI icon of the current **active window** surrounded by a thick red border.

- You can pop up on a subVI and select **Show VI Hierarchy**n the **Hierarchy Window** with the selected subVI surrounded by a thick red border.

- If the **Hierarchy Window** is already open, you can bring it to the front by selecting it from the list of open windows under the **Window** menu.

The **Hierarchy Window** for the Thermometer Scale subVI.vi is shown in Figure 4.20. The window displays the dependencies of VIs by providing information on VI callers and subVIs. As you move the **Operating** tool over objects in the window, the name of the VI is shown below the VI icon. This window also contains a toolbar, as shown in Figure 4.20, that you can use to configure several types of settings for displayed items.

You can switch the **Hierarchy Window** display mode between horizontal and vertical display by pressing the **Horizontal Layout** or **Vertical Layout** button hierarchy toolbar. In a horizontal display, subVIs are shown to the right of their calling VIs; in a vertical display, they are shown below their calling VIs. In either case, the subVIs are always connected with lines to their calling VIs. The window shown in the Figure 4.20 is displayed vertically.

Arrow buttons and arrows beside nodes indicate what is displayed and what is hidden according to the following rules:

- A red arrow button pointing towards the node indicates some or all subVIs are hidden, and clicking the button will display the hidden subVIs.

- A black arrow button pointing towards the subVIs of the node indicates all immediate subVIs are shown.

- A blue arrow pointing towards the callers of the node indicates the node has additional callers in this VI hierarchy but they are not shown at the present time. If you show all subVIs, the blue arrow will disappear. If a node has no subVIs, no red or black arrow buttons are shown.

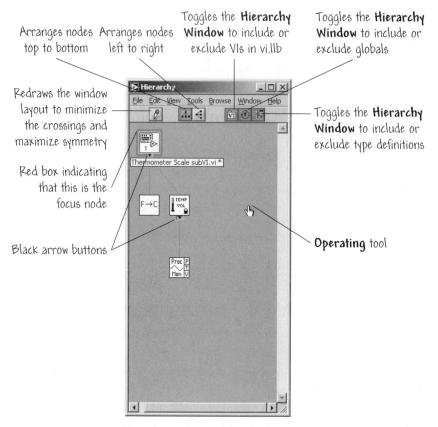

Arranges nodes top to bottom

Arranges nodes left to right

Toggles the **Hierarchy Window** to include or exclude VIs in vi.llb

Toggles the **Hierarchy Window** to include or exclude globals

Redraws the window layout to minimize the crossings and maximize symmetry

Toggles the **Hierarchy Window** to include or exclude type definitions

Red box indicating that this is the focus node

Black arrow buttons

Operating tool

FIGURE 4.20
The **Hierarchy Window** for Thermometer Scale subVI.vi.

Double clicking on a VI or subVI opens the front panel of that node. You also can pop up on a VI or subVI node to access a menu with options, such as showing or hiding all subVIs, opening the VI or subVI front panel, editing the VI icon, and so on.

You can initiate a search for a given VI by simply typing the name of the node directly onto the **Hierarchy Window**. When you begin to type in text from the keyboard a small search window will automatically appear displaying the text that has been typed and allowing you to continue adding text. The search for the desired VI commences immediately. The search window is illustrated in Figure 4.21. If the characters currently displayed in the search window do not match any node names in the search path, the system beeps, and no more characters can be typed. You can then use the <Backspace> or <Delete> key to delete one or more characters to resume typing. The search window disappears automatically if no keys are pressed for a certain amount of time, or you can press the <Esc> key to remove the search window immediately. When a match is found you can use the right or down arrow key, or the <Enter> key on Windows

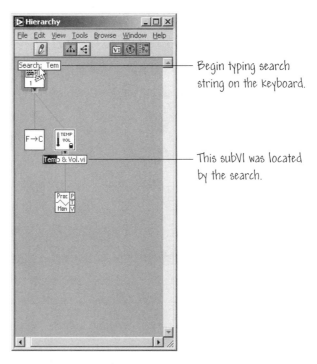

FIGURE 4.21
Searching the **Hierarchy Window** for VIs.

and the <Return> key on Macintosh to find the next node that matches the search string. To find the previous matching node, press the left or up arrow key, or the <Shift-Enter> on Windows and <Shift-Return> on Macintosh.

BUILDING BLOCK

4.9   BUILDING BLOCKS: MEASURING VOLUME

In this exercise you will continue the work on the VI Volume1 created in Chapter 2. The goal is to make the necessary modifications so that you can use the VI as a subVI. This means that you need to edit the icon and assign the connector terminals. The first step is to create the icon and connector. You can use Figure 4.22 as a guide.

When you have finished readying the subVI, save it in the Users Stuff folder as VolumeSubVI.vi. In many cases you will develop subVIs for general use; hence you will want them accessible from the **All Functions** palette. To accomplish this task for the VolumeSubVI VI, you need to save it in the folder User.lib. After you have saved the subVI in User.lib, you will need to exit LabVIEW and then open it again. Once this is done, open a new VI, switch to the block diagram,

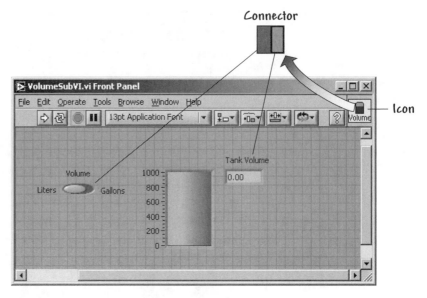

FIGURE 4.22
Wiring the connector of Volume subVI.

and show the Functions≫All Functions palette. Click on the **User Libraries** palette, and you should find the icon for VolumeSubVI, as shown in Figure 4.23.

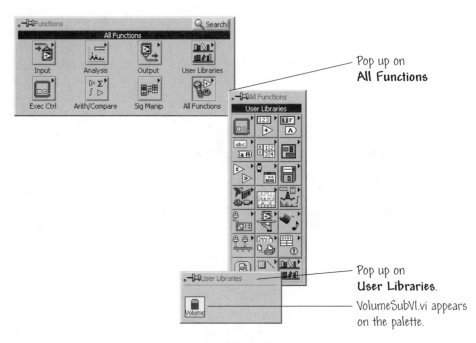

FIGURE 4.23
Locating the Volume subVI on the **User Library** palette.

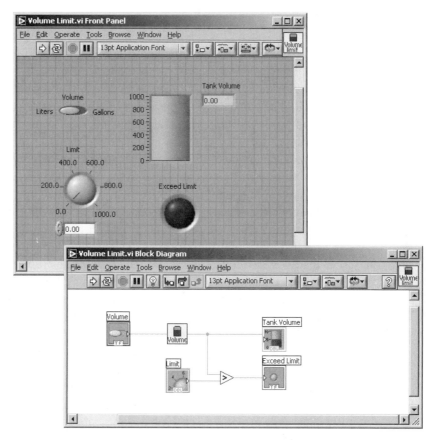

FIGURE 4.24
Using the VolumeSubVI.

The subVI is now readily accessible to you in your future programming endeavors. You could also have accessed the subVI through Functions≫Select a VI. . . by navigating to Users Stuff where you previously saved the program.

Using the VolumeSubVI, construct a VI that indicates (using a large LED) whether a volume limit has been exceeded. The volume limit should be a user input parameter. You can use the VI shown in Figure 4.24 as a guide. When you have completed construction of the program and have verified that it is working properly (using the various debugging tools, such as **Highlight Execution** and probes), you can save the VI as Volume Limit in the Users Stuff folder in the Learning directory.

4.10   RELAXED READING: LABVIEW REAL-TIME AT BIOSPHERE

The Biosphere 2 research facility has been operating since the 1980s, The facility was described as "an engineering marvel" and recently voted by the Discovery

Channel as the third most advanced engineering project of the twentieth century. Biosphere 2 was a 3.2-hectare footprint, closed-system facility, containing seven terrestrial biomes, including an ocean system, a desert biome and a tropical rainforest. From the outset, this experimental facility provided novel scientific insights into the complex natural processes that operate on Earth at the planetary scale. The core design team of Biosphere 2 is now performing advanced research in the rapidly developing fields of biospherics, which includes controlled ecological life support systems and advanced life support systems. The current initiative is the development of a series of small-scale modular biospheres.

Researchers at the Biosphere Foundation were challenged with constructing an advanced prototype Laboratory Biosphere to perform closed-system experiments requiring the acquisition of data from a wide range of sensors, as well as the archiving, retrieval, and graphical presentation of the results. They also needed to remotely control critical mechanical systems in real time. The solution that they developed employed the National Instruments FieldPoint I/O system to collect data and provide control capabilities, as well as LabVIEW Real-Time and the LabVIEW Datalogging and Supervisory Control Module to manipulate and archive data and output data over the Internet for controlled remote access.

The Laboratory Biosphere was designed to perform experiments on sustainable, soil-based, agricultural systems and evaluate interactions along multiple vectors, including light, water, atmospheric composition, temperature, and humidity to determine optimal growing conditions for a variety of selected crops within closed-system environments. The entire range of modular biosphere systems will provide essential base-line data for many developing technologies. In addition, this data, together with the practical know-how developed in the operation of these complex systems, will be critical to a wide range of future applications, which potentially includes the successful operation of a sustainable research base on Mars.

Because of the pioneering nature of the development of the biosphere system, no single off-the-shelf hardware or software systems exist to provide immediate solutions to the complex specifications of the modular biosphere project. Initially, the use of microprocessors to acquire data from a large number of sensors and provide feedback and control systems using workstations was considered. However, such an approach would have demanded long hours dedicated to designing, programming, testing, and tweaking the systems before achieving even simple initial objectives. Instead, the National Instruments LabVIEW Real-Time and the LabVIEW Datalogging and Supervisory Control Module running under Windows 2000 and NI FieldPoint I/O hardware were selected as the basis of the overall solution. The system included a range of eight-channel analog input modules to collect and transfer data from the sensors and modules to perform manual and automated system control. With LabVIEW Real-Time and FieldPoint I/O, a low-cost and highly-integrated solution was achieved. The system is flexible enough for future development and eliminates the need for custom-designed system components.

In the summer of 2002, a significant amount of interesting data was collected during a closed experiment in growing a crop of dwarf soybeans. The ability of the NI-based system to integrate with the Internet was key for rapid publication and dissemination of real-time data. Eventually, it will be possible for students of all levels to track and participate in real-time discussions of closed-system experiments as they develop from start to finish.

New features currently being added to the system include integrating cameras into the test module that will use the vision software functionality to provide feedback to better understand and control of the conditions of the sealed Laboratory Biosphere.

For more information, contact:

Gerard Houghton or Mark Van Thicco
Biosphere Foundation
9 Silver Hills Road
Santa Fe, NM 87505

4.11 SUMMARY

Constructing subVIs was the main topic of this chapter. One of the keys to constructing successful LabVIEW programs is understanding how to build and use subVIs. SubVIs are the primary building blocks of modular programs, which are easy to debug, understand, and maintain. They are analogous to subroutines in text-based programming languages like C or Fortran. Two methods of creating subVIs were discussed—creating subVIs from VIs and creating subVIs from VI selections. A subVI is represented by an icon in the block diagram of a calling VI and must have a connector with terminals properly wired to pass data to and from the calling VI. The **Icon Editor** was discussed as the way to personalize the subVI icon so the information about the function of the subVI is readily apparent from visual inspection of the icon. The editor has a tool set similar to that of most common paint programs. The **Hierarchy Window** was introduced as a helpful tool for managing the hierarchical nature of your programs.

Active window: Window that is currently set to accept user input, usually the frontmost window. You make a window active by clicking on it or by selecting it from the **Windows** menu.

Control: Front panel object for entering data to a VI interactively or to a subVI programmatically.

Description box: Online documentation for VIs.

Dialog box: An interactive screen with prompts in which you describe additional information needed to complete a command.

Function: A built-in execution element, comparable to an operator or statement in a conventional programming language.

Hierarchy Window: Window that graphically displays the hierarchy of VIs and subVIs.

Icon Editor: Interface similar to that of a paint program for creating VI icons.

Indicator: Front panel object that displays output.

Marquee: Moving, dashed border surrounding selected objects.

Pseudocode: Simplified, language-independent representation of programming code.

SubVI: A VI used in the block diagram of another VI—comparable to a subroutine.

Top-level VI: The VI at the top of the VI hierarchy. This term distinguishes a VI from its subVIs.

VI library: Special file that contains a collection of related VIs.

EXERCISES

E4.1   Open a new VI and switch to the block diagram window. Select Write Characters to File.vi from the Functions≫All Functions≫File I/O palette and place it on the block diagram. Notice that the **Run** button breaks when this subVI is placed on the block diagram. Click the **Broken Run** button to access the **Error List**. The one listed error states that the VI has a required input. As you know from the chapter, each input terminal can be categorized as required, recommended or optional. You must wire values to required inputs before you can run a VI. Recommended and optional inputs do not prevent the VI from running and use the default value for that input. Default values are shown when you open (in this case, double-click on the VI).

Select the error from the **Error List** dialog box by clicking on it and then choose **Show Error**. Notice that a terminal on the middle left side of the Write Characters to File.vi icon is highlighted by a black box, which quickly transitions to a marquee around the icon. Open **Context Help** (Help≫Show Context Help or <Ctrl+H>) to read about the VI, paying special attention to the inputs listed as **Required**. You should see that the character string terminal is a required input. Right-click on the terminal on the icon and select **Create Control**. Notice that once the waveform type is attached to the icon, the error disappears and you can run the VI.

Navigate to the Functions≫All Functions≫Analyze≫Mathematics≫Calculus palette and select the Integration.vi and place it on the block diagram. Notice that the **Run** button does not break. Use **Context Help** again and notice that all of the inputs to this VI are **Recommended**, but not **Required**, and you will be able to run the VI without any inputs wired to it.

E4.2   Open a new VI and switch to the block diagram window. Navigate to the Functions≫All Functions≫Analyze≫Signal Processing≫Signal Generation palette, select the Signal Generator by Duration.vi and place it on the block diagram. Determine the errors that cause the **Broken Run** to appear. Determine a fix to the problem so that the VI can run properly.

E4.3   Open Slope.vi, which calculates the slope between two points, (X1, X2) and (Y1, Y2). The Slope.vi can be found in the Exercises&Problems\Chapter 4 folder in the Learning directory. The **Run** button is broken; hence, the VI is not executable. Click the **Run** button to access the Error List and you will find that the error is found within the subVI. You must open the subVI by double-clicking on it and correct the errors within it before the main VI will run. From within the SlopeSub.vi, go to the block diagram and wire the output of the **Subtract** terminal to the input of the **Divide** terminal. Save the subVI, return to the main VI, and notice that the **Run** button is not broken. Enter values into the four numeric controls in Slope.vi and run the VI to vary the slope between two points. Stop the VI.

Leave the SlopeSub.vi open and go to File≫New VI to open a new VI. Go to the block diagram of the new VI to practice the various ways to insert subVIs into a main VI. First, click on the icon of the SlopeSub.vi and drag it onto the block diagram on the new VI. Place the **Wiring** tool over the icon and see that all the input and output terminals are present. Second, from the block diagram, go to Functions≫Select a VI… and browse to the location on your computer where you saved SlopeSub.vi. Click the **Open** button and place the icon of the SlopeSub.vi on the block diagram. Remember that both methods of placing subVIs on block diagrams generate the same result.

E4.4 Open a new VI and go to Help≫Find Examples…. Click on the **Search** tab, type **Express** into the string labeled **Enter keyword(s)**, and press the **Search** button. This will locate shipping examples that predominantly use Express VIs to accomplish various tasks. Open the Lissajous2.vi by double-clicking on it and run the VI changing the controls on the front panel to alter the curve. Now go to the block diagram and see that there are not many Express VIs needed to perform this program.

From the block diagram, go to Browse≫Show VI Hierarchy and you will see that there are many additional subVIs used than you have seen from the block diagram. Use the tools you learned in this chapter to become familiar with the VI Hierarchy Window and explore the various subVIs used in Lissajous2.vi.

PROBLEMS

P4.1 Construct a VI that computes the average of three numbers input by the user using native LabVIEW functions. One computation in your program should be summing the three input numbers followed by a division by 3. The resulting average should be displayed on the front panel. Also, add a piece of code that multiplies the computed average by a random number in the range $[0, \ldots, 1]$. Create a subVI by grouping the parts of the code that compute the average. Remember to edit the icon so that it represents the function of the subVI, namely, the average of three numbers. Figure P4.1 can be used as a guide.

P4.2 Open Check Limit.vi, which generates a sine wave and plots it on a waveform graph. Check Limit.vi can be found in the Exercises&Problems\Chapter 4 folder in the Learning directory. The values of the sine wave are compared with a numeric control, therefore evaluating whether the sine wave exceeds the set limit. It also uses a While Loop that we will learn in the next chapter.

Edit the icon so that it only has the necessary number of terminals and connect them to the appropriate controls and indicators so that all of the front panels objects can be accessed if this VI were used in the future as a subVI. Also, change the VI so the toolbar, menu bar, and scroll bars are not visible when the VI executes. **Hint:** Navigate to File≫VI Properties… and choose **Window Appearance**, and then choose **Customize**.

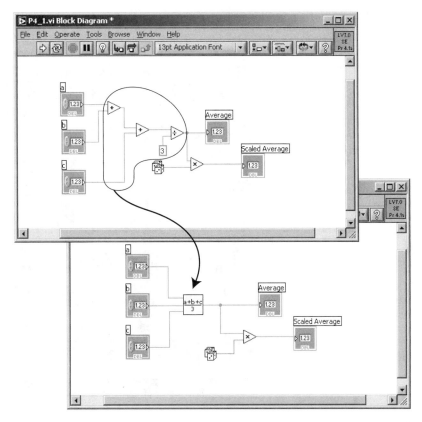

FIGURE P4.1
Computing the average of three numbers using a subVI to compute the average.

P4.3 Create a subVI that multiplexes four inputs to a single output. The subVI should have four floating-point numeric controls (denoted **In1** thru **In4**), one floating-point numeric indicator (denoted by **Out**) and one unsigned 8-bit integer (denoted by **Select**). If **Select** = 1, then **Out** = **In1**; if **Select** = 2, then **Out** = **In2**; if **Select** = 3, then **Out** = **In3**; and if **Select** = 4, then **Out** = **In4**. **Hint:** The **Select VI** from the **Comparison** palette may be useful.

P4.4 Create a VI that executes the **Quit LabVIEW VI** from Functions≫All Functions≫Application Control palette. Open a new VI and place the **Quit LabVIEW VI** on the block diagram. To edit the VI properties, select File≫VI Properties... and choose the **Execution** category. Check the box next to **Run when opened**. Save the VI in Users Stuff. Close the VI and then open it again. What happens? Try to figure out how you can edit the VI. (**Hint:** A subVI may be useful.)

P4.5 Complete the crossword puzzle.

Across

1. Interface similar to that of a paint program for creating VI icons.
5. A built-in execution element.
6. Window that is currently set to accept user input.
9. A VI used in the block diagram of another VI.
11. The VI whose front panel, block diagram, or Icon Editor is the active window.
12. Front panel object that displays output.
13. An interactive screen that prompts you for additional information.
14. Window that graphically displays the hierarchy of VIs and subVIs.

Down

2. The VI at the top of the VI hierarchy. This term distinguishes a VI from subVIs.
3. Front panel object for entering data to a VI interactively or subVI programmatically.
4. Special file that contains a collection of related VIs.
7. Online documentation for VI objects.
8. Simplified language-independent representation of programming code.
10. Moving, dashed border surrounding selected objects.

CHAPTER 5

Structures

Structures govern the execution flow in a VI. This chapter introduces you to five structures in LabVIEW: the For Loop, the While Loop, the Case structure, and the Flat Sequence and Stacked Sequence structures. Formula Nodes will be introduced as an effective way to implement mathematical equations. It is shown how script nodes can be used to execute MATLAB code within LabVIEW. We will also discuss the subject of controlling the execution timing of VIs.

GOALS

1. Study For Loops, While Loops, Case structures, and Flat Sequence and Stacked Sequence structures.

2. Understand how to use timing functions in LabVIEW.

3. Understand the use of shift registers and feedback nodes.

4. Become familiar with Formula Nodes and MATLAB Script Nodes.

5. Appreciate common wiring errors with structures.

5.1 THE FOR LOOP

For Loops and **While Loops** control repetitive operations in a VI, either until a specified number of iterations completes (i.e., the For Loop) or while a specified condition is true (i.e., the While Loop). A difference between the For Loop and the While Loop is that the For Loop executes a predetermined number of times, and a While Loop executes until a certain conditional input becomes false. For Loops and While Loops are found on the **Structures** palette of the Functions≫All Functions menu—see Figure 5.1. In this section, we concentrate on For Loops. While Loops are discussed in the next section.

A For Loop executes the code (known as the **subdiagram**) inside its borders a total of N times, where the N equals the value in the count terminal, as illustrated in Figure 5.2.

The For Loop has two terminals: the **count terminal** (an input terminal) and the **iteration terminal** (an output terminal). You can set the count explicitly by wiring a value from outside the loop to the count terminal, or you can set the count implicitly with auto-indexing (a topic discussed in Chapter 6). The bottom and right edges of the iteration terminal are exposed to the inside of the loop, allowing you access to the current value of the count. The iteration terminal (see Figure 5.2) contains the current number of completed loop iterations, where 0 represents the first iteration, 1 represents the second iteration, and continuing up to $N - 1$ to represent the Nth iteration.

Both the count and iteration terminals are long integers with a range of 0 through $2^{31} - 1$. You can wire a floating-point number to the count terminal, but it will be rounded off and **coerced** to lie within the range 0 through $2^{31} - 1$. The For Loop does not execute if you wire 0 to the count terminal!

The For Loop is located on the Functions≫All Functions≫Structures palette. Unlike many other **objects**, the For Loop is not dropped on the block diagram immediately. Instead, a small icon representing the For Loop appears in the block diagram, giving you the opportunity to size and position the loop. To do so, first click in an area above and to the left of all the objects that you want to execute within the For Loop, as illustrated in Figure 5.3. While holding down the mouse button, drag out a rectangle that encompasses the objects you want to place inside the For Loop. A For Loop is created upon release of the mouse button. The For Loop is a resizable box—use the **Positioning** tool for resizing by grabbing a corner of the For Loop and stretching to the desired dimensions. You can add additional block diagram elements to the For Loop by dragging and dropping them inside the loop boundary. The For Loop border will highlight as objects move inside the boundaries of the loop, and the block diagram border will highlight as you drag objects out of the loop.

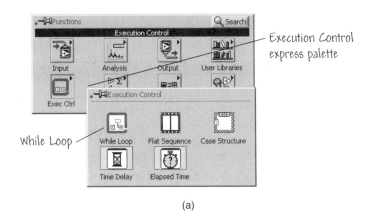

(a)

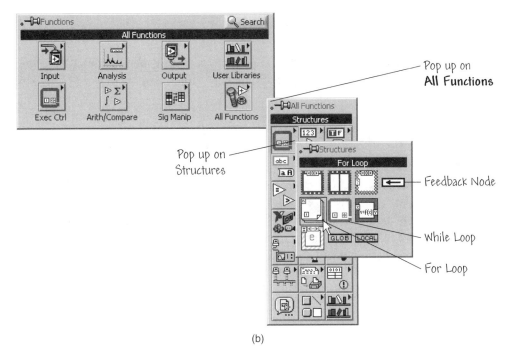

(b)

FIGURE 5.1
(a) The While Loop is found in the **Exec Ctrl** express palette. (b) The For Loop and While Loop are found on the **Structures** palette.

If you move an existing **structure** (e.g., a For Loop) so that it overlaps another object on the block diagram, the partially covered object will be visible above one

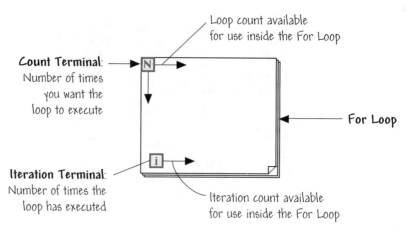

FIGURE 5.2
The For Loop.

edge of the structure. If you drag an existing structure completely over another object, the covered object will display a thick shadow to warn you that the object is underneath.

5.1.1   Numeric Conversion

Most of the numeric controls and indicators you have used so far have been double precision, floating-point numbers. Numbers can be represented as integers (byte [I8], word [I16], or long [I32]) or floating-point numbers (single, double, or extended precision). If you wire together two terminals that are of different data types, LabVIEW converts one of the terminals to the same representation as the other terminal. As a reminder, a **coercion dot** is placed on the terminal where the conversion takes place.

For example, consider the For Loop count terminal shown in Figure 5.4. The terminal representation is long integer. If you wire a double-precision, floating-point number to the count terminal, the number is converted to a long integer. Notice the small gray dot in the count terminal of the first For Loop—that is the coercion dot. To change the representation of the count terminal input, pop up on the terminal and select **Representation**, as illustrated in Figure 5.5. A palette will appear from which you can select the desired representation. In the case of the For Loop, you can change the count terminal input from double-precision, floating-point to long integer.

When the VI converts floating-point numbers to integers, it rounds to the nearest integer. If a number is exactly halfway between two integers, it is rounded to the nearest even integer. For example, the VI rounds 6.5 to 6, but rounds 7.5 to 8. This is an IEEE standard method for rounding numbers.

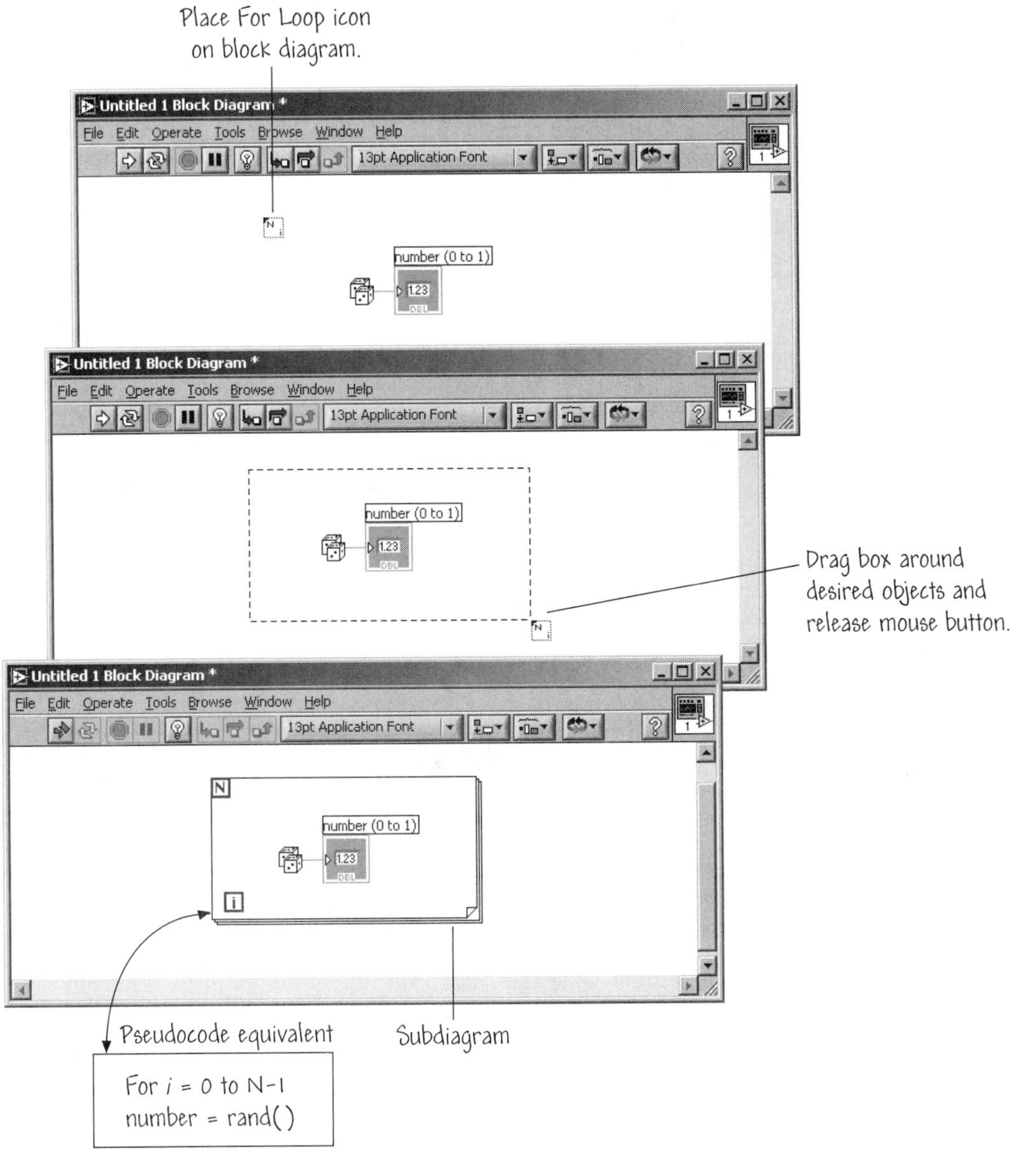

Place For Loop icon
on block diagram.

Drag box around
desired objects and
release mouse button.

Pseudocode equivalent   Subdiagram

For i = 0 to N-1
number = rand( )

FIGURE 5.3
Placing a For Loop on the block diagram.

A
For Loop
Example

In this exercise we will place a random number object inside a For Loop and display the random numbers and For Loop counter on the front panel. The VI shown in Figure 5.6 can be constructed by following these steps:

1. Place a random number function on the block diagram. The random number function can be found in the Functions≫Arithmetic & Comparisons≫

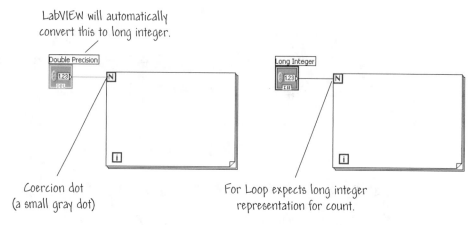

FIGURE 5.4
Converting double-precision, floating-point numbers at the count terminal.

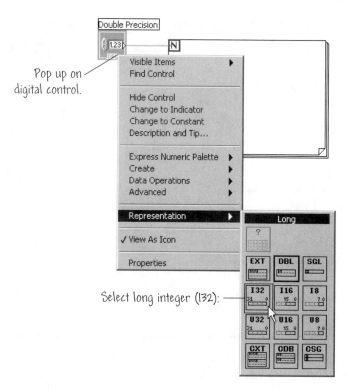

FIGURE 5.5
Changing the representation of a front panel numeric object.

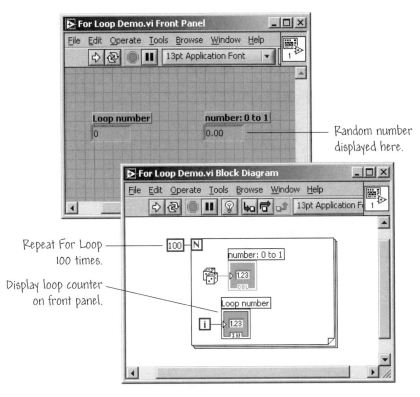

FIGURE 5.6
Displaying a series of random numbers using the For Loop.

Numeric palette. Create an indicator on the random number function and label it number: 0 to 1.

2. Place the For Loop on the block diagram so that the random number is enclosed within the loop, as shown in Figure 5.6.

3. Create a constant to the For Loop by popping up on the count terminal and selecting **Constant** from the **Create** menu. Set the value of the constant (which will be zero by default) to 100. This will let the For Loop execute one hundred times.

4. Create an indicator on the iteration terminal and label it Loop number.

5. Debug and run the program. One suggestion is to run the program with **Highlight Execution**, otherwise the program may run too fast to observe the loop execution.

6. On the front panel you will see the loop counter increment from 0 to 99 (that is, 100 iterations), and the random number between 0 and 1 should be displayed each iteration. Notice that the numeric indicator counts from 0 to 99, and *not* from 1 to 100!

7. Save the VI as For Loop Demo.vi in the Users Stuff folder in the Learning directory.

A working version of the VI called For Loop Demo.vi *can be found in the* Chapter 5 *folder in the* Learning *directory.*   ◆

5.2   THE WHILE LOOP

A While Loop is a structure that repeats a section of code until a certain condition is met. It is comparable to a Do Loop or a Repeat-Until loop in traditional programming languages. The While Loop, shown in Figure 5.7, executes the subdiagram inside its borders until a certain condition is satisfied. The While Loop

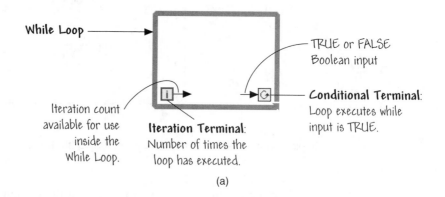

(a)

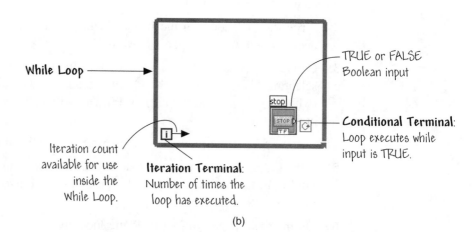

(b)

FIGURE 5.7
(a) The While Loop from the **Structures** palette. (b) The While Loop from the **Exec Ctrl** express palette.

has two terminals: the **conditional terminal** (an input terminal) and the **iteration terminal** (an output terminal). The iteration terminal of the While Loop behaves exactly like the For Loop iteration terminal. It is an output numeric terminal that outputs the number of times the loop has executed. The conditional terminal input is a Boolean variable: TRUE or FALSE. The While Loop executes until the Boolean value wired to its conditional terminal is either True of False, depending on whether the conditional terminal is set to **Stop if True** or **Continue if True**.

The VI checks the conditional terminal at the *end* of each iteration; therefore, the While Loop always executes at least once. If the value at the conditional terminal is TRUE, another iteration is performed; otherwise the loop terminates. The default value of the conditional terminal is FALSE, so it follows that the While Loop iterates only once if you leave the conditional terminal unwired.

You place the While Loop in the block diagram by (a) selecting it from the **All Functions≫Structures** subpalette of the **Functions** palette or (b) selecting it from the **Execution Control** Express palette, as illustrated in Figure 5.1. Similar to the For Loop, the While Loop is not dropped on the block diagram immediately. Instead, a small icon representing the While Loop appears in the block diagram, giving you the opportunity to size and position the loop, as shown in Figure 5.8. To do so, first click with the While Loop icon in an area above and to the left of all the objects that you want to execute within the While Loop, as shown in Figure 5.8. While holding down the mouse button, drag out a rectangle that encompasses the objects you want to place inside the While Loop. A While Loop is created upon release of the mouse button.

The completed While Loop is a resizable box—once the While Loop is placed on the block diagram, you can use the **Positioning** tool to resize the box by grabbing a corner and stretching the box to the desired dimensions. Additional block diagram elements can be added to the While Loop by dragging and dropping the desired objects inside the While Loop box. As with For Loops, the While Loop border will highlight as the object moves inside, and the block diagram border will highlight when you drag an object out of the While Loop. Once the While Loop (or For Loop) is on the block diagram, you cannot place an object inside the structure by dragging the structure over the object! Doing this will simply cause the structure to be placed over the object (that is, the structure will be the frontmost object). The correct procedure is to drag and drop objects inside existing While Loops (or other structures, such as For Loops).

You can change the way the conditional terminal functions by right clicking on the While Loop and choosing **Stop if True**. Then, rather than having the While Loop **Continue if True**, it will now **Stop if True**. An example is shown in Figure 5.9. In Fig. 5.9a, the While Loop continues if the value of x is greater than 0.5 and the Enable Boolean is pushed (TRUE). Conversely, in Fig. 5.9b the While Loop will stop if the value of x is greater than 0.5 and the Enable Boolean is pushed (TRUE).

Place While Loop icon
on block diagram.

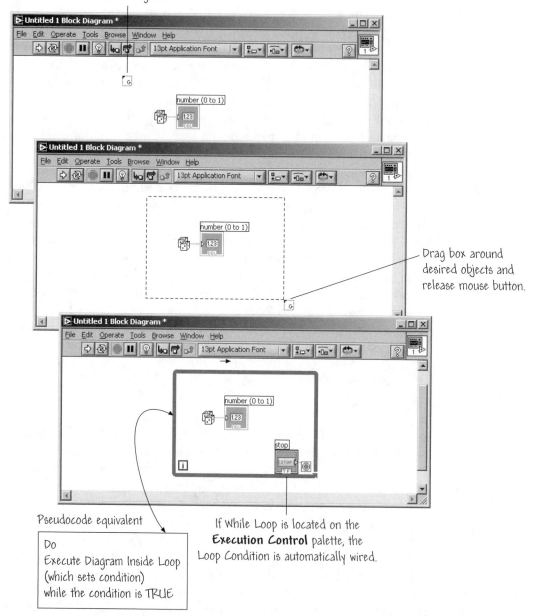

Drag box around
desired objects and
release mouse button.

Pseudocode equivalent

> Do
> Execute Diagram Inside Loop
> (which sets condition)
> while the condition is TRUE

If While Loop is located on the
Execution Control palette, the
Loop Condition is automatically wired.

FIGURE 5.8
Placing a While Loop in the block diagram.

Continue if True   Stop if True

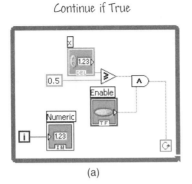
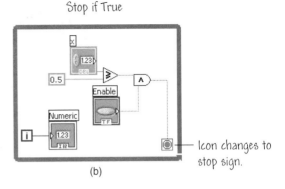

Icon changes to stop sign.

(a)   (b)

FIGURE 5.9
Continue if True or Stop if True.

The first time through a For Loop or a While Loop, the iteration count is zero. If you want to register how many times the loop has actually executed, you must add 1 to the count.

A While Loop Example

In this exercise we will place a random number object inside a While Loop and display the random numbers and While Loop counter on the front panel. This is very similar to the For Loop exercise in the previous section. The VI shown in Figure 5.10 can be constructed by following these steps:

1. Place a random number function on the block diagram (found in the Functions≫Arithmetic & Comparison≫Numeric palette). Create an indicator on the random number function and label it number: 0 to 1.

2. Place the While Loop on the block diagram so that the random number is enclosed within the loop, as shown in Figure 5.10. Use the Functions≫All Functions≫Structures palette to locate the While Loop.

3. Create a control to the conditional terminal by popping up on the terminal and selecting **Create Control**. A Boolean variable will appear on the block diagram, and simultaneously an on/off button will appear on the front panel. This will be used to stop the While Loop iterations while in the **Run** mode.

4. Create an indicator on the iteration terminal and label it Loop number.

5. To run the program, use the **Operating** tool and set the button on the front panel to the "on" position. Click on the **Run** button to start the program execution, and run the program with **Highlight Execution** to observe the program data flow.

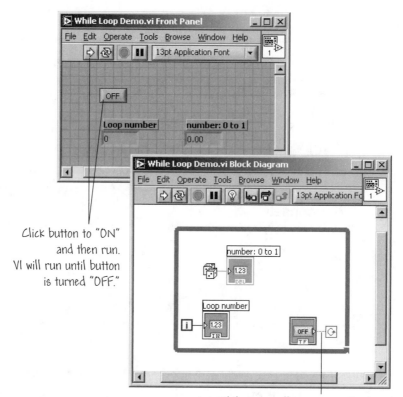

Click button to "ON"
and then run.
VI will run until button
is turned "OFF."

While Loop will execute until value
is FALSE, which occurs when button is "OFF."

FIGURE 5.10
Displaying a series of random numbers using the While Loop.

6. On the front panel you will see the loop counter continue to increment until you push the button to the "off" position. This causes the conditional terminal to change to FALSE, and the While Loop iterations cease.

*If you want to run a VI multiple times, place all the code in a loop rather than using the **Run Continuously** button. It is better to stop a VI with a Boolean control than with the **Abort Execution** button.*

7. Save the VI as While Loop Demo.vi in Learning\Users Stuff.

*A working version of the VI called **While Loop Demo.vi** can be found in the **Chapter 5** folder in the **Learning** directory.*

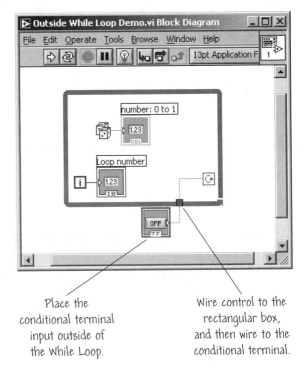

Place the
conditional terminal
input outside of
the While Loop.

Wire control to the
rectangular box,
and then wire to the
conditional terminal.

FIGURE 5.11
Placing the conditional terminal input outside of the While Loop.

If you place the terminal of the Boolean control outside the While Loop, as shown in Figure 5.11, you create either an infinite loop or a loop that executes only once, depending on the initial value. Why? Because the Boolean input data value is read before it enters the loop—remember data flow programming?—and not within the loop or after completion of the loop.

To experiment with this, modify your VI by placing the Boolean terminal outside of the While Loop. Connect the Boolean terminal to the While Loop border, as illustrated in Figure 5.11. A green rectangle will appear on the loop border. Then wire the green rectangle to the conditional terminal.

Conduct the following two numerical experiments: First, set the button state on the front panel to "on" and then run the VI (use **Highlight Execution** to watch the data flow). After a reasonable period of time, press the button to "off." What happens? The VI does not stop running—it is in an infinite loop. Can you explain this behavior using the notion of data flow programming? Since LabVIEW operates under data flow principles, inputs to the While Loop must pass their data before the loop executes. A While Loop passes data out only after the loop completes all iterations. The infinite loop experiment can be stopped by clicking on the **Abort Execution** button on the front panel toolbar.

For a second experiment, set the button state on the front panel to "off" and then run the VI (use **Highlight Execution** to watch the data flow). What happens in this case? You should see that the VI will execute once through the While Loop.

Save the VI with the conditional terminal outside the While Loop as Outside While Loop Demo.vi in the Users Stuff folder in the Learning directory.

A working version of Outside While Loop Demo.vi *can be found in the* Chapter 5 *folder in the* Learning *directory.*   ◆

5.3   SHIFT REGISTERS AND FEEDBACK NODES

When programming with loops, you often need to access data from previous iterations of the loop. Two ways of accessing the data from previous iterations of the loop are by utilizing the shift register and the Feedback Node. Both methods are discussed in this section.

5.3.1   Shift Registers

Shift registers transfer values from one iteration of a For Loop or While Loop to the next. The shift register is comprised of a pair of terminals directly opposite each other on the vertical sides of the loop border, as shown in Figure 5.12. You create a shift register by popping up on the left or right loop border and selecting **Add Shift Register** from the short cut menu.

The right terminal (the rectangle with the up arrow) stores the data as each iteration finishes. The stored data from the previous iteration is shifted and appears at the left terminal (the rectangle with the down arrow) at the beginning of

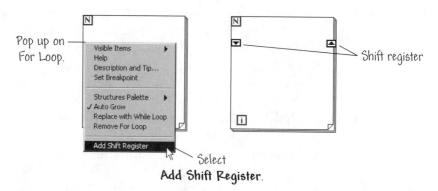

FIGURE 5.12
Shift register containing a pair of terminals.

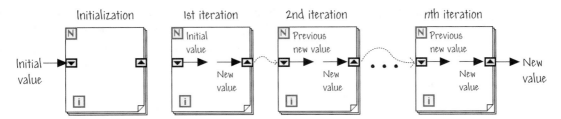

FIGURE 5.13
Passing data from one loop iteration to next using shift registers.

the next iteration, as shown in Figure 5.13. A shift register can hold any data type, including numeric, Boolean, strings (see Chapter 9), and arrays (see Chapter 6), but the data wired to the terminals of each register must be of the same type. The shift register conforms to the data type of the first object that is wired to one of its terminals.

Consider a simple illustrative example. Suppose we have the two situations depicted in Figure 5.14. Both cases look very similar—but the code in (b) contains a shift register. The code in (b) computes a running sum of the iteration count within the For Loop. Each time through the loop, the new sum is saved in the shift register. At the end of the loop, the total sum of 10 is passed out to the

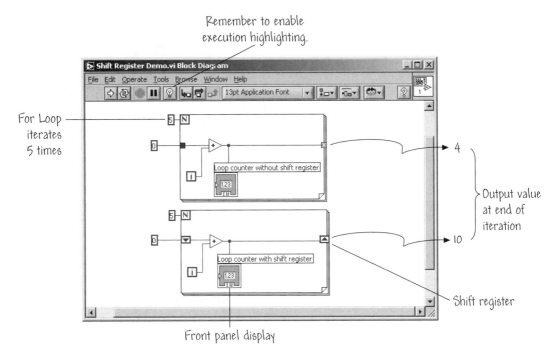

FIGURE 5.14
A simple example showing the effect of adding a shift register to a For Loop.

numeric indicator. Why 10? Because the sum of the iteration count is 10 ($=$ $0 + 1 + 2 + 3 + 4$). On the other hand, the code in (a) that does not contain the shift register does not save values between iterations. Instead, a zero is added to the current iteration count each time, and only the last value of the iteration counter ($= 4$) will be passed out of the loop.

You can run the simulation shown in Figure 5.14 by opening the VI titled **Shift Register Demo.vi** *which is in* **Learning\ Chapter 5** *directory—enable* **Highlight Execution** *before running and watch the data flow through the code!*

5.3.2 Using Shift Registers to Remember Data Values from Previous Loop Iterations

You can configure the shift register to store data values from previous iterations. To prepare the loop to store previous values, first create additional terminals on the loop border, as illustrated in Figure 5.15. If, for example, you add four elements to the left terminal, you can store and access values from the previous four iterations. You add the additional elements by popping up on the left terminal of the shift register and choosing **Add Element** from the short cut menu.

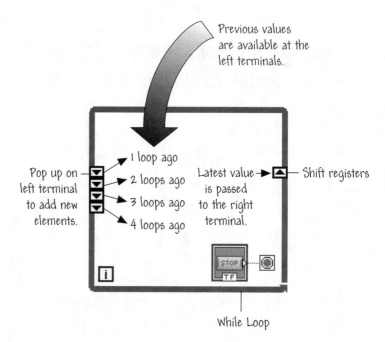

FIGURE 5.15
Adding elements to the shift register to access values from previous loop iterations.

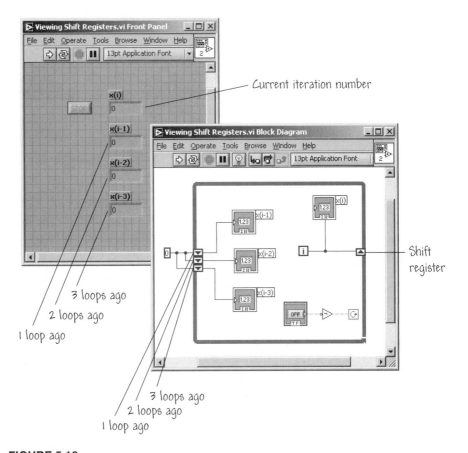

FIGURE 5.16
A demonstration of the use of shift registers to access data from previous iterations of a While Loop.

**Using
Shift
Registers**

In this example we will open an existing VI and use it to watch the data flow in a While Loop containing shift registers. Begin by opening Viewing Shift Registers.vi located in Learning\Chapter 5.

The front panel has four numeric indicators, as shown in Figure 5.16. The $X(i)$ indicator will display the current value, which will shift to the left terminal at the beginning of the next iteration. The $X(i-1)$ indicator will display the value one iteration ago, the $X(i-2)$ indicator will display the value two iterations ago, and the $X(i-3)$ indicator will display the value three iterations ago. The shift register is initialized to zero.

Before running the program, make sure that **Highlight Execution** is enabled. This will allow you to view the data flow and to watch the shift registers access data from previous iterations of the While Loop. Run the VI and watch the bubbles indicating the data flow. Notice that in each iteration of the While Loop, the VI "funnels" the previous values through the left terminals of the shift register.

$X(i)$ shifts to the left terminal, $X(i - 1)$, at the beginning of the next iteration. The values at the left terminal funnel downward through the terminals. In this example, the VI retains the last three values. To retain more values, you would need to add more elements to the left terminal of the shift register. When you're finished observing the VI operation, close it and do not save any changes.   ◆

5.3.3 Initializing Shift Registers

Shift registers are initialized by wiring a constant or control to the left terminal of the shift register from outside the loop. Shift register initialization is demonstrated in Figure 5.17. On the first execution the final value of the shift register is 12. Why 12? Because the final value is a sum of the iteration count ($= 0 + 1 + 2 + 3 + 4$) plus the initial value ($= 2$). On the second execution of the code that contains the initialized shift register, the result is exactly the same as on the first execution. In fact, the results on all subsequent executions are identical to the first execution: the final value of the shift register is 12. That is, the final value of the shift register from the first execution does not play a role in the second run—this is as it should be!

Unless the shift register is explicitly initialized, the first time the VI is executed the initial value of the shift register will be the default value for the shift register data type—if the shift register data type is Boolean, the initial value will be FALSE. Similarly, if the shift register data type is numeric, the initial value will be zero. The bottom row of Figure 5.17 illustrates what happens if you execute code twice with uninitialized shift registers. On the first execution the final value of the shift register is 10. Why 10? Can you explain this result? Running

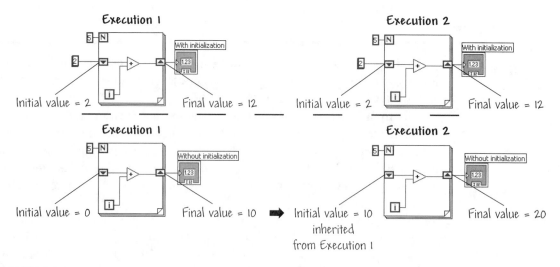

FIGURE 5.17
Initializing shift registers.

the code again, without closing the VI first, results in a final value of the shift register equal to 20. This result is due to the fact that on the second execution the initial value of the shift register is equal to 10 (left over from the previous run), and that is added to the sum of the iteration counter on the second run, which is equal to 10. When the shift register is not explicitly initialized, on the second run the shift register will take on the last value of the first run. Always use initialized shift registers for consistent results!

Data values stored in the shift register are stored until you close the VI and remove it from memory. That is, if you run a VI containing uninitialized shift registers, the initial values taken by the shift registers during subsequent executions will be the last values from previous executions. This can make debugging your VI a difficult process because this situation is hard to detect.

The shift register initialization demonstration depicted in Figure 5.17 can be found in **Learning\Chapter 5** *and is called* **Shift Register Init Demo.vi.**

Computing a Running Average

In this example, we will use shift registers to compute the running average of a sequence of random numbers. Since the random number function in LabVIEW provides random numbers from 0 to 1, we expect that the average will be 0.5. How many random numbers will it take to obtain an average near 0.5?

Figure 5.18 shows a VI that computes the running average of several random numbers. The length of the random number sequence is input via the slide control, as depicted on the front panel in Figure 5.18. Using the block diagram in the figure as a model, develop your own VI to compute the running average of a sequence of random numbers. The formula that is coded to compute the average is

$$Ave_i = \frac{i}{i+1} Ave_{i-1} + \frac{1}{i+1} RN_i,$$

where $i = 0, 1, \cdots, N - 1$, Ave_i is the computed average at the ith iteration, and RN_i is the current random number from the random number function. If you aren't familiar with computing a running average, just concentrate on the programming aspects of the VI shown in Figure 5.18 and don't worry about the formula. The main point is to understand how to use the shift registers in conjunction with For Loops and While Loops!

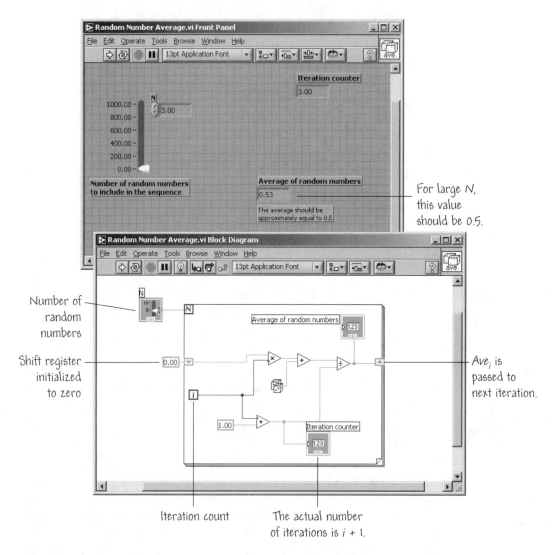

FIGURE 5.18
Computing a running average of a sequence of random numbers.

In this example, the shift register is used to pass the value of the variable Ave_{i-1} from one iteration to the next. You should notice as you run the code that for small values of N (e.g., $N = 3$) the average is generally not close to the expected value of $Ave_N = 0.5$; however, as you increase N, the average gets closer and closer to the expected value. Try it out!

The VI depicted in Figure 5.18 can be found in **Learning\Chapter 5** *and is called* **Random Number Average.vi***. Check it out if you can't get yours to work or if you want to compare your results.* ◆

5.3.4   Feedback Nodes

The Feedback Node can be utilized in a For Loop or While Loop if you wire the output of a subVI, function, or group of subVIs and functions to the input of that same VI, function, or group—that is, if you create a *feedback* path. Like a shift register, the Feedback Node stores data when the loop completes an iteration, sends that value to the next iteration of the loop, and transfers any data type. The Feedback Node is illustrated in Figure 5.19 inside of a While Loop.

Once the feedback is wired inside the loop, both the Feedback Node arrow and the Initializer Terminal will automatically appear. The Feedback Node arrow indicates the direction of data flow along the wire. Initializing a Feedback Node resets the initial value the Feedback Node passes the first time the loop executes when the VI runs. If you leave the input of the Initializer Terminal unwired, each time the VI runs, the initial input of the Feedback Node is the last value from the previous execution or the default value for the data type if the loop has never executed.

You also can select the Feedback Node on the **Structures** palette and place it inside a For Loop or While Loop (see Figure 5.1 for location on the palette). If you place the Feedback Node on the wire before you branch the wire that connects the data to the tunnel, the Feedback Node passes each value to the tunnel. If you place the Feedback Node on the wire after you branch the wire that connects data

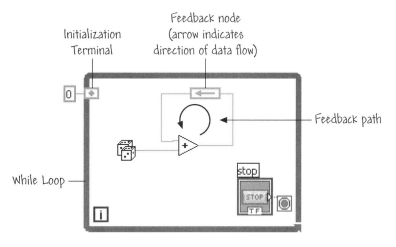

FIGURE 5.19
The Feedback Node.

to the tunnel, the Feedback Node passes each value back to the input of the VI or function and then passes the last value to the tunnel.

Use the Feedback Node to avoid unnecessarily long wires in loops.

Practice with Feedback Nodes

In this example we compare the use of the shift register with the Feedback Node and illustrate the differences in placing the Feedback Node in the feedback path and the feedforward path.

Open the Feedback Node Demo VI located in the **Chapter 5** folder of the **Learning** directory. The block diagram is shown in Figure 5.20. The block diagram has three parts:

1. In the top section of the code, the Feedback Node is placed after the branch.

2. In the middle section of the code, the Feedback Node is placed before the branch.

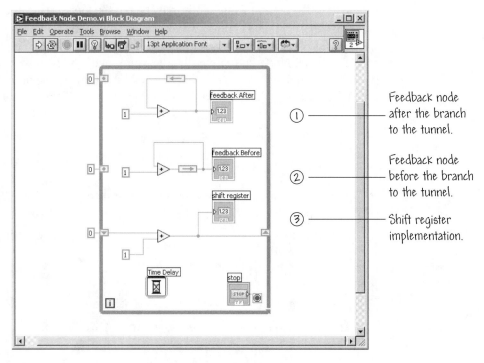

FIGURE 5.20
Different implementations of the Feedback Node.

3. In the bottom section of the code, shift registers are used instead of Feedback Nodes.

Wiring a 0 to the Initialization Terminal or to the shift register initializes each diagram. The Time Delay function slows the operation of the code for better viewing. Run the VI and observe the outputs on the front panel.

The top section of the code reads the initialized Feedback Node and passes this value to the Add function.

The middle section of the code reads the initialized Feedback Node and passes this value to the indicator. This Add function does not execute until the next iteration of the loop. This section of code will output a number always one less than the first section of code.

The Feedback Node implementation in the top section of the code and the shift register implementation in the bottom section of the code have the same functionality.

If you want to replace a Feedback Node with a shift register, right click the Feedback Node and select Replace with Shift Register.   ◆

5.4   CASE STRUCTURES

A **Case structure** is a method of executing conditional text. This is analogous to the common If...Then...Else statements in conventional, text-based programming languages. You place the Case structure on the block diagram by selecting it from the **Structures** subpalette of the **Functions≫All Functions** palette or from the **Execution Control** Express palette, as shown in Figure 5.21. As with the For Loop and While Loop structures, you can either drag the Case structure icon to the block diagram and enclose the desired objects within its borders, or you can place the Case structure on the block diagram, resize it as necessary, and drag objects inside the structure.

The Case structure can have multiple subdiagrams. The subdiagrams are configured like a deck of cards of which only one card is visible at a time. At the top of the Case structure border is the Selector label. The diagram identifier can be numeric, Boolean, string, or enumerated type control. An enumerated type control is unsigned byte, unsigned word, or unsigned long and is selectable from the decrement and increment buttons, as depicted in Figure 5.22(a) and (b). The Selector label displays the values that cause the corresponding subdiagrams to execute. The Sequence Structures are similar in form to the Case structure and are depicted in Figure 5.22(c) and (d). If the Sequence structure Selector label is numeric, the Selector label value is followed by a Selector label range, which

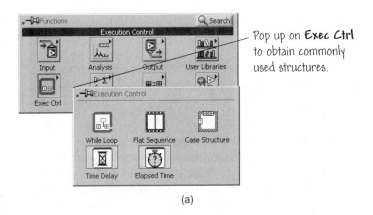

Pop up on **Exec Ctrl** to obtain commonly used structures.

(a)

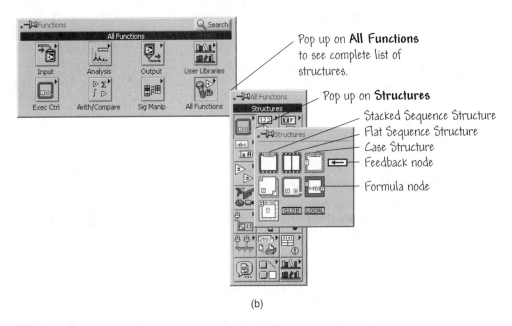

Pop up on **All Functions** to see complete list of structures.

Pop up on **Structures**

Stacked Sequence Structure
Flat Sequence Structure
Case Structure
Feedback node

Formula node

(b)

FIGURE 5.21
(a) Selecting the Case structure from the **Execution Control** Express palette. (b) Select Case Structure from the **All Functions≫Structures** palette.

shows the minimum and maximum values for which the structure contains a subdiagram. To view other subdiagrams (that is, to see the subdiagrams within the "stack of cards"), click the decrement (left) or increment (right) button to display the previous or next subdiagram, respectively. Decrementing from the first subdiagram displays the last, and incrementing from the last subdiagram displays the first subdiagram—it wraps around!

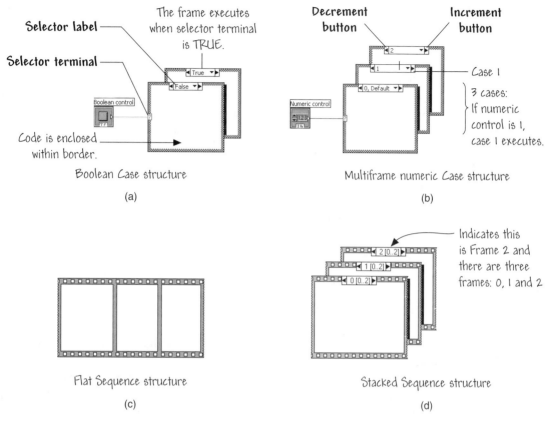

FIGURE 5.22
Overview of Case and Sequence structures.

You can position the selector terminal anywhere on the Case structure along the left border. The selector label automatically adjusts to the input data type. For example, if you change the value wired to the selector from a numeric to a Boolean, then cases 0 and 1 change to FALSE and TRUE, respectively. Here is a point to consider in changing the data type: if the Case structure selector originally received numeric input, then n cases $0, 1, 2, \cdots, n$ may exist in the code. Upon changing the selector input data from numeric to Boolean, cases 0 and 1 change automatically to FALSE and TRUE. However, cases $2, 3, \cdots, n$ are not discarded! You must explicitly delete these extra cases before the Case structure can execute.

You can also type in and edit values directly into the selector label using the **Labeling** tool. The selector values are specified as a single value, as a list or as a range of values. A list of values is separated by commas, such as $-1, 0, 5,$ 10. A range is typed in as 10..20, which indicates all numbers from 10 to 20, inclusively. You also can use open-ended ranges. For example, all numbers less than or equal to 0 are represented by ..0. Similarly, 100.. represents all numbers greater than or equal to 100.

The case selector can also use string values that display in quotes, such as "red," and "green." You don't need to type in the quotes when entering the values unless the string contains a comma or the symbol "..". In a case selector using strings, you can use special backslash codes (such as \r, \n, and \t) for nonalphanumeric characters (carriage return, line feed, tab, respectively).

If you type in a selector value that is not the same type as the object wired to the selector terminal, then the selector value displays in red, and your VI is broken. Also, because of the possible round-off error inherent in floating-point arithmetic, you cannot use floating-point numbers in case selector labels. If you wire a floating-point type to the case terminal, the type is rounded to the nearest integer, and a coercion dot appears. If you try to type in a selector value that is a floating-point number, then the selector value displays in red, and your VI is broken.

5.4.1   Adding and Deleting Cases

If you pop up on the Case structure border, the resulting menu gives you the many options shown in Figure 5.23. You choose **Add Case After** to add a case after the case that is currently visible. **Delete This Case**, or **Add Case Before** to add a case before the currently visible case. You can also choose to copy the currently shown case by selecting **Duplicate Case** and to delete the current visible case by selecting **Delete This Case**. When you add or remove cases (i.e., subdiagrams) in a case structure, the diagram identifiers are automatically updated to reflect the inserted or deleted subdiagrams.

Sometimes you may want to rearrange the listed order of the cases in the structure. For example, rather than have the cases listed $(0, 1, 2, 3, \cdots)$, you might want them listed in the order $(0, 2, 1, 3, \cdots)$. Resorting the order in which the cases appear in the case structure on the block diagram does not affect the run-time behavior of the Case structure! It is merely a matter of programming preference. You can change the order in which the cases are listed in the structure by selecting **Rearrange Cases**... from the short cut menu. When you do so, the dialog box shown in Figure 5.23 appears. The **Sort** button sorts the case selector values based on the first selector value. To change the location of a selector, click the selector value you want to move (it will highlight when selected) and drag it to the desired location in the stack. In the **Rearrange Cases** dialog box, the section entitled "Complete selector string" shows the selected case selector name in its entirety in case it is too long to fit in the "Case List" box located in the top portion of the dialog box. Other on-line assistance can be found in the context-sensitive help in the **Help** menu.

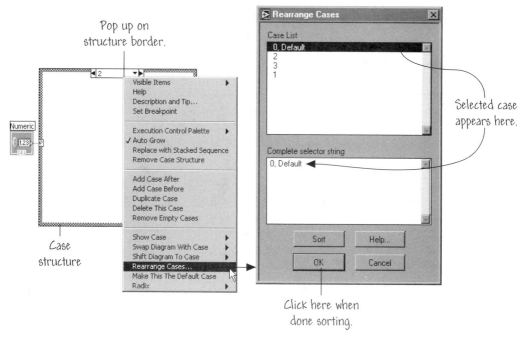

FIGURE 5.23
Adding and deleting cases in Case structures.

The **Make This The Default Case** item in the short cut menu specifies the case to execute if the value that appears at the selector terminal is not listed as one of the possible choices in the selector label. Case statements in other programming languages generally do not execute any case if a case is out of range, but in LabVIEW you must either include a default case that handles out-of-range values or explicitly list every possible input value. For the default case that you define, the word "Default" will be listed in the selector label at the top of the Case structure.

◆ **A Simple Case Structure Example**

In this exercise you will build a VI that uses a Boolean Case structure, shown in Figure 5.24. The input numbers from the front panel will pass through **tunnels** to the Case structure, and there they are either added or subtracted, depending on the Boolean value wired to the selector terminal.

What is a tunnel? A tunnel is a data entry point or exit point on a structure. You can wire a terminal from outside the Case structure to a terminal within the structure. When you do this a rectangular box will appear on the structure border—this represents the tunnel. You can also wire an external terminal to the

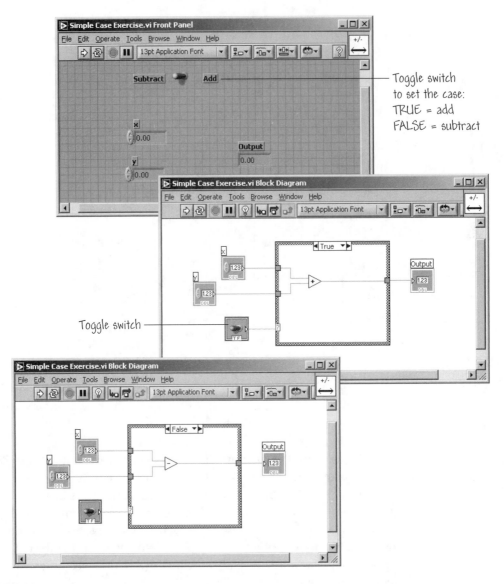

FIGURE 5.24
A Boolean Case structure example to add or subtract two input numbers.

structure border to create the tunnel, and then wire the terminal to an internal terminal in a second step. Tunnels can be found on other structures, such as Sequence structures, While Loops, and For Loops. The data at all input tunnels is available to all cases.

In this exercise you will need to create two tunnels to get data into the Case structure and one tunnel to pass data out of the structure. Using Figure 5.24 as a guide, construct a VI that utilizes the Case structure to perform the following task: if the Boolean wired to the selector terminal is TRUE, the VI will add the numbers; otherwise, the VI will subtract the numbers. The Boolean value of the selector terminal is toggled using a horizontal toggle switch found on the front panel.

The VI depicted in Figure 5.24 can be found in **Learning\ Chapter 5** *and is called* **Simple Case Exercise.vi.** ◆

5.4.2   Wiring Inputs and Outputs

As previously mentioned, the data at all input terminals (tunnels and selection terminal) is available to all cases. Cases are not required to use input data or to supply output data, but if any one case supplies output data, all must do so. Forgetting this fact can lead to coding errors. Correct and incorrect wire situations are depicted in Figure 5.25, where for the FALSE case two input numbers are added and sent (through the tunnel) out of the structure, and for the TRUE case the computer system beeps.

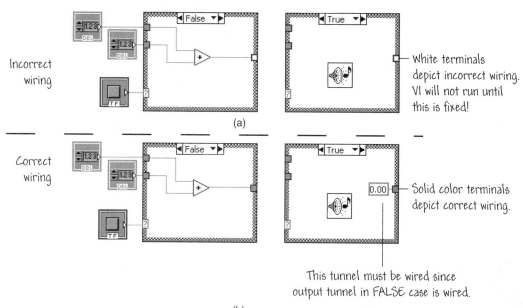

FIGURE 5.25
Wiring Case structure inputs and outputs.

When you create an output tunnel in one case, tunnels appear at the same position on the structure border in the other cases. You must define the output tunnel for each case. Unwired tunnels look like white squares, and when they occur, the **Broken Run** button will also appear, so be sure to wire to the output tunnel for each unwired case. You can wire constants or controls to unwired cases by popping up on the white square and selecting **Constant** or **Control** from the **Create** menu. When all cases supply data to the tunnel, it takes on a solid color consistent with the data type of the supplied data, and the **Run** button appears.

Using Case Structures

In this exercise, you will build a VI that computes the ratio of two numbers. If the denominator is zero, the VI outputs ∞ to the front panel and causes the system to make a beep sound. If the denominator is not zero, the ratio is computed and displayed on the front panel.

Begin by opening a new VI. Build the simple front panel shown in Figure 5.26. The numeric control **x Numerator** supplies the numerator x, and the numeric control **y Denominator** supplies the denominator y. The **x/y** numeric indicator displays the ratio of the two input numbers. Switch to the block diagram and place the Case structure in the window.

By default, the Case structure selector terminal is Boolean—this is the format we desire in this situation. You can display only one case at a time, and we'll start by considering the TRUE case. Remember that to change cases, you need to click on the increment and/or decrement arrows in the top border of the Case structure.

Place the other diagram objects and wire them as shown in Figure 5.26. The main objects are:

1. The Not Equal to 0? function (found in the Functions≫Arithmetic & Comparison≫Express Comparison menu) checks whether the denominator is zero. The function returns a TRUE if the denominator is not equal to 0.

2. The Divide function (found in the Functions≫Arithmetic & Comparison≫Express Numeric menu) returns the ratio of the numerator and denominator numbers.

3. The Positive Infinity constant indicates that a divide by zero has been attempted (see the Functions≫Arithmetic & Comparison≫Express Numeric≫Express Numeric Constants menu).

4. The Beep.vi function (found in the library vi.lib\platform\system.lib) causes the system to issue an audible tone when division by zero has been attempted. On Windows platforms, all input parameters to the beep function are ignored, while on the Macintosh, you can specify the tone frequency in Hertz, the du-

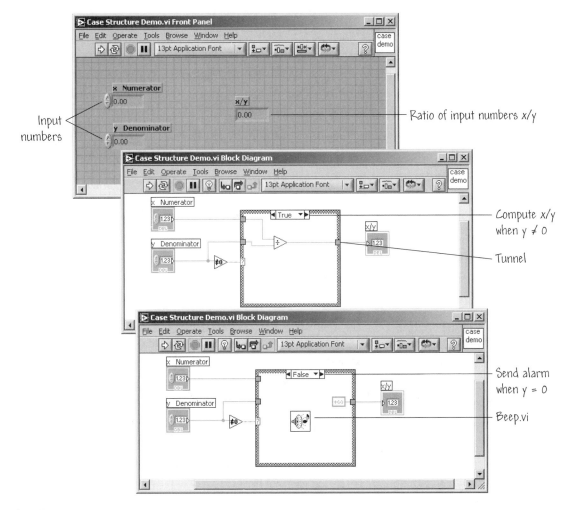

FIGURE 5.26
A VI that computes the ratio of two numbers and uses Case structures to handle division by zero.

ration in milliseconds, and the intensity as a value from 0 to 255, with 255 being the loudest. Double click on the icon to see this VI's front panel.

You must define the output tunnel for each case. When you create an output tunnel in the TRUE case, an output tunnel will appear at the same position in the FALSE case. Unwired tunnels look like white squares. In this and all other VI development, be sure to wire to the output tunnel for each unwired case. In Figure 5.26, the constant ∞ is wired to the output tunnel in the FALSE case.

The VI will execute either the TRUE case or the FALSE case. If the denominator number is not equal to zero, the VI will execute the TRUE case and return the ratio of the two input numbers. If the denominator number is equal to zero, the VI will execute the FALSE case and output ∞ to the digital indicator and make the system beep.

Once all objects are in place and wired properly, return to the front panel and experiment with running the VI. Change the input numbers and compute the ratio. Make the denominator input equal to zero and listen for the system beep. When you are finished, save the VI as **Case Structure Demo.vi** in the **Users Stuff** folder in the **Learning** directory. Close the VI.

A working version of this VI called **Case Structure Demo.vi** *can be found in* **Learning\Chapter 5**.   ◆

5.5 FLAT AND STACKED SEQUENCE STRUCTURES

The **Sequence structure** executes subdiagrams sequentially. The subdiagrams look like a frame of film; hence they are known as **frames**. There are two classes of sequence structures: the Flat Sequence and the Stacked Sequence. Although the two classes of Sequence structures are similar, there are key differences that will be pointed out in the forthcoming discussions. The Flat Sequence structure and the Stacked Sequence structure are depicted in Figure 5.22(c) and (d), respectively.

Determining the execution order of a program by arranging its elements in a certain sequence is called **control flow**. Most text-based programming languages (such as Fortran and C) have inherent control flow because statements execute in the order in which they appear in the program. In data flow programming, a node executes when data is available at all of the node inputs (this is known as **data dependency**), but sometimes you cannot connect one node to another. When data dependencies are not sufficient to control the data flow, the Sequence structure is a way of controlling the order in which nodes execute. You use the Sequence structure to control the order of execution of nodes that are not connected with wires. Within each frame, as in the rest of the block diagram, data dependency determines the execution order of nodes.

On Stacked Sequence structures, the code that you want to execute first is placed inside the border of frame 0(0..x), the code to be executed second is placed inside the border of frame 1(0..x), and so on. The interval (0..x) represents the range of frames in the Stacked Sequence structure. Only when the last frame completes does data leave the structure. As with the Case structure, only one frame of a Stacked Sequence structure is visible at a time. Clicking on the arrows

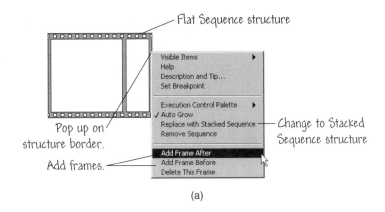

(a)

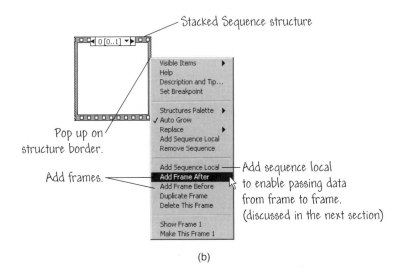

(b)

FIGURE 5.27
Removing and adding frames to the Sequence structure. (a) Flat Sequence.
(b) Stacked Sequence.

at the top of the structure allows you to flip through the other frames. New frames are created by popping up on the structure border and selecting **Add Frame After** or **Add Frame Before**, as illustrated in Figure 5.27.

Recall that the outputs on Case structures can have one data source per case. In contrast, on Stacked Sequence structures, output tunnels can have only one data source. The output can originate from any frame, but data leaves the Stacked Sequence structure only when it completes execution entirely, not when the individual frames finish. Sequence locals are used to pass data between frames in Stacked Sequence structures. Data at input tunnels is available to all frames, just as with Case structures.

Data flow for the Flat Sequence structure differs from data flow for other structures. Frames in a Flat Sequence structure execute in order and when all data wired to the frame are available. New frames are created by popping up on the structure border and selecting **Add Frame After** or **Add Frame Before**, as illustrated in Figure 5.27. Data is passed between frames using tunnels in Flat Sequence structures. The data leaves each frame as the frame finishes executing.

Use the Stacked Sequence structure if you want to conserve space on the block diagram. Use the Flat Sequence structure to avoid using sequence locals and to better document the block diagram.

5.5.1   Sequence Locals

Sequence locals are variables that pass data between frames of a Stacked Sequence structure. You create sequence locals on the border of a frame—the data wired to a sequence local is then available in subsequent frames. The data, however, is not available in frames preceding the frame where the sequence was created.

To obtain a sequence local, choose **Add Sequence Local** from the structure border short cut menu (see Figure 5.27). You cannot add another sequence local if you pop up too close to an existing sequence local or over the subdiagram display window. Once a sequence local terminal is placed on the border, you can drag it to any other unoccupied location on the border. At first, a sequence local terminal is just a small yellow box, but once a data source is wired to the terminal, an outward-pointing arrow appears in the frame of the terminal. The sequence local terminals in subsequent frames contain inward-pointing arrows to show that they are a data source for that frame. You cannot use the sequence local in frames preceding the source frame since it hasn't been assigned a value yet! To remind you that the sequence local does not contain a value in preceding frames (hence is not available for use), it appears as a dimmed rectangle. To remove a sequence local terminal, pop up on the terminal and select **Remove**.

The example in Figure 5.28 shows a four-frame Stacked Sequence structure. A sequence local in frame 1 passes the value of a random number function to subsequent frames. The random number value is available in frame 2—as indicated by the arrow pointing into frame 2. In this simple example, the random number in frame 2 is being displayed by a digital indicator. In frame 3 we attempt to wire the random number to a digital control. The result is a broken wire since we cannot wire a digital control to a sequence local that is a source of data. Also, notice that the random number value is not available in frame 0, as indicated by the dimmed yellow square. Remember that data is not available in frames preceding the frame in which the sequence local is created!

5.5.2   Evaluate and Control Timing in a Sequence Structure

It is useful to be able to control and time the execution of VIs. You can use Express VIs or traditional VIs to accomplish the timing task. The Wait (ms) and Tick Count (ms) functions (located in the **Time&Dialog** palette of the Functions≫All Functions menu) in conjunction with a sequence structure let you accomplish these tasks using traditional VIs. The units of time are given in milliseconds.

You can also accomplish the timing task using Express VIs. The Time Delay Express VI and the Elapsed Time Express VI are located on the **Functions≫Execution Control** Express palette, as shown in Figure 5.29. When the Time Delay Express VI is placed on the block diagram, a dialog box appears in which you set the time delay value. A similar dialog box opens with the Elapsed Time Express VI. All units are given in seconds.

The Wait (ms) function causes your VI to wait a specified number of milliseconds before it continues execution. The function waits the specified number of milliseconds and then returns the millisecond timer's end value. The Tick Count (ms) function returns the value of the millisecond timer and is commonly used to calculate elapsed time. The base reference time (that is, zero milliseconds) is undefined. Therefore, you cannot convert the millisecond timer output value to a real-world time or date. Note that the value of the millisecond timer wraps from $2^{32} - 1$ to 0.

The internal clock does not have high resolution—about 1 ms on Windows XP/2000/NT/98 and Macintosh. The resolutions are driven by operating system limitations and not by LabVIEW.

A simple example illustrating the use of the Time Delay Express VI and the Elapsed Time Express VIs is shown in Figure 5.30. Open the VI Timing with Sound Demo located in Learning\Chapter 5. Set the time between beeps on the slider control and run the VI in **Run Continuously** mode. You should hear a sound approximately every n seconds according to the setting on the slider control. The actual time between beeps (to the resolution of the clock) is displayed on the front panel to verify that the desired timing has been achieved. On Windows XP/2000/NT/98 and Macintosh, if you reduce the time between beeps to 0.001 second (using the **Labeling** tool to type in the value in the digital display of the slider control), you will find that the actual time between beeps is about 1 ms. However, if you continue to reduce the time below 0.001, the actual time between beeps will not reduce accordingly—you have reached the clock resolution of your system.

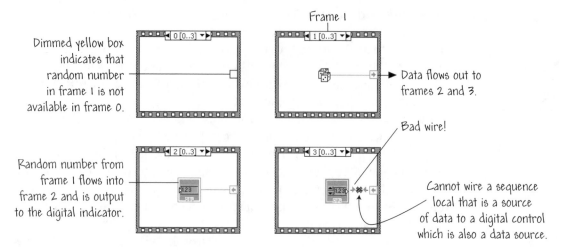

Dimmed yellow box indicates that random number in frame 1 is not available in frame 0.

Frame 1

Data flows out to frames 2 and 3.

Bad wire!

Random number from frame 1 flows into frame 2 and is output to the digital indicator.

Cannot wire a sequence local that is a source of data to a digital control which is also a data source.

FIGURE 5.28
Many forms of the sequence local.

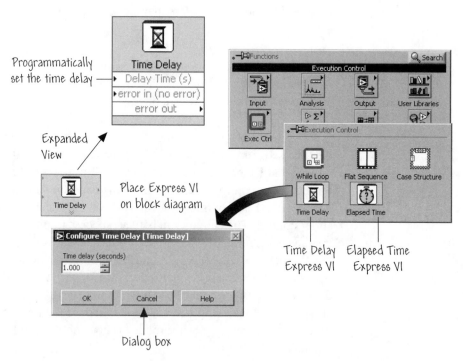

Programmatically set the time delay

Expanded View

Place Express VI on block diagram

Dialog box

Time Delay Express VI   Elapsed Time Express VI

FIGURE 5.29
The Time Delay Express VI and the Elapsed Time Express VI.

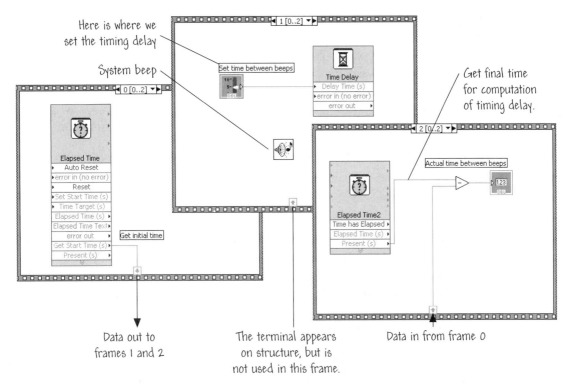

FIGURE 5.30
Controlling the timing of a VI using the Stacked Sequence structure.

5.5.3   Avoid the Overuse of Sequence Structures

In general, VIs can operate with a great deal of inherent parallelism. Sequence structures tend to hide parts of the program and to interfere with the natural data flow. The use of Sequence structures prohibits parallel operations, but does guarantee the order of execution. Asynchronous tasks that use I/O devices can run concurrently with other operations if Sequence structures do not prevent them from doing so.

We will be discussing I/O devices (such as GPIB, serial ports, and data acquisition boards) more in subsequent chapters (see Chapters 8 and 10), and you will understand better how the use of Sequence structures can inhibit the performance of such devices. The objective here is to alert you to the idea that you should avoid the overuse of Sequence structures. The use of Sequence structures does not negatively impact the computational performance of your program, but it does interrupt the data flow. You should write programs that take advantage of the concept of data flow programming!

5.6   THE FORMULA NODE

The **Formula Node** is a structure that allows you to program one or more algebraic formulas using a syntax similar to most text-based programming languages. It is useful when the equations have many variables or otherwise would require a complex block diagram model for implementation. The Formula Node itself is a resizable box (similar to the Sequence structure, Case structure, For Loop, and While Loop) in which you enter formulas directly in the code, in lieu of creating block diagram subsections.

Consider the equation

$$y = x - e \sin x,$$

where $0 \le e \le 1$. This is a famous equation in astrodynamics, known as Kepler's equation. If you implement this equation using regular LabVIEW arithmetic functions, the block diagram looks like the one in Figure 5.31. You can implement the same equation using a Formula Node as shown in the same figure.

5.6.1   Formula Node Input and Output Variables

You place the Formula Node on the block diagram by selecting it from the **Structures** subpalette of the **Functions≫All Functions** palette (see Figure 5.21). You create the input and output terminals of the Formula Node by popping up on the border of the node and choosing **Add Input** or **Add Output** from the short cut menu, as shown in Figure 5.32. Output variables have a thicker border than input

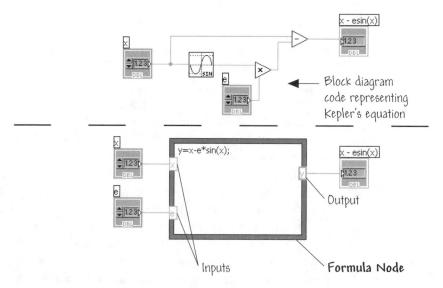

FIGURE 5.31
Implementing formulas in a Formula Node can often simplify the coding.

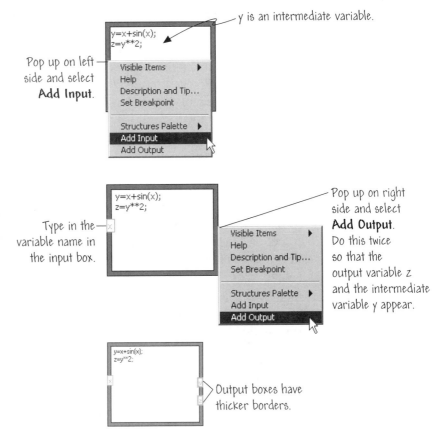

FIGURE 5.32
Formula Node input and output variables.

variables. You can change an input to an output by selecting **Change to Output** from the short cut menu, and you can change an output to an input by selecting **Change to Input** from the short cut menu.

Once the necessary input and/or output terminals are on the Formula Node, enter the input and output variable names in their respective boxes using the **Labeling** tool. Every variable used in the Formula Node must be declared as an input or an output with no two inputs and no two outputs possessing the same name. An output can, however, have the same name as an input. Intermediate variables (that is, variables used in internal calculations) must be declared as outputs, although it is not necessary for them to be wired to external nodes. In Figure 5.32, the y variable and the z variable are both declared as outputs, although the intermediate variable y does not have to be wired to an external node.

5.6.2 Formula Statements

Formula statements use a syntax similar to most text-based programming languages for arithmetic expressions. You can add comments by enclosing them inside a slash-asterisk pair (/*comment*/). Figure 5.33 shows the operators and functions that are available inside the Formula Node. You can access the same list of available operators and functions by choosing Help≫Show Context Help and moving the cursor over the Formula Node in the block diagram.

You enter the formulas inside the Formula Node using the **Labeling** tool. Each formula statement must terminate with a semicolon, and variable names are case-sensitive. There is no limit to the number of variables or formulas in a Formula Node. If you have a large number of formulas, you can either enlarge the Formula Node using the **Positioning** tool, or you can pop up in the Formula Node (not on the border) and choose **Scrollbar**. The latter method will put a scrollbar in the Formula Node, and with the **Operating** tool you can scroll down through the list of formulas for viewing.

The following example shows how you can perform conditional branching inside a Formula Node. Consider the following code fragment that computes the ratio of two numbers, x/y:

$$\text{if } (y \neq 0) \text{ then}$$
$$z = x/y$$
$$\text{else}$$

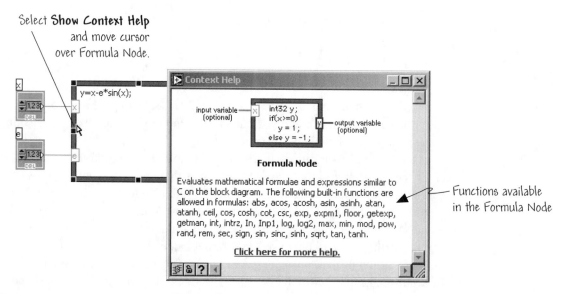

FIGURE 5.33
Functions available in Formula Nodes.

215

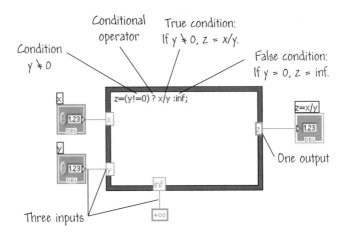

FIGURE 5.34
Implementing the formula $z = x/y$ in a Formula Node.

$z = +\infty$
end if

When $y = 0$, the result is set to ∞. You can implement the code fragment given above using a Formula Node, as illustrated in Figure 5.34.

The Formula Node demonstration depicted in Figure 5.34 can be found in **Learning\Chapter 5** *and is called* **Formula Code.vi.** *Open the VI and try it out!*

5.7   MATLAB SCRIPT NODES

LabVIEW has script nodes that allow you to execute external scripts. In particular, you can execute MATLAB scripts using a script node. MATLAB[1] is a popular interactive program for scientific and engineering calculations and visualization. Since many students have developed and used MATLAB scripts (known as m-files), in this section we focus on using script nodes in LabVIEW to execute MATLAB scripts.

LabVIEW uses ActiveX technology to implement MATLAB script nodes, therefore MATLAB script nodes are available only on Windows platforms.

---

1. MATLAB is a registered trademark of The Mathworks, Inc.

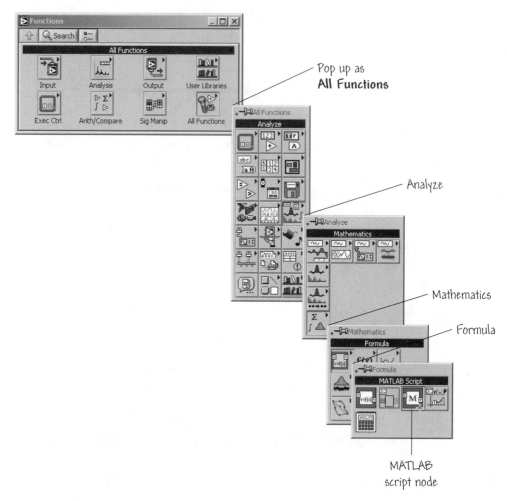

FIGURE 5.35
Accessing the MATLAB script node.

5.7.1   Accessing the MATLAB Script Node

The MATLAB script node is accessed on the **All Functions** palette, as illustrated in Figure 5.35. The process of placing the script node on the block diagram is similar to For Loops and While Loops. First select the MATLAB script node from the palette and place it on the block diagram. Using the **Positioning** tool, the script node is extended to the desired size, as illustrated in Figure 5.36.

5.7.2   Entering Scripts into the MATLAB Script Node

There are two ways to place a MATLAB script in the script node. You can use the **Operating** or **Labeling** tool to enter the script in the MATLAB script node, or if

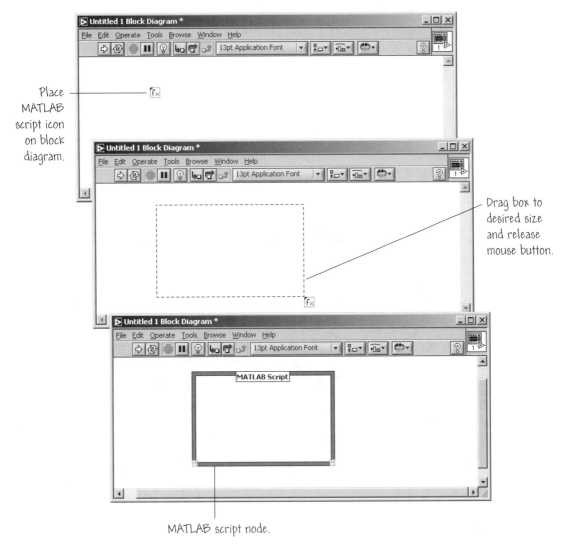

Place MATLAB script icon on block diagram.

Drag box to desired size and release mouse button.

MATLAB script node.

FIGURE 5.36
Placing a MATLAB script node on the block diagram.

you already have a script written, you can import it. To import a script, right-click the MATLAB script node and select **Import** from the short cut menu to display the **Choose a script** dialog box, as shown in Figure 5.37. When the dialog box appears, select the file you want to import and click **Open**. The MATLAB script text will then appear in the script node. It is suggested that you write your script and run it within the MATLAB environment for testing and debugging purposes before you import it into LabVIEW.

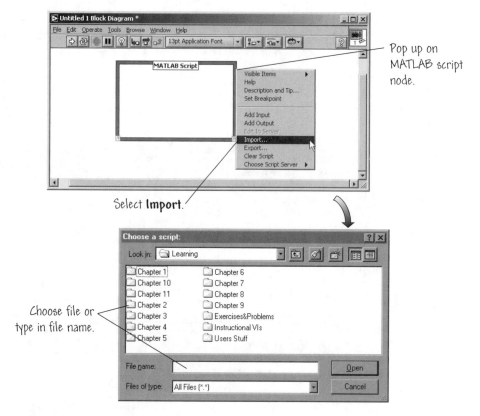

Pop up on MATLAB script node.

Select **Import**.

Choose file or type in file name.

FIGURE 5.37
Importing a script into a MATLAB script node.

You must have MATLAB installed on your computer to use MATLAB script nodes because script nodes invoke the MATLAB script server to execute MATLAB scripts.

5.7.3 Input and Output Variables

As with Formula Nodes, you need to add inputs and outputs for variables to the MATLAB script node. To add an output variable, right-click the MATLAB script node frame and select **Add Output** from the short cut menu, as shown in Figure 5.38. Similarly, to add an input variable, right-click the MATLAB script node frame and select **Add Input** from the short cut menu. When the input and output variables appear on the node, you can add their names. You can also use the **Labeling** tool to edit the variable names at any time.

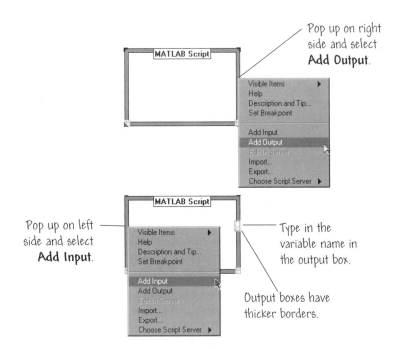

FIGURE 5.38
Adding input and output variables to the MATLAB script node.

By default the MATLAB script node includes one input and one output for the error in and error out parameters. To take advantage of the error-checking parameters for debugging information, it is suggested that you create an indicator for the error out terminal on the MATLAB script node before you run the VI. This allows you to view the error information generated at runtime.

 Just as with a regular formula node, you can display a scrollbar within your script node by popping up on the node and selecting Visible Items≫Scrollbar.

You can create controls and indicators for each input and output on the MATLAB script node. For example, to create an indicator on an output terminal, right-click the output terminal select Create≫Indicator from the short cut menu. LabVIEW will create an indicator on the front panel and wire terminals to the output on the block diagram, as shown in Figure 5.39.

5.7.4   Saving MATLAB Scripts

You may want to save your MATLAB script to a text file. In this way, you can later open this text file in LabVIEW, thus importing the MATLAB script into LabVIEW. To save a MATLAB script, right-click the MATLAB script node and select **Export** from the short cut menu to display the **Name the script** dialog

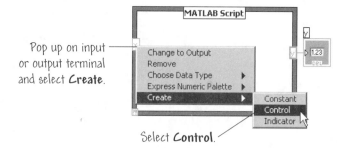

Pop up on input
or output terminal
and select **Create**.

Select **Control**.

FIGURE 5.39
Adding control and indicators to the input and output variables.

box, as illustrated in Figure 5.40. Enter the desired new file name or select the file you want to overwrite and click **Save**. MATLAB script files are text files, and although text files usually have a .txt extension, MATLAB files have a .m extension. This is consistent with MATLAB m-file naming conventions.

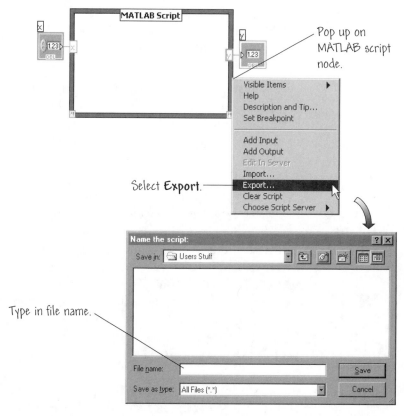

Pop up on
MATLAB script
node.

Select **Export**.

Type in file name.

FIGURE 5.40
Saving MATLAB scripts to a text file.

TABLE 5.1   LabVIEW and MATLAB Data Types.

	LabVIEW Data Types	MATLAB Data Types
`I32`	Signed 32-bit integer numeric	N/A
	Double-precision floating-point numeric	Real
`abc`	String	N/A
`[I32]`	1D array signed 32-bit integer numeric	N/A
`DBL`	1D array double-precision floating-point numeric	Real Vector
`[I32]`	Multidimensional array signed 32-bit integer numeric	N/A
`[DBL]`	Multidimensional array double-precision floating point numeric	Real Matrix
`[CDB]`	Complex double	Complex
`CDB`	1D array complex double	Complex Vector
`[CDB]`	Multidimensional array complex double	Complex Matrix

5.7.5   MATLAB Data Types in LabVIEW

LabVIEW recognizes the MATLAB datatypes. Table 5.1 shows the LabVIEW data types and the corresponding data types in MATLAB. You can change the data type of an input or output terminal on a script node. MATLAB is a loosely typed script language, hence the datatype of a variable is not determined until after the script executes. Therefore, LabVIEW cannot determine the variable type in Edit mode. LabVIEW queries the script server to find out possible datatypes, and lets you choose which LabVIEW datatype each terminal should be.

If you incorrectly configure a variable's datatype, LabVIEW produces either an error or incorrect information at runtime.

You should always verify the data type of the script node inputs and outputs. In MATLAB, the default data type for any new input or output in Real. To change the datatype of an input or output terminal on a script node, first right-click the

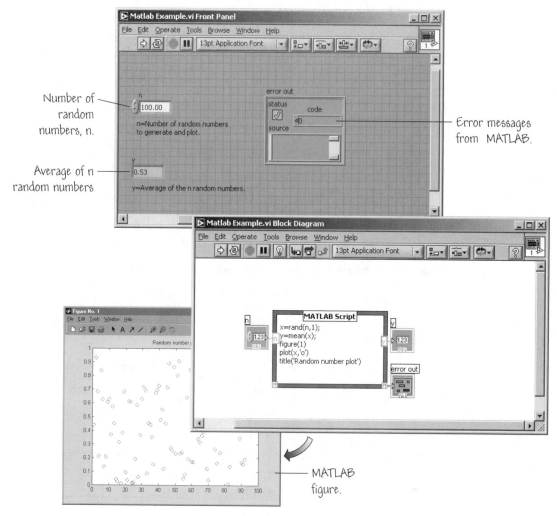

FIGURE 5.41
Using a script node to generate a MATLAB plot.

terminal of the input or output and select **Choose Data Type** on the short cut menu. A list of the available datatypes appears from which you can choose the preferred data type.

A MATLAB Example

In this exercise you will construct a VI that utilizes a MATLAB script node. The purpose of the VI is to generate and plot a given number of random numbers in MATLAB. The script will also compute the average of the random numbers for output. The VI is shown in Figure 5.41. Open a new VI and begin by placing a MATLAB script node on the block diagram. Size it large enough to allow for the required lines of code (refer to the block diagram in Figure 5.41).

Add an input and an output to the script node. The input variable is the number of random numbers to generate, and the output variable is the average of the random numbers. Make the input and output variables double-precision floating-point numerics. Also, place a string indicator on the script node to show an error out messages generated in MATLAB. Run the VI when ready. LabVIEW will launch the MATLAB application if it is not already opened. Vary the number of random numbers in the set and see how the average varies correspondingly. The MATLAB script node will generate a plot of the random numbers using MATLAB graphics. When you are finished experimenting, save the VI as **Matlab Example.vi** in the **Users Stuff** folder.

A working version of **Matlab Example.vi** *can be found in the* **Chapter 5** *folder in the* **Learning** *directory.*   ◆

5.8   SOME COMMON PROBLEMS IN WIRING STRUCTURES

When wiring the structures presented in this chapter, you may encounter wiring problems. In this section, we discuss some of the more common wiring errors and present suggestions on how to avoid them. Five common problems with wiring structures are:

- Assigning more than one value to a Sequence local.
- Wiring from multiple frames of a Sequence structure.
- Failing to wire a tunnel in all cases of a Case structure.
- Overlapping tunnels.
- Wiring underneath rather than through a structure.

The first two problems are actually variations of the multiple sources error.

5.8.1   Assigning More Than One Value to a Sequence Local

A sequence local can be assigned a value in only one frame. Figure 5.42 shows the value π assigned to the sequence local in frame 0 and then another attempt to assign a value to this same local variable in frame 1—this results in a bad wire. You can use the value of the sequence local in all frames that follow the frame in which the assignment took place.

5.8.2   Wiring from Multiple Frames of a Sequence Structure

Figure 5.43 depicts two Sequence structure frames attempting to assign values to the same tunnel. The tunnel turns white to signal this error, which is just another variation of the multiple sources error.

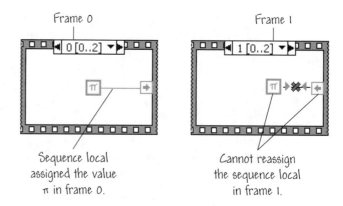

FIGURE 5.42
Local variables in a Sequence structure can be assigned a value in only one frame.

5.8.3   Failing to Wire a Tunnel in All Cases of a Case Structure

When you wire from a Case structure to an object outside the structure, you must connect output data from all cases to the object. Failure to do so will result in a bad tunnel, as illustrated in Figure 5.44. This problem is a variation of the no source error. Why? Because at least one case would not provide data to the object outside the structure when that case executed. The problem is easily solved by wiring to the tunnel in all cases. Can you explain why this is not a multiple sources violation? It seems like one object is receiving data from multiple sources. The answer is that only one case executes at a time and produces only one output value per execution of the Case structure. If each Case did not output a value, then data flow execution would stop on the cases that did not output a value.

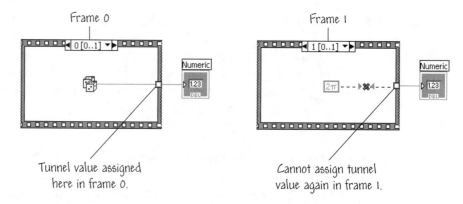

FIGURE 5.43
Two Sequence structure frames cannot assign values to the same tunnel.

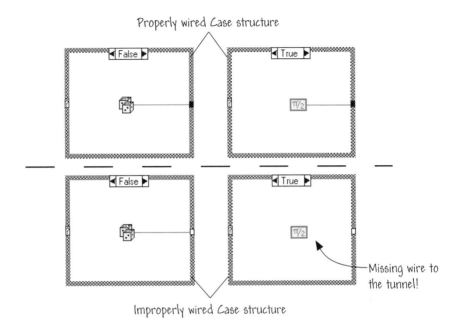

FIGURE 5.44
Failing to wire a tunnel in all cases of a Case structure leads to problems.

5.8.4 Overlapping Tunnels

Tunnels are created as you wire, resulting occasionally in tunnels that overlap each other. Overlapping tunnels do not affect the execution of the diagram, so this is not really an error condition. But overlapping tunnels make editing and debugging of the VI more difficult. You should avoid creating overlapping tunnels! The fix is easy—drag one tunnel away with the **Positioning** tool to expose the other tunnels. A problematic situation is illustrated in Figure 5.45.

5.8.5 Wiring Underneath Rather Than Through a Structure

Suppose that you want to pass a variable through a structure. You accomplish this by clicking either in the interior or on the border of the structure as you are wiring. This process is illustrated in Figure 5.46. If you fail to click in the interior or on the border, the wire will pass underneath the structure, and some segments of the wire may be hidden. This condition is not an error per se, but hidden wires are visually confusing and should be avoided.

As you wire through the structure, you are provided with visual cues for guidance during the wiring process. For example, when the **Wiring** tool crosses the left border of the structure, a highlighted tunnel appears. This lets you know that a tunnel will be created at that location as soon as you click the mouse button. You should click the mouse button! However, if you continue to drag the tool

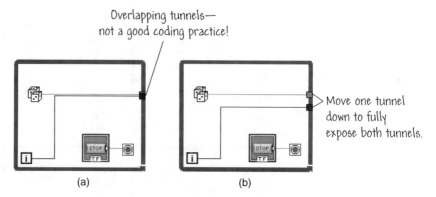

FIGURE 5.45
Sometimes overlapping tunnels occur when wiring structures.

through the structure (without clicking the mouse on the left-hand-side border) until the tool touches the right border of the structure, a second highlighted tunnel appears. If you continue to drag the **Wiring** tool past the right border of the structure without clicking, both tunnels disappear. When this occurs, the wire passes underneath the structure rather than through it, as depicted in Figure 5.46.

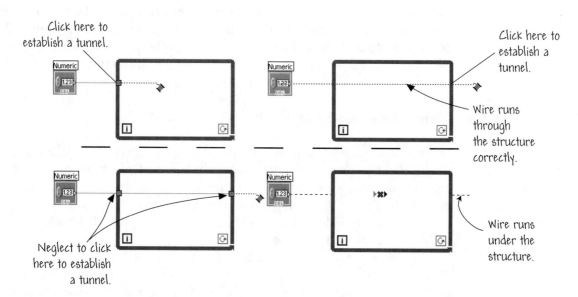

FIGURE 5.46
Wiring underneath rather than through a structure is a common problem.

BUILDING BLOCK

5.9 BUILDING BLOCKS: MEASURING VOLUME

In this exercise you will continue the work on the building blocks VI Volume Limit created in Chapter 4. The goal is to make the necessary modifications so that you can acquire multiple volume measurements using a For Loop and control the timing of the loop. You should use Figure 5.47 as a guide.

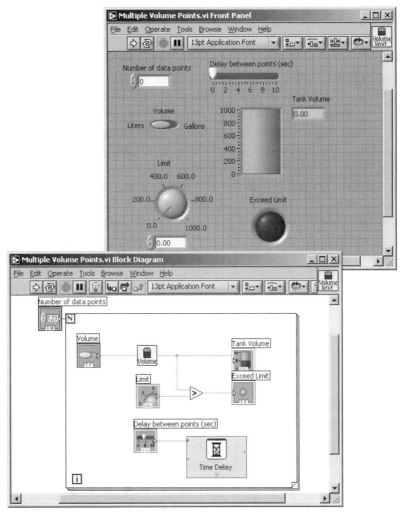

FIGURE 5.47
Wiring the connector of the Volume subVI.

When you have finished editing and debugging the VI, save it in the Users Stuff folder as Multiple Volume Points.vi.

Run the VI for various delays between data points—try 5 sec and count the interval to see if the delay really is 5 seconds (actually, the timing will only be approximate!). You will see in Chapter 7 that a better way to control the For Loop timing is to use the Wait Until Next ms Multiple function.

5.10   RELAXED READING: INVESTIGATING GLOBAL WARMING WITH LABVIEW

The Antarctic ice sheet plays a crucial role in the behavior of the Earth's ice-ocean-atmosphere system because it influences the distribution of heat and mass within that system. Recently, there has been concern that global atmospheric warming could lead to a substantial rise in sea level, and the greatest uncertainty lies in the contribution of melting ice sheets. Improved understanding of this process requires integration of direct measurements of the ice sheet mass balance from satellite measurements and modeling of ice sheet dynamics.

Understanding of the fundamentals of ice flow is not currently sufficiently advanced to make quantitative predictions. Because field measurements cannot define the state of stress when modeling the flow of glaciers and ice sheets, a control system was required to simulate the stress of the flow of glaciers and ice sheets. The objective was to be able to measure the mechanical properties of anisotropic ice from the full depth of a borehole in Antarctica. The control system needed to be flexible and possess a high level of adaptability to measure a large range of physical parameters.

With LabVIEW, it was possible to construct the required multi-axis control system at a lower cost solution than typical commercial servo-hydraulic turnkey solutions. The system configuration is illustrated in Figure 5.48. A 12-bit PCI-6025 board performs the data acquisition. This board also provides the stepper motor control and timing for the ultrasonic transducer tomography system. It was possible to switch ultrasonic transducer transmitters and receivers to achieve the required pattern of data. In this way, the physical parameter changes in the ice specimen could be measured. Two PCI-7030 RT Series boards and 6030E 16-bit DAQ boards, each with 16 single-ended analog inputs and two analog outputs, made up the multi-axis digital servo hydraulic controller. A LabVIEW Real-Time application provided the error amplifier and valve drive voltage via the analog outputs to drive a precision hydraulic servo valve attached to each of the 60-kN double-acting hydraulic actuators.

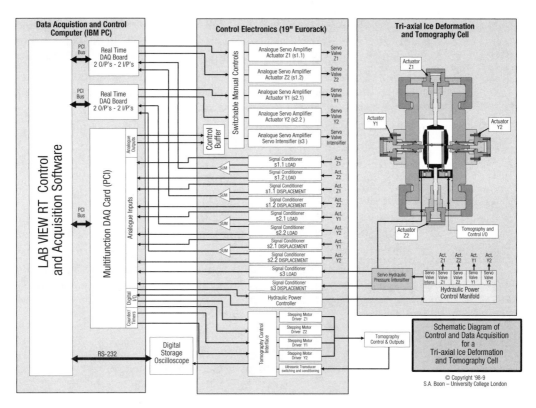

FIGURE 5.48
Schematic of the Control System.

The RT Series PCI hardware enabled an interactive solution. The LabVIEW RT application uses a series of layered windows for viewing and setting up of:

- Load or displacement increments, ramp rate, and tomography transducer selection

- Overall test progress of tomography with load and displacement monitoring

- Control loop function (PID and other outer loop controls) for changes in ice sample stiffness

- Mode of control (load or displacement) for individual or paired actuators

- Graphs for stress and strain of each axis

- Ultrasonic transducer pulsing levels and digital oscilloscope controls

- Transducer calibration

- Test data capture and replay

The application loads the LabVIEW RT program onto the PCI-7030 control boards after the test parameters are entered. It then awaits the commands from the main PC-based application. With the analog inputs on the RT Series acquisition board, multiple signals can be input from other axes into the closed-loop control system to ensure a stable environment for the tomography component of the experiments. The handshake system between the main control program and the RT Series board control program, combined with soft test limits, ensures a safe shutdown in the event of equipment failures. The LabVIEW RT control program effectively closed the control loop by taking data from the test setup and modifying the valve drive voltage to the actuator systems. The system provides feedback directly to the RT Series board by the conditioned transducer signals, which are connected to the board's analog input channels.

With LabVIEW RT, a low-cost multi-axis control system was constructed in a familiar programming environment with a high level of quality technical support. This project represents a successful application of LabVIEW to the study of the Antarctic ice sheets flows-an important area of research for planetary scientists.

For more information:

Dr. Steve Boon
The Rock and Ice Physics Laboratory
Department of Geological Sciences
University College of London
Gower Street, London, WC1E 6BT
s.boon@ucl.ac.uk

5.11   SUMMARY

In this chapter we studied four structures (For Loops, While Loops, Flat and Stacked Sequence structures, and Case structures). The While Loop and the For Loop are used to repeat execution of a subdiagram placed inside the border of the resizable loop structure. The While Loop executes as long as the value at the conditional terminal is TRUE. The For Loop executes a specified number of times. Shift registers are variables that transfer values from previous iterations to the beginning of the next iteration.

The Case structure and the Sequence structures are utilized to control data flow. Case structures are used to branch to different subdiagrams depending on the values of the selection terminal of the Case structure. Sequence structures are used to execute diagram functions in a specific order. The portion of the diagram to be executed first is placed in the first frame of the Stacked Sequence structure, the diagram to be executed second is placed in the second frame, and so on. Sequence locals pass values between Stacked Sequence structure frames. Use the Stacked Sequence structure if you want to conserve space on the block diagram. Use the Flat Sequence structure to avoid using sequence locals and to better document the block diagram.

You can directly enter formulas on the block diagram in the Formula Node. The Formula Node is useful when the function equation has many variables or otherwise would require a complex block diagram model for implementation.

LabVIEW has script nodes that allow you to execute external scripts. In this chapter, we discussed the MATLAB script node, and showed how to input external MATLAB code for execution in LabVIEW.

We also discussed the matter of controlling and timing the execution of structures. The Wait (ms) and Tick Count (ms) functions in conjunction with a Sequence structure let you accomplish these tasks using traditional VIs. You can also accomplish the timing task using the Time Delay Express VI and the Elapsed Time Express.

KEY TERMS

Case structure: A conditional branching control structure that executes one and only one of its subdiagrams, based on specific inputs. It is similar to If-Then-Else and Case statements in conventional programming languages.

Coercion: The automatic conversion performed in LabVIEW to change the numeric representation of a data element.

Coercion dot: Dot that appears where LabVIEW is forced to convert a numeric representation of one terminal to match the numeric representation of another terminal.

Conditional terminal: The terminal of a While Loop that contains a Boolean value that determines whether the loop performs another iteration.

Count terminal: The terminal of a For Loop whose value determines the number of times the For Loop executes its subdiagram.

Control flow: Determining the execution order of a program by arranging its elements in a certain sequence.

Data dependency: The concept that block diagram nodes do not execute until data is available at all the node inputs.

For Loop: Iterative loop structure that executes its subdiagram a set number of times.

Formula Node: Node that executes formulas that you enter as text. Especially useful for lengthy formulas too cumbersome to build in the block diagram.

Frames: Diagrams associated with Sequence structures that look like frames of film and control the execution order of the program.

Iteration terminal: The terminal of a For Loop or While Loop that contains the current number of completed iterations.

Object: Generic term for any item on the front panel or block diagram, including controls, nodes, and wires.

Sequence local: Terminal that passes data between the frames of a Sequence structure.

Sequence structure: Program control structure that executes its subdiagram in numeric order. Commonly used to force nodes that are not data-dependent to execute in a desired order.

Shift register: Optional mechanism in loop structures used to pass the value of a variable from one iteration of a loop to a subsequent iteration.

Structure: Program control element, such as a Sequence, Case, For Loop, or While Loop.

Subdiagram: Block diagram within the border of a structure.

Tunnels: Relocatable connection points for wires from outside and inside the structures.

While Loop: Loop structure that repeats a section of code until a Boolean condition is False.

E5.1 Open the block diagram of Rearrange Cases.vi found in the Exercises&Problems\Chapter 5 folder in the Learning directory. Analyze all the possible temperature values listed in the Case Structure. Notice that the case values are not in ascending numerical order. In order to change the order of case frames, right-click on the **Case** border and select **Rearrange Cases**. This will bring up the **Rearrange Cases** wizard, which displays the current order of the cases. Click the **Sort** button to arrange the cases in ascending numerical order. Click **OK** if you want to leave them in ascending order. Otherwise, if you want the case frames to display in any other order, click and drag the values in the Case List.

E5.2 Open the block diagram of Gray Dots.vi found in the Exercises&Problems\Chapter 5 folder in the Learning directory. Notice there are two gray dots to signify numeric conversion. In general, you should try to eliminate as many gray dots in your VIs as possible because they slow down execution and use more memory for mathematical operations. To eliminate gray dots, right-click on the terminal with the gray dot and select **Representation**. In this VI, right-click on the indicator terminal labeled Data Points and choose the same representation as the input value—which is DBL. To eliminate the other gray dot on the N terminal of the For Loop, change the representation on the control terminal titled Number of Data Points to match the N terminal—which is I32.

E5.3 Open Case Errors.vi found in the Exercises&Problems\Chapter 5 folder in the Learning directory and click on the **Broken Run** arrow. This will display a list of the errors in this VI. Use the error list to determine how to fix the VI so you can run it.

E5.4 Using a single While Loop, construct a VI that executes a loop N times or until the user presses a stop button. Be sure to include the Time Delay Express VI so the user has time to press the **Stop** button.

E5.5 Open the block diagram of the Avoid Sequence.vi found in the Exercises&Problems\Chapter 5 folder in the Learning directory, and gain an understanding of the program. Rewrite this VI without using the Flat sequence structure or the Stacked sequence structure. The VI can execute in the correct order without the sequence structures using data flow. Remember that it is important to use the sequence structures only when it is not possible to wire nodes together to ensure that they execute in the proper order.

E5.6   Open the block diagram of the Loan Calculator.vi found in the Exercises&Problems\Chapter 5 folder in the Learning directory. This program will compute monthly payments required to pay off a loan based on the inputs on the front panel. The user has three choices for how the interest is compounded and two choices for the payment frequency. Click on the **Broken Run** arrow to display the list of errors. You will need to program the following equations into the four Formula Express VIs:

Interest Compounded Monthly	$i = \text{APR}/\text{PymtFreq}$
Interest Compounded Daily	$i = (1 + (\text{APR}/365))^{\text{PymtFreq}} - 1$
Interest Compounded Continuously	$i = e^{(\text{APR}/\text{PymrFreq})} - 1$
Payment Amount	$A = P[i(1+i)^N]/[(1+i)^N - 1]$

E5.7   Open the block diagram of the Statistics.vi found in the Exercises&Problems\Chapter 5 folder in the Learning directory. This program will compute the number of combinations possible if you are to select r out of n possible objects. Add Case structures to the code to utilize the inputs **Does Order Matter?** and **Can Samples Repeat?** inputs and display the appropriate result on the front panel. The table below outlines which equation should be used in which case:

	Order Matters	**Order Doesn't Matter**
Samples Repeat	n^r	$(n-1+r)!/(n-1)!r!$
Samples Cannot Repeat	$n!/(n-r)!$	$n!/r!(n-r)!$

E5.8   Open the block diagram of the Compute Equilibrium.vi found in the Exercises&Problems\Chapter 5 folder in the Learning directory. This VI computes the equilibrium force required to keep a beam in equilibrium. Add a Case structure to the program so the user can choose whether the native LabVIEW functions or the Formula Express VIs will be used to calculate the result. The Formula Express VIs required are already programmed for you; you just need to add the case structure, move them into the appropriate case, and wire them correctly.

   Review: In order for a beam to be at equilibrium, the sum of the moments about any point must be zero. This example uses the sum of the moments about one end of the beam and the sum of the moments about the point where force A is located to determine the C and z values associated with the equilibrium force.

PROBLEMS

P5.1 Complete the crossword puzzle.

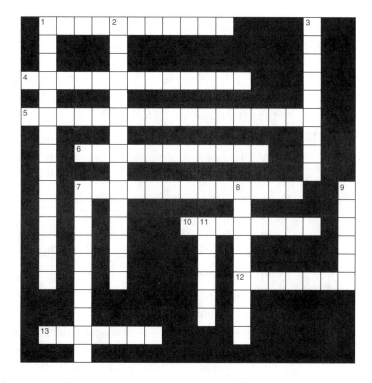

Across

1. Dot that appears where LabVIEW converts numeric representations of terminals to match.
4. Terminal that passes data between the frames of a Sequence structure.
5. The terminal of a For Loop or While Loop that contains the current number of completed iterations.
6. Node that executes formulas that you enter as text.
7. Optional mechanism in loop structures used to pass the value of a variable from iteration to iteration.
10. The automatic conversion performed to change the numeric representation of a data element.
12. Relocatable connection points for wires from outside and inside the structures.
13. Iterative loop structure that executes its subdiagram a set number of times.

Down

1. A conditional branching structure that executes one of its subdiagrams based on specific inputs.
2. The terminal of a For Loop whose value determine the number of times the For Loop executes.
3. Loop structure that repeats a section of code until a Boolean condition is False.
7. Block diagram within the border of a structure.
8. Program control element, such as a Sequence, Case, For Loop, or While Loop.
9. Diagrams associated with Sequence structures that control the execution order of the program.
11. Generic term for any item on the front panel or block diagram, including controls, nodes and wires.

P5.2   Use a For Loop to generate 100 random numbers. Determine the most current maximum and minimum number as the random numbers are being generated. This is sometimes referred to as a "running" maximum and minimum. Display the running maximum and minimum values as well as the current random number on the front panel. Be sure to include the Time Delay Express VI so the user is able to watch the values update as the For Loop executes.

P5.3   Construct a VI that displays a random number between 0 and 1 once every second. Also, compute and display the average of the last four random numbers generated. Display the average only after four numbers have been generated; otherwise display a 0. Each time the random number exceeds 0.5, generate a beep sound using the **Beep.vi**.

P5.4   Use the MATLAB script node to obtain a plot of a sine wave on a MATLAB figure. The sine wave frequency should be an input to the MATLAB Script Node.
     Note: Skip this problem if you are unfamiliar with programming in MATLAB.

P5.5   Construct a VI that has three Round LEDs on the front panel. When you run the program, the first LED should turn on and stay on. After one second, the second LED should turn on and stay on. After two more seconds, the third LED should turn on and stay on. All LEDs should be on for three seconds, and then the program should end.

P5.6   Create a time trial program to compare the average execution times of the Formula Node and the native LabVIEW Math Functions. This program will require a For Loop, a Flat or Stacked Sequence Structure, and a Case Structure. The For Loop is required to run the time trial N times and then the results can be averaged using the Statistics Express VI. The Sequence Structure is required to sample the Tick Count before and after the code executes. The Case Structure is required to determine whether the user would like to execute the Formula Node or the native LabVIEW Math Functions. Run the time trial for each of the cases. Which method has the fastest execution time? Which method is the easiest to program? Which method is the easiest to understand if someone else were to look at your code?

CHAPTER 6

Arrays and Clusters

The array data type and the cluster data type are presented in this chapter. An array is a variable-sized collection (or grouping) of data elements that are all the same type, such as a group of floating-point numbers or a group of strings. A cluster is a fixed-sized collection of data elements of mixed types, such as a group containing floating-point numbers and strings. You will learn how to use built-in functions to manipulate arrays and clusters. The important concept of polymorphism is introduced. Polymorphism is the ability of a function to adjust to input data of different types, dimensions, or representations.

GOALS

1. Understand how to create and use arrays.

2. Learn to use a variety of built-in array functions.

3. Understand the concept of polymorphism.

4. Become familiar with creating and using clusters.

5. Learn to manipulate clusters using built-in functions.

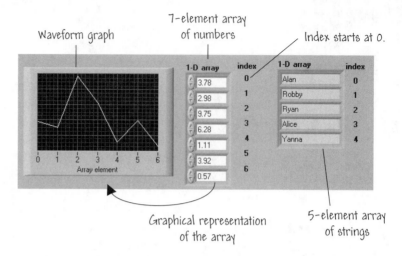

FIGURE 6.1
One-dimensional array examples.

6.1 ARRAYS

An **array** is a *variable-sized* collection of data elements that are all the *same type*. In contrast, a **cluster** is a *fixed-sized* collection of data elements of *mixed types*. In what situations might you use arrays? One scenario that benefits from the use of arrays and which you will encounter frequently involves working with a collection of data for plotting purposes. Arrays are quite useful as a mechanism for organizing data for plotting. Arrays are also helpful when you perform repetitive computations or when you are solving problems that are naturally formulated in matrix-vector notation, such as solving systems of linear equations. Using arrays in your VIs leads to compact block diagram codes that are easier to develop because of the large number of built-in array functions and VIs.

Arrays can have one or more dimensions with up to 2^{31} elements per dimension. The maximum number of elements depends on the available memory. The individual elements of an array can be any type with the exceptions that you cannot have an array of arrays, an array of charts, or an array of graphs. You access an individual array element by its index. The index is zero-based, implying that the array index is in the range 0 to $n - 1$, where n is the number of elements in the array. A one-dimensional (1D) array is shown in Figure 6.1. The first element of the 1D array has index 0, the second element has index 1, and so on.

Another way to display an array is with a waveform graph. In Figure 6.1, the waveform graph is used to display the numeric array in which each successive element of the array is plotted on the graph. To generate an *x* versus *y* graph, you could use a two-dimensional array (2D) with one column containing the *x* data points and the other column containing the *y* data points. To learn more about graphs, refer to Chapter 7.

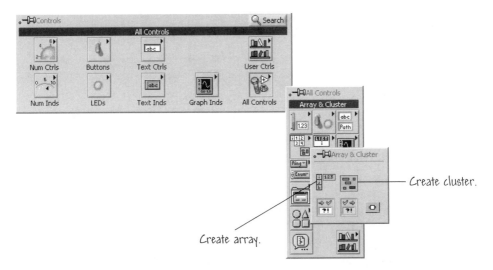

FIGURE 6.2
Creating an array control from the **Array & Cluster** subpalette of the **Controls** palette.

6.1.1   Creating Array Controls and Indicators

It takes two steps to create an array control or indicator. The two steps involve combining an **array shell** from the **Array & Cluster** subpalette of the **Controls≫All Controls** palette, as shown in Figure 6.2, with a valid element, which can be a numeric, Boolean, or cluster. Actually, valid elements also include strings—we will discuss strings in Chapter 9. In any case, charts, graphs, or other arrays are not valid elements to combine with the array control or indicator.

The two steps that lead to the creation of an array control or indicator follow:

- Select an empty array shell from the **Array & Cluster** subpalette of the **Controls≫All Controls** palette and drop it onto the front panel, as illustrated in Figure 6.3.

- Drag a valid data object (such as a numeric, Boolean, or string) into the array shell, as shown in Figure 6.4. The array shell resizes to accommodate its new type. You can also place the valid data object directly into the shell using the array shell's short cut menu.

To display more elements of the array, use the **Positioning** tool to grab a resizing handle on the corner of the array window and stretch the object to the desired number of visible array elements. The index value shown in the box on the left side of the array corresponds to the first visible element in the array. You can move through the array by clicking on the up and down arrows in the index display.

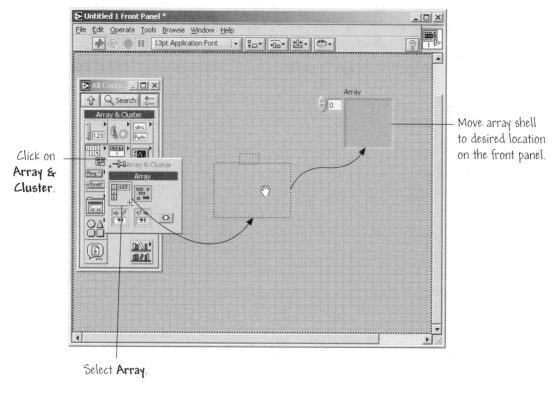

Click on **Array & Cluster**.

Select **Array**.

Move array shell to desired location on the front panel.

FIGURE 6.3
Placing an empty array shell on the front panel.

The block diagram terminal of an array shell is black when first dropped on the front panel. This indicates that the data type is undefined. The terminal contains brackets that denote array structures, as illustrated in Figure 6.5. When you drop a valid data object (such as a numeric, Boolean, or string) into the array shell, the array terminal on the block diagram turns from black to a color reflecting the data type. In Figure 6.5 the array block diagram terminal is pink, indicating that the array contains strings. When you wire arrays in the block diagram, you will find that array wires are thicker than wires carrying a single value.

Remember that you must assign a data object to the empty array shell before using the array on the block diagram. If you do not assign a data type, the array terminal will appear black with an empty bracket.

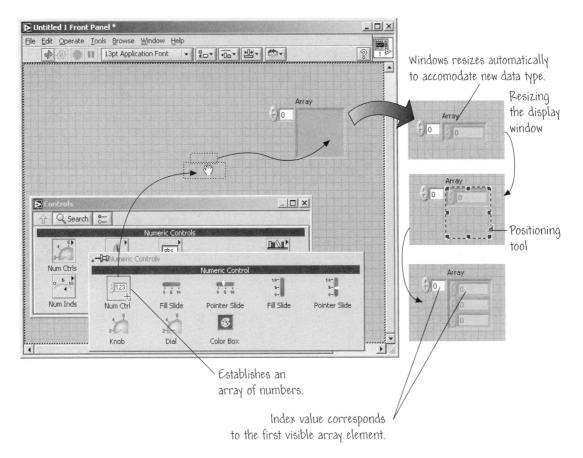

FIGURE 6.4
Drag a valid data object into the array shell to establish the array type.

6.1.2   Multidimensional Arrays

A two-dimensional (2D) array requires two indices—a row index and a column index—to locate an element. A three-dimensional array requires three indices, and in general, a *n*-dimensional array requires *n* indices. You add dimensions to the array in one of two ways: (a) by using the **Positioning** tool to resize the index display, or (b) by popping up on the array index display and choosing **Add Dimension** from the short cut menu (see Figure 6.6). An additional index display appears for each dimension you add. The example in Figure 6.6 shows a 2D digital control array.

You can reduce the dimension of an array by resizing the index display appropriately or by selecting **Remove Dimension** from the pop-up menu.

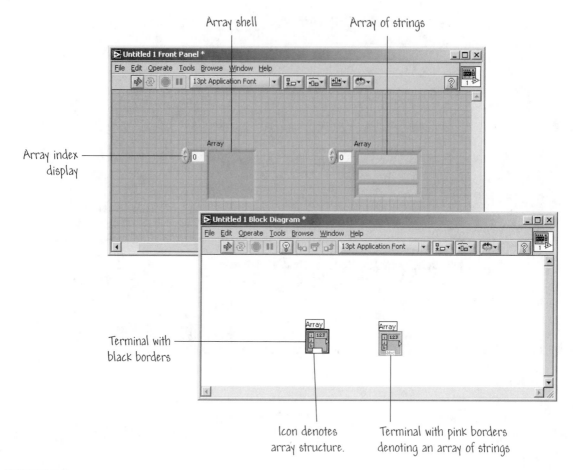

FIGURE 6.5
The terminal of an array shell is black, denoting an undefined data type.

6.2 CREATING ARRAYS WITH LOOPS

With the For Loop and the While Loop you can create arrays automatically with a process known as **auto-indexing**. Figure 6.7a shows a For Loop creating a 10-element array using auto-indexing. On each iteration of the For Loop, the next element of the array is created. In this case the loop counter is set to 10; hence a 10-element array is created. If the loop counter was set to 20, then a 20-element array would be created. The array passes out of the loop to the indicator after the loop iterations are complete.

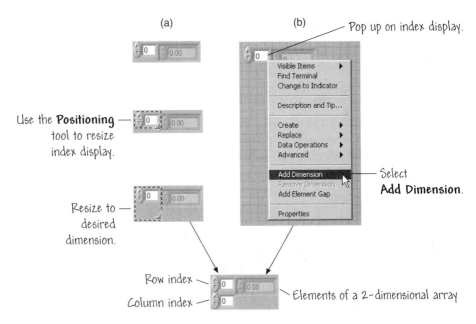

FIGURE 6.6
Adding dimensions to an array.

It is possible to pass a value out of a loop without creating an array. This requires that you disable auto-indexing by popping up on the tunnel (the square on the loop border through which the data passes) and selecting **Disable Indexing** from the short cut menu. In Figure 6.7b, auto-indexing is disabled; thus when the VI is executed, only the last value returned from the Random Number function passes out of the loop.

Open and run **Array Auto Index Demo.vi** *shown in Figure 6.7. It can be found in the* **Chapter 6** *folder in the* **Learning** *directory. After running the program, you should find that with auto-indexing enabled, the indicator array will have 10 elements (indexed 0, 1, · · · , 9), and with auto-indexing disabled only the last value of the random number function is passed out of the For Loop.*

You can also pass arrays into a loop one element at a time or the entire array at once. With auto-indexing enabled, when you wire an array (of any dimension) from an external node to an input tunnel on the loop border, then elements of the array enter the loop one at a time, starting with the first component. This is illustrated in Figure 6.8a. With auto-indexing disabled, the entire array passes into the loop at once. This is illustrated in Figure 6.8b.

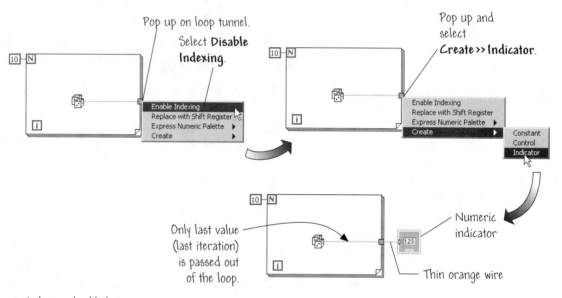

(a) Indexing enabled

(b) Indexing disabled

FIGURE 6.7
Auto-indexing is the ability of For Loops and While Loops to automatically index and accumulate arrays at their boundaries.

 *With For Loops auto-indexing is enabled by default. In contrast, with While Loops auto-indexing is disabled by default. If you desire auto-indexing, you need to pop up on the While Loop tunnel and choose **Enable Indexing** from the short cut menu.*

With a For Loop with auto-indexing enabled, an array entering the loop automatically sets the loop count to the number of elements in the array, thereby eliminating the need to wire a value to the loop count, N. What happens if you explicitly wire a value to the loop count that is different than an array size entering

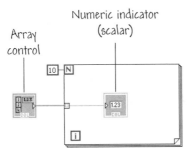

(a) Indexing enabled

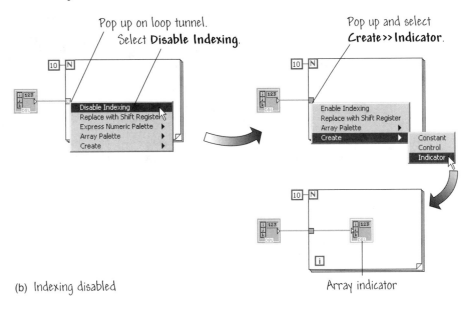

(b) Indexing disabled   Array indicator

FIGURE 6.8
Auto-indexing applies when you are wiring arrays into loops.

the loop? Or what if you wire two arrays with a different number of elements to a For Loop? The answer is that if you enable auto-indexing for more than one array entering a For Loop, or if you set the loop count by wiring a value to N with auto-indexing enabled, the actual loop count becomes the smaller of the two. For example, in Figure 6.9, the array size ($= 4$), and not the loop count N ($= 5$), sets the For Loop count—because the array size is the smaller of the two. The Array Size function used in the VI shown in Figure 6.9 will be discussed in the next section along with other common built-in array functions provided.

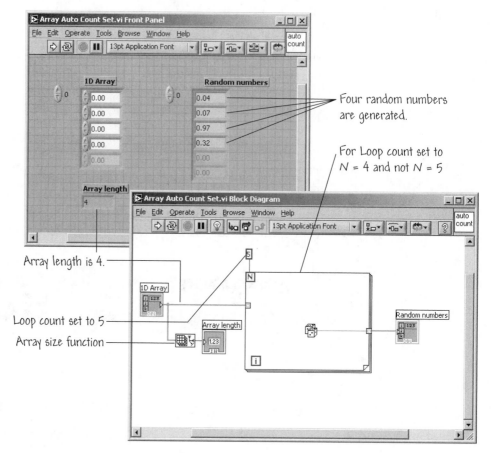

FIGURE 6.9
Automatically setting the For Loop count to the array size.

Open and run Array Auto Count Set.vi shown in Figure 6.9. It can be found in the Chapter 6 folder in the Learning directory. After running the program, you should find that only four random numbers are generated, despite the fact that the loop count N has been set to five!

6.2.1 Creating Two-Dimensional Arrays

You can use two nested For Loops (that is, one loop inside the other) to create a 2D array. The outer For Loop creates the row elements, and the inner For Loop creates the columns of the 2D array. Figure 6.10 shows two For Loops creating a 2D array of random numbers using auto-indexing.

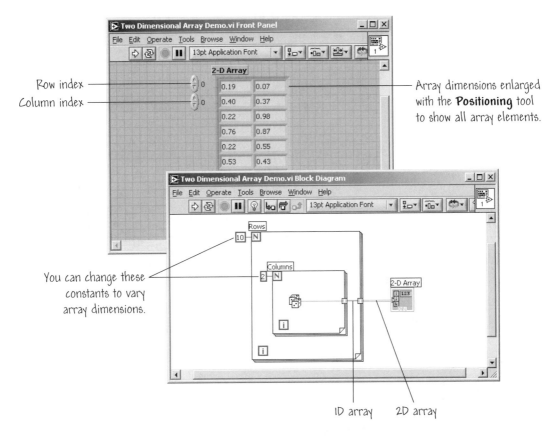

FIGURE 6.10
Creating a 2D array using two For Loops and auto-indexing.

    Open and run Two-Dimensional Array Demo.vi shown in Figure 6.10. It can be found in the Chapter 6 folder in the Learning directory. After running the program, you should find that a 2D array has been created.

6.3 ARRAY FUNCTIONS

Many built-in functions are available to manipulate arrays. Most of the common array functions are found in the **Array** subpalette of the **Functions≫All Functions** palette, as illustrated in Figure 6.11. Several of the heavily used functions are pointed out in Figure 6.11 and discussed in this section.

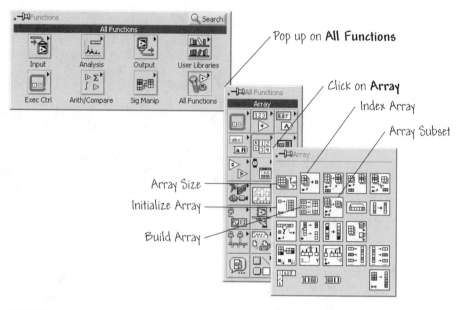

FIGURE 6.11
The array functions palette.

6.3.1 Array Size

The function Array Size returns the number of elements in the input array, as illustrated in Figure 6.12. If the input is an n-dimensional array, the output of the array function Array Size is a one-dimensional array with n elements—each element containing a dimension size.

You can run the array size demonstration shown in Figure 6.12 by opening Array Size Demo.vi in the Chapter 6 folder in the Learning directory. After executing the VI once to verify that the 1D array has length 4 and the 2D array has 2 rows and 4 columns, add an element to the 1D array and run the VI again. What happens? The 1D Array size value increments by 1.

6.3.2 Initialize Array

The function Initialize Array creates an n-dimensional array with elements containing the values that you specify. All elements of the array are initialized to the same value. To create and initialize an array that has more than one dimension, pop up on the lower-left side of the Initialize Array node and select **Add Dimension** or use the **Positioning** tool to grab a resizing handle to enlarge the node. You can remove dimensions by selecting **Remove Dimension** from the function

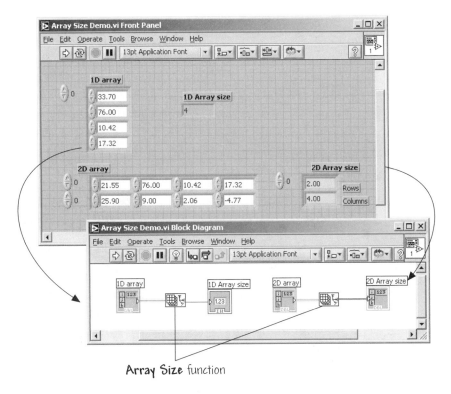

FIGURE 6.12
The function Array Size returns the number of elements in the input array.

short cut menu or with the **Resizing** cursor. The Initialize Array is useful for allocating memory for arrays. For instance, if you are using shift registers to pass an array from one iteration to another, you can initialize the shift register using the Initialize Array function.

The input determines the data type and the value of each element of the initialized array. The dimension size input determines the length of the array (see Figure 6.13a). For example, if the input value **element** is a double-precision floating-point number with the value of 5.7, and dimension size has a value of 100, the result is a 1D array of 100 double-precision floating-point numbers all set to 5.7, as illustrated in Figure 6.13b. You can wire inputs to the Initialize Array function from front panel control terminals, from block diagram constants, or from calculations in other parts of the block diagram. One method of obtaining the output array is to pop up on the Initialize Array function and choose Create≫ Indicator. Also, you can pop up and choose Create≫ Control and Create≫ Constant to create inputs for the element and dimension size. Figure 6.13c depicts a 2D array with 3 rows and 2 columns initialized with the long integer value of 0.

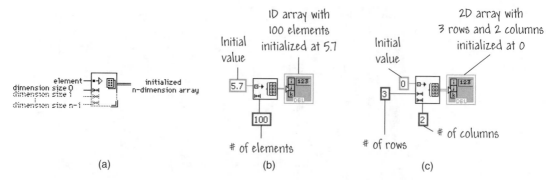

(a)   (b)   (c)

FIGURE 6.13
The function Initialize Array creates an array of a certain dimension containing the same value for each element.

*You can run the array initialization demonstration shown in Figure 6.13 by opening **Array Initialization Demo.vi** located in the **Chapter 6** folder in the **Learning** directory. Change the initial value of the two arrays on the block diagram, run the VI and examine the result.*

6.3.3   Build Array

The function Build Array concatenates multiple arrays or adds extra elements to an array. The function accepts two types of inputs—scalars and arrays—therefore it can accommodate both arrays and single-valued elements.

The Build Array function appears with one scalar input when initially placed in the block diagram window, as shown in Figure 6.14.

Array inputs have brackets, while element inputs do not. Array inputs and scalar inputs are not interchangeable! Pay special attention to the inputs of a Build Array function, otherwise you may generate wiring errors that are difficult to detect.

You can add as many inputs as you need to the Build Array function. Each input to the function can be either a scalar or an array. To add more inputs, pop up on the left side of the function and select **Add Input**. You also can enlarge the Build Array node by placing the **Positioning** tool at one corner of an object to grab and drag the resizing handle. You can remove inputs by shrinking the node with the resizing handle or by selecting **Remove Input** from the short cut window.

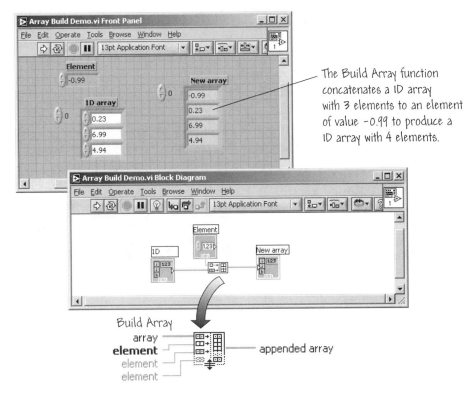

FIGURE 6.14
Build Array concatenates multiple arrays or appends elements to an array.

The input type (element or array) is automatically configured when wired to the Build Array function. The Build Array function shown in Figure 6.14 is configured to concatenate an array and one scalar element into a new array.

The Build Array function concatenates the elements or arrays in the order in which they appear in the function, top to bottom. If the inputs are of the same dimension, you can right-click on the function and select **Concatenate Inputs** to concatenate the inputs into a longer array of the same dimension. If you leave the **Concatenate Inputs** option unselected, you will add a dimension to the array.

The Array Build demonstration shown in Figure 6.14 is located in the **Chapter 6** *folder in the* **Learning** *directory and is called* **Array Build Demo.vi.** *Open, run and experiment with the VI.*

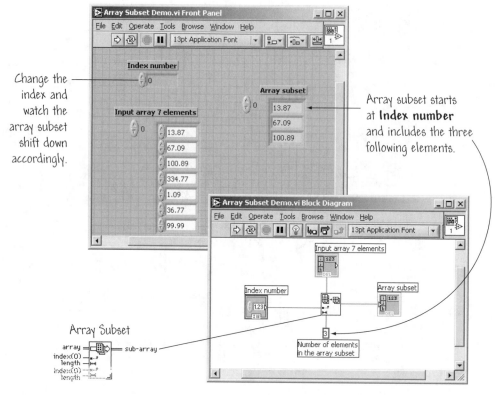

FIGURE 6.15
The function Array Subset returns a portion of an array starting at **index** and containing **length** elements.

6.3.4 Array Subset

The function Array Subset returns a portion of an array starting at **index** and containing **length** elements. In the example shown in Figure 6.15 you'll find that the array index begins with 0.

The array subset demonstration shown in Figure 6.15 can be found in the Chapter 6 *folder in the* Learning *directory and is called* Array Subset Demo.vi. *Open and run the VI in the* **Run Continuously** *mode. Using the* **Operating** *tool, change the value of* **Index Number** *and watch the Array Subset function return a portion of the array, starting at the index that you specify and including the three subsequent elements.*

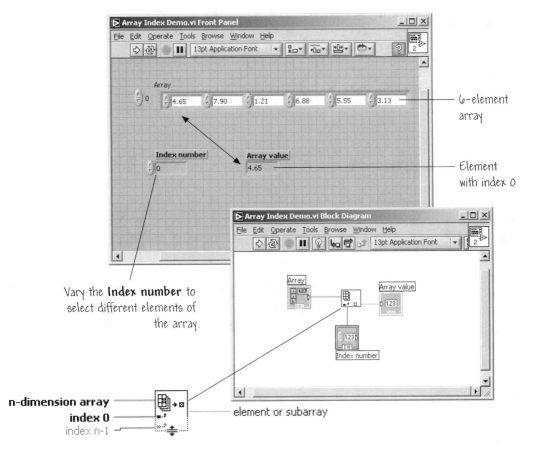

FIGURE 6.16
The function Index Array accesses an element of an array.

6.3.5   Index Array

The function Index Array accesses an element of an array. The example shown in Figure 6.16 uses the input **Index number** to specify which element of the array to access. Remember that the index number of the first element is zero.

The index array demonstration shown in Figure 6.16 can be found in the Chapter 6 *folder of the* Learning *directory and is called* Array Index Demo.vi. *Open and run the VI in* **Run Continuously** *mode. Using the* **Operating** *tool, change the value of* **Index number** *and watch the Index Array function return a different element based on the index value.*

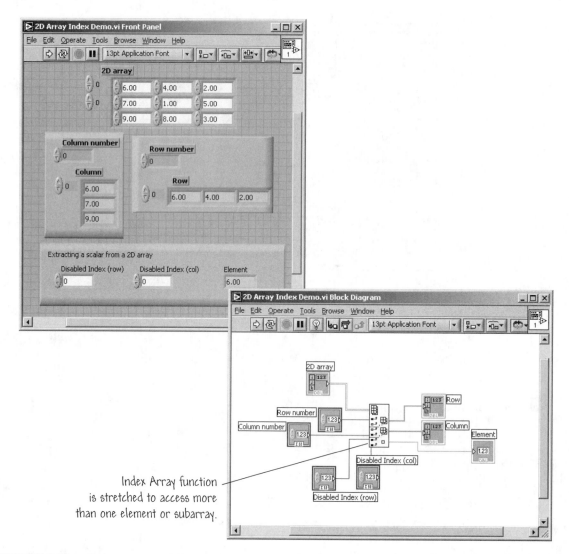

FIGURE 6.17
Using Index Array to extract a row or column of an array.

The Index Array function automatically resizes to match the dimensions of the wired input array. For example, if you wire a 1D array to the Index Array function, a single index input will show. Similarly, if you wire a 2D array the Index Array function, two index inputs will show—one for the row and one for the column.

Once you have wired an input array to the Index Array function, you can access more than one element, or subarray (e.g., a row or a column) using the **Positioning** tool to manually resize the function once placed on the block diagram. When you expand the Index Array function it will expand in increments determined by the dimensions of the array wired to the function. In Figure 6.17, the Index Array function is expanded so that three subarrays can be extracted: a row, a column, and a single element. The index inputs you wire determine the shape of the subarray you want to access or modify. For example, if the input to an Index Array function is a 2D array and you wire only the row input, you extract a complete 1D row of the array. If you wire only the column input, you extract a complete 1D column of the array. If you wire the row input and the column input, you extract a single element of the array. These three cases are illustrated in Figure 6.17. Each input group is independent and can access any portion of any dimension of the array.

The 2D index array demonstration shown in Figure 6.17 can be found in the **Chapter 6** *folder of the* **Learning** *directory and is called* **2D Array Index Demo.vi**. *Open and run the VI in* **Run Continuously** *mode. Using the* **Operating** *tool, change the* **Column number** *and* **Row number** *values and verify the numbers returned in* **Column** *and* **Row**.

Practice with Arrays

In this exercise you will get more practice using array functions. Open Practice with Arrays.vi located in the Chapter 6 folder of the Learning directory. The front panel and block diagram of the incomplete VI are shown in Figure 6.18. This VI is not completed—you will finish wiring and debugging the VI for practice.

The front panel contains four arrays and one numeric control. The completed VI concatenates the input arrays and the numeric control values to form a new array. Using the Array Size function and the Array Initialize function, the VI creates a new array of appropriate dimension and with all elements of the array initialized to 1. In the final computation of the VI, the difference between the two new arrays is computed, and the result is displayed on the front panel. When you compute the difference of two arrays with the same number of elements, the differencing operation subtracts the array values element by element.

The completed block diagram should resemble the one shown in Figure 6.19. Remember to save and close the VI in the Users Stuff folder.

The completed VI for this exercise can be found in the **Chapter 6** *folder of the* **Learning** *directory and is called* **Practice with Arrays Done.vi**.   ◆

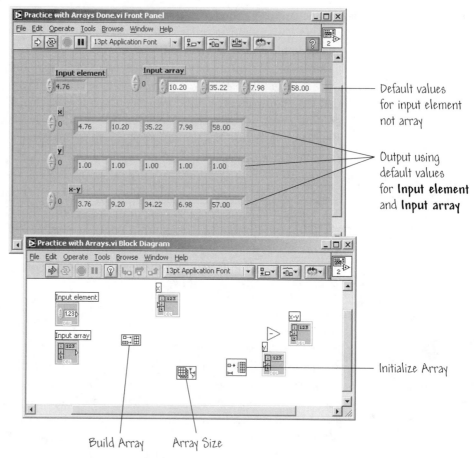

FIGURE 6.18
Practice with array functions.

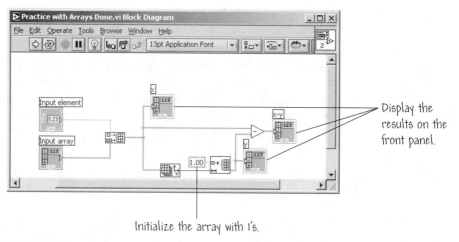

FIGURE 6.19
The completed block diagram for the array practice exercise.

6.4   POLYMORPHISM

Polymorphism is the ability of certain LabVIEW functions (such as Add, Multiply, and Divide) to accept as inputs of different dimensions and representations. Arithmetic functions that possess this capability are polymorphic functions. For example, you can add a scalar to an array or add together two arrays of different lengths. Figure 6.20 shows some of the polymorphic combinations of the Add function.

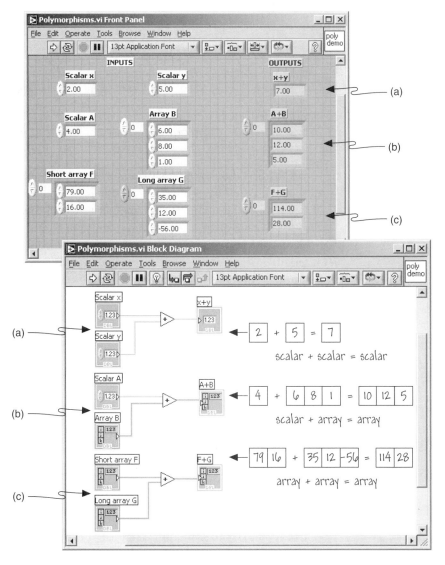

FIGURE 6.20
Polymorphic combinations of the Add function.

Array with 10 elements

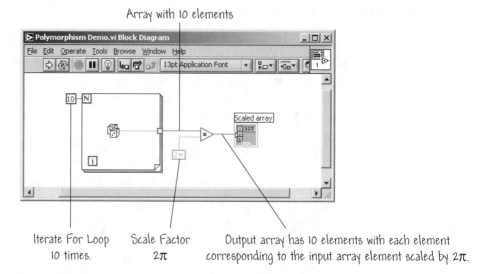

Iterate For Loop
10 times.

Scale Factor
2π

Output array has 10 elements with each element
corresponding to the input array element scaled by 2π.

FIGURE 6.21
Demonstrating the polymorphic capability of the Multiply function.

In the first combination shown in Figure 6.20a, the result of adding a scalar and a scalar is another scalar. In the second combination shown in 6.20b, the result of adding a scalar to an array is another array. In this situation, the scalar input is added to each element of the input array. In the third combination depicted in 6.20c, an array of length 2 is added to an array of length 3, resulting in an array of length 2 (the length of the shorter of the two input arrays). Array addition is performed component-wise; that is, each element of one array is added to the corresponding element of the other array.

When two input arrays have different lengths, the output array resulting from some arithmetic operation (such as adding the two arrays) will be the same size as the smaller of the two input arrays. The arithmetic operation applies to corresponding elements in the two input arrays until the shorter array runs out of elements—then the remaining elements in the longer array are ignored!

Consider the VI depicted in Figure 6.21. Each iteration of the For Loop generates a random number that is stored in the array and readied for output once the loop iterations are finished. After the loop finishes execution (after 10 iterations), the Multiply function multiplies each element in the array by the scaling factor 2π. Notice that the Multiply function has two inputs: an array and a scalar. The polymorphic capability of the Multiply function allows the function to take inputs of differing dimension (in this instance, an array of length 10 and a scalar) and produce a sensible output. What happens when you have two array inputs for a Multiply function? In that case, corresponding elements of each array are multiplied.

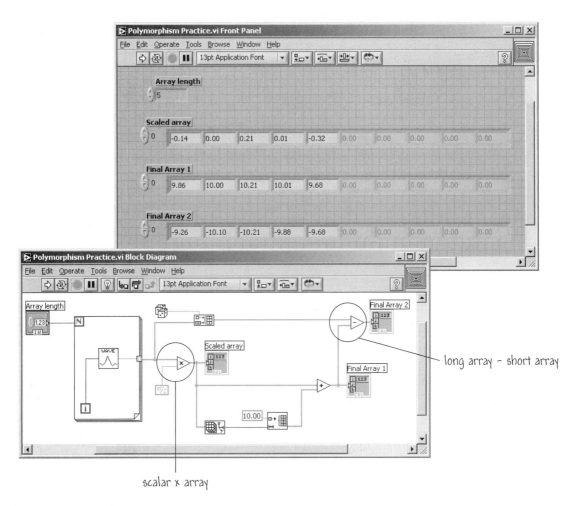

FIGURE 6.22
Practicing using polymorphism.

The two demonstrations of polymorphism shown in Figures 6.20 and 6.21 can be found in the **Chapter 6** *folder in the* **Learning** *directory and are called* **Polymorphisms.vi** *and* **Polymorphism Demo.vi**, *respectively. Open and run each VI. Vary the values of the array elements and examine the results. You can also edit the array lengths and test your knowledge of polymorphisms by first predicting the result and then verifying your prediction by running the VI.*

Practice Using Polymorphism

Open a new VI and recreate the front panel and block diagram shown in Figure 6.22. When you have finished the construction, experiment with the VI. A suggested way to run the VI is in **Run Continuously** mode. Increment and decrement the variable **Array length** and watch the array length change.

This exercise demonstrates two applications of polymorphism. Referring to Figure 6.22, it can be seen that in one case the Multiply function operates on two inputs of different dimension: an array and a scalar $\pi/2$. In the second illustration of polymorphism, the Subtract function has two array inputs of differing lengths. The resulting array **Final Array 2** has length equal to the length of the shorter array (the length is equal to length of **Scaled array**). Three array functions are also utilized: Build Array, Array Size, and Initialize Array. Save the VI as Polymorphism Practice.vi and place it in the Users Stuff folder.

The completed polymorphism practice VI shown in Figure 6.22 can be found in the Chapter 6 folder in the Learning directory and is called Polymorphism Practice.vi. ◆

6.5   CLUSTERS

A **cluster** is a data structure that, like arrays, groups data. However, clusters and arrays have important differences. One important difference is that clusters can group data of different types, whereas arrays can group only like data types. For example, an array might contain ten numeric indicators, whereas a cluster might contain a numeric control, a toggle switch and string control. And although cluster and array elements are both ordered, you access cluster elements by **unbundling** some or all the elements at once rather than indexing one element at a time. Clusters are also different from arrays in that they are of a fixed size.

One similarity between arrays and clusters is that both are made up of either controls or indicators. In other words, clusters cannot contain a mixture of controls and indicators. An example of a cluster is shown in Figure 6.23. The cluster shown has four elements: a numeric control, a horizontal toggle switch, a string control (more on strings in Chapter 9), and a knob. The cluster data type appears frequently when graphing data (as we will see in the next chapter).

Clusters are generally used to group related data elements that appear in multiple places on the block diagram. Because a cluster is represented by only one wire in the block diagram, its use has the positive effect of reducing wire clutter and the number of connector terminals needed by subVIs. A cluster may be thought of as a bundle of wires wherein each wire in the cable represents a different element of the cluster. On the block diagram, you can wire cluster terminals only if they have the same type, the same number of elements, and the same order of elements. Polymorphism applies to clusters as long as the clusters have the same number of elements in the same order.

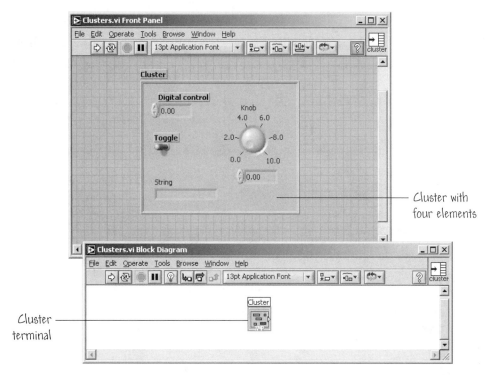

Cluster terminal

Cluster with four elements

FIGURE 6.23
A cluster with four elements: numeric control, toggle switch, string control, and a knob.

6.6   CREATING CLUSTER CONTROLS AND INDICATORS

Cluster controls and indicators are created by placing a **cluster shell** on the front panel, as shown in Figure 6.24. A new cluster shell has a resizable border and an (optional) label. When you pop up in the empty element area of the cluster shell, the **Controls≫All Controls** palette appears, as illustrated in Figure 6.25. You create a cluster by placing any combination of numerics, Booleans, strings, charts, graphs, arrays, or even other clusters from the **Controls** palette into the cluster shell. Remember that a cluster can contain controls *or* indicators, but not both. The cluster becomes a control cluster or indicator cluster depending on the first element you place in the cluster. For example, if the first element placed in a cluster shell is a numeric control, then the cluster becomes a control cluster.

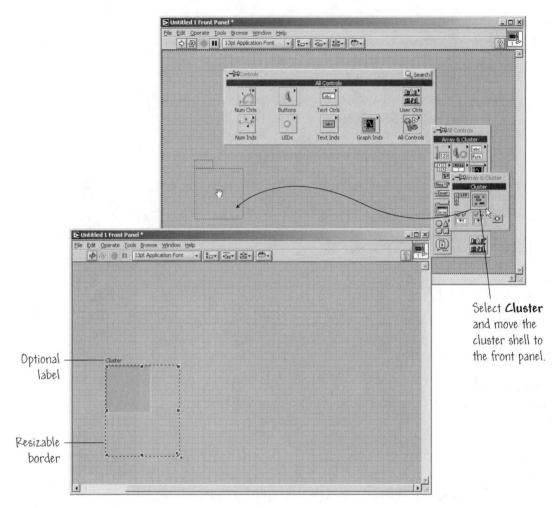

FIGURE 6.24
Creating and resizing a cluster.

Any objects added to the cluster afterwards become control objects. Selecting **Change To Control** or **Change To Indicator** from the short cut menu of any cluster element changes the cluster and all its elements from indicators to controls or from controls to indicators, respectively. You can also drag existing objects from the front panel into the cluster shell.

6.6.1   Cluster Order

The elements of a cluster are ordered according to when they were placed in the cluster rather than according to their physical position within the cluster

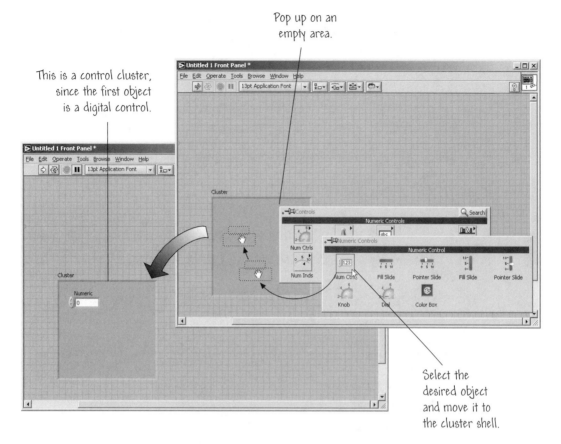

FIGURE 6.25
Creating a cluster by placing objects in the cluster shell.

shell. The first object placed in the cluster shell is labeled element 0, the second object inserted is element 1, and so on. The order of the remaining elements is automatically adjusted when an element is removed from the cluster. The cluster order determines the order in which the elements appear as terminals on the Bundle and Unbundle functions in the block diagram (more on these cluster functions in the next sections). You must keep track of your cluster order if you want to access individual elements in the cluster, because individual elements in the cluster are accessed by order.

You can examine and change the order of the elements within a cluster by popping up on the cluster border on the front panel and choosing the item **Reorder Controls in Cluster**. . . from the short cut menu, as shown in Figure 6.26.

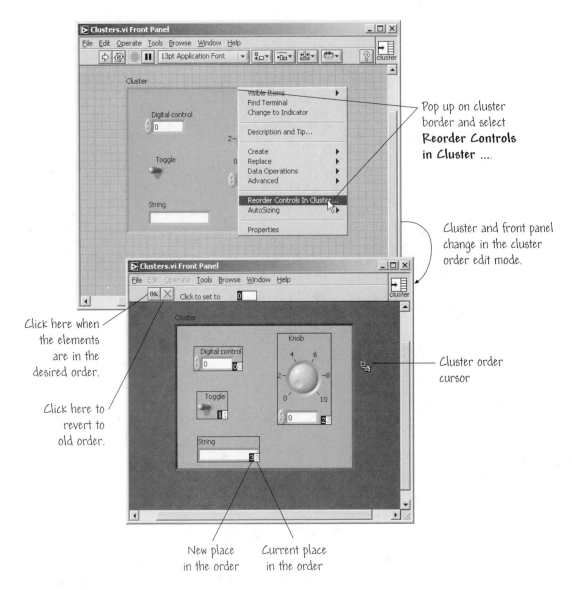

FIGURE 6.26
Examining and changing the cluster order.

Notice in the figure that a new set of buttons replaces the toolbar, and the cluster appearance changes noticeably. Even the cursor changes to a hand with a # sign above—this is the cluster order cursor.

Two boxes appear side-by-side in the bottom right corner of each element in the cluster. The white boxes indicate the current place of that element in the cluster order. The black boxes indicates the new location in the order, in case you have changed the order. The numbers in the white and black boxes are identical until you make a change in the order. Clicking on an element with the cluster order cursor changes the current place of the element in the cluster order to the number displayed in the toolbar at the top of the front panel. When you have completed arranging the elements within the cluster to your satisfaction, click on the OK button, or you can revert to the old order by clicking on the X button.

A simple example shows the importance of the cluster order. Consider the situation depicted in Figure 6.27a, where the front panel contains two clusters. In the first cluster, the first component is a numeric control, while in the second cluster, the first component is a numeric indicator. In the associated block diagram, the cluster control properly wires to the cluster indicator. Now, in Figure 6.27b, the cluster order has been changed where the string control is the first component of the cluster control. The numeric indicator is still the first component of the cluster indicator. The wire connecting is now broken, and if you attempted to run the VI you would get an error message stating that there is a type conflict.

6.6.2   Using Clusters to Pass Data to and from SubVIs

A VI connector panel can have a maximum of 28 terminals (as discussed in Section 4.3). In general, you do not want to pass information individually to all 28 terminals when calling a VI or a subVI. A good rule-of-thumb is to use connector passes with 14 or fewer terminals. Otherwise, the terminals are very small and may be difficult to wire. Using clusters, you can group related controls together. One cluster control uses one terminal on the connector, but it can contain may controls. Similarly, you can use a cluster to group indicators.

This allows the subVI to pass multiple outputs using only one terminal. Figure 6.28, illustrates the benefit of using clusters to pass data to and from subVIs.

6.7   CLUSTER FUNCTIONS

Two of the more important cluster functions are the Bundle and Unbundle functions. These functions (and others) are found in the **Cluster** subpalette of the **All Functions** palette, as illustrated in Figure 6.29.

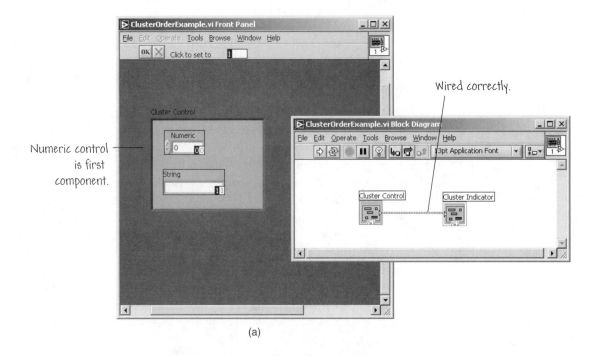

(a)

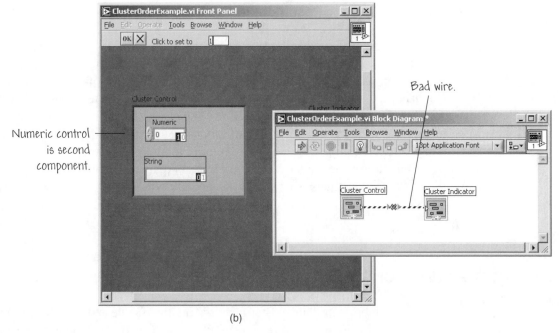

(b)

FIGURE 6.27
Example showing the importance of cluster order: (a) Wiring is correct. (b) Wiring is incorrect.

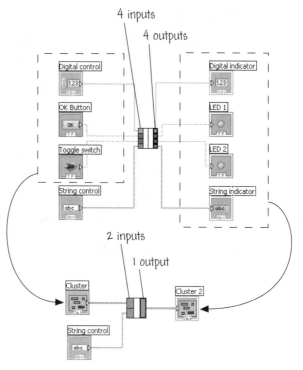

FIGURE 6.28
Using clusters to pass data to and from subVIs.

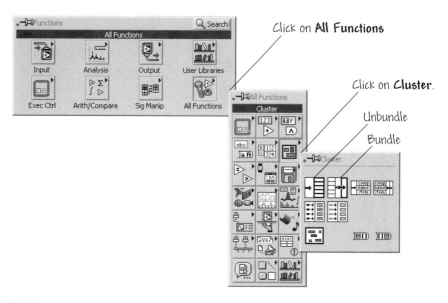

FIGURE 6.29
The **Cluster** palette.

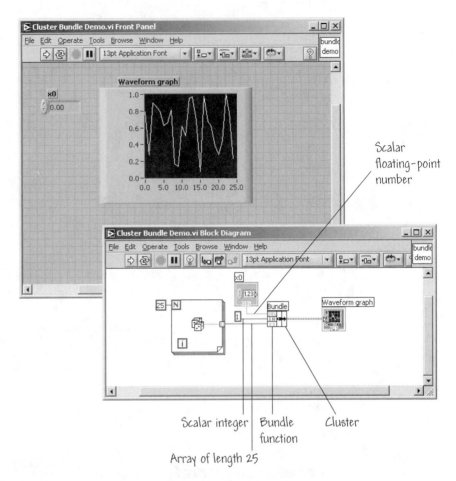

FIGURE 6.30
The Bundle function.

6.7.1 The Bundle Function

The Bundle function is used to assemble individual elements into a single new cluster or to replace elements in an existing cluster. When placed on the block diagram, the Bundle function appears with two element input terminals on the left side. You can increase the number of inputs by enlarging the function icon (with the resizing handles) vertically to create as many terminals as you need. You can also increase the number of inputs by popping up on the left side of the function and choosing **Add input** from the menu. Since you must wire all the inputs you create, only enlarge the icon to show the exact desired number of element inputs. When you wire to each input terminal, a symbol representing the data type of the wired element appears in the originally empty terminal, as illustrated in Figure 6.30. The order of the elements in the cluster is equal to the order of inputs to the Bundle function. The order is defined top to bottom, which means that the element you wire to the top terminal becomes element 0, the element you wire to the second terminal becomes element 1, and so on.

The example depicted in Figure 6.30 shows the Bundle function creating a cluster from three inputs: a floating-point real number, an integer, and an array of numbers generated by the For Loop. The output of the new cluster is wired to a waveform graph that displays the random numbers on a graph. Graphs (such as the waveform graph) are covered in Chapter 7.

The example shown in Figure 6.30 can be found in the **Chapter 6** *folder in the* **Learning** *directory and is called* **Cluster Bundle Demo.vi**. *When you open and run the VI you will see that varying the input x0 changes the x-axis origin.*

In addition to input terminals for elements on the left side, the Bundle function also has an input terminal for clusters in the middle, as shown in Figure 6.31. Sometimes you want to replace or change the value of one or two elements of a cluster without affecting the others. The example in Figure 6.31 shows a convenient way to change the value of two elements of a cluster. Because the cluster input terminal (that is, the middle terminal) of the Bundle function is wired to an existing cluster named **Cluster**, the only element input terminals that must be wired are those that are associated with cluster elements that you want to replace—in the illustration in Figure 6.31 the second and fourth element values are being replaced. To replace an element in the cluster using the Bundle function, you first place the Bundle function on the block diagram and then resize the function to show exactly the same number of input terminals as there are elements in the existing cluster that you want to modify. The next step is to wire the existing cluster to the middle input terminal of the Bundle function. Afterwards, any other input terminals on the left side that are wired will replace the corresponding elements in the existing cluster. Remember that if the objective is to create a new cluster rather than modify an existing one, you do not need to wire an input to the center cluster input of the Bundle function.

The example depicted in Figure 6.31 shows the Bundle function being used to change the value of the numeric control and the knob. The corresponding VI can be found in the Chapter 6 folder in the Learning directory and is called Cluster Element Replacement.vi. Open and run the VI in **Run Continuously** mode. **Cluster** is a control cluster and **Cluster Output** is an indicator cluster. Toggle the horizontal toggle switch (using the **Operating** tool). You should see that the toggle switch in **Cluster Output** also switches. Now vary the **Knob** value away from the default value of 4.8. Notice that the **Knob** in **Cluster Output** does not change. This phenomenon occurs because the **Knob** is the fourth element of the cluster and it's value is being replaced by the value set in the numeric control **Knob input**. To change the value of the **Knob** in **Cluster Output** you change the value of **Knob input**. Try it out!

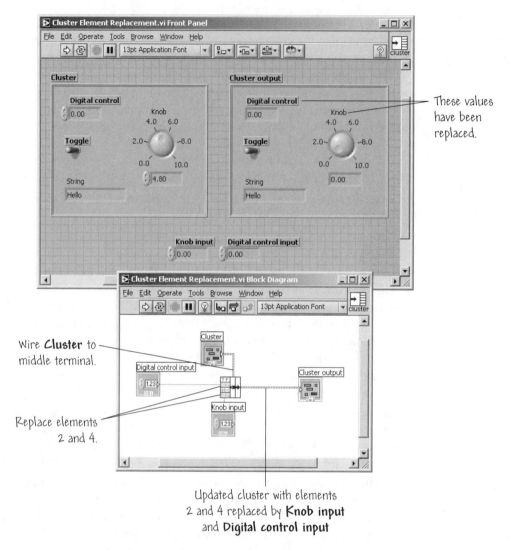

FIGURE 6.31
Using the Bundle function to replace elements of an existing cluster.

6.7.2   The Unbundle Function

The Unbundle function extracts the individual components of a cluster. The output components are arranged from top to bottom in the same order as in the cluster. When placed on the block diagram, the Unbundle function appears with two element output terminals on the right side. You adjust the number of terminals with the resizing handles following the same method as with the Bundle function, or you can pop up on the right side of the function and choose **Add Output** from the menu. Element 0 in the cluster order passes to the top output terminal, element

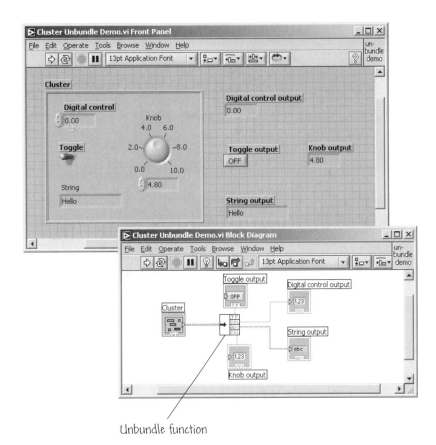

Unbundle function

FIGURE 6.32
The Unbundle function.

1 passes to the second terminal, and on down the line. The cluster wire remains broken until you create the correct number of output terminals, at which point the wire becomes solid. When you wire an input cluster to the correctly sized Unbundle, the previously blank output terminals will assume the symbols of the data types in the cluster.

*While all elements are unbundled using the **Unbundle** function, the **Unbundle By Name** function can access one or more elements in a cluster.*

The example depicted in Figure 6.32 shows the Unbundle function being used to unpack the elements of the data structure **Cluster**. The cluster has four elements, and each element is split from the cluster and wired to its own individual indicator for viewing on the front panel.

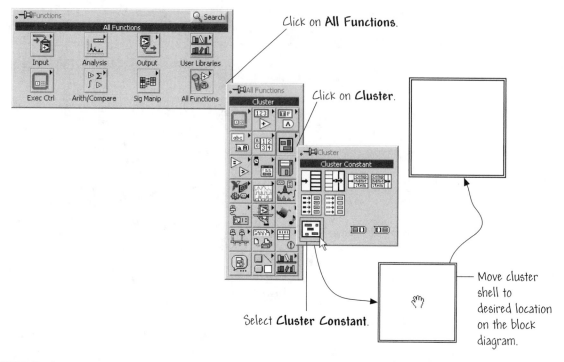

FIGURE 6.33
Creating a cluster constant on the block diagram.

*The example shown in Figure 6.32 can be found in the **Chapter 6** folder in the **Learning** directory and is called **Cluster Unbundle Demo.vi**. Open and run the VI in **Run Continuously** mode and notice that varying any values in the control cluster immediately changes the values of the various indicators.*

6.7.3   Creating Cluster Constants on the Block Diagram

You can use the same technique you used on the front panel to create a cluster constant on the block diagram. On the block diagram, choose Cluster≫Cluster Constant from the **Functions** palette to create the cluster shell as illustrated in Figure 6.33. Click in the block diagram to place the cluster shell, and place other constants of the appropriate data type within the cluster shell.

If you have a cluster control or indicator on the front panel and want to create a cluster constant containing the same components in the block diagram, you can either drag that cluster from the panel to the diagram or select Create≫Constant from the short cut menu, as depicted in Figure 6.34.

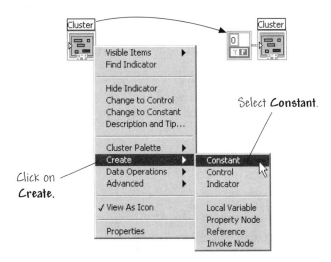

FIGURE 6.34
Create a cluster constant using a cluster on the front panel.

6.7.4   Using Polymorphism with Clusters

Since the arithmetic functions are polymorphic, they can be used to perform computations on clusters of numbers. As shown in Figure 6.35, you use the arithmetic functions with clusters in the same way you use them with arrays of numerics. You can also use the string-to-number functions to convert a cluster of numerics to a cluster of strings (strings are covered in Chapter 9).

> *The VI in Figure 6.35 can be found in* **Chapter 6** *of the* **Learning** *directory. The VI is called* **ClusterScaling.vi.**

BUILDING BLOCK

6.8   BUILDING BLOCKS: MEASURING VOLUME

In the Chapter 5 Building Block exercise, you constructed a VI that could acquire multiple volume measurements using a For Loop, and then you controlled the timing of the loop. The output was displayed on a tank indicator along with an LED to indicate when a volume limit had been exceeded. In this exercise you will modify the VI Multiple Volume Points that you developed in Chapter 5 and saved in the Users Stuff folder.

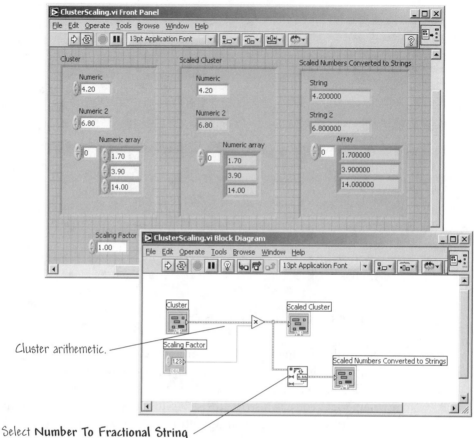

FIGURE 6.35
Using polymorphism with clusters.

The objective is to use the Build Array function to organize the volume data in a manner suitable for graphing. The target front panel and block diagram are shown in Figure 6.36. Open Multiple Volume Points.vi and edit it using Figure 6.36 as a guide.

There are two new objects used in this VI: the waveform graph and the Curve Fit Express VI. Waveform graphs are the subject of Chapter 7 and are used here as a way to motivate the upcoming discussions on graphs and charts. The subject of curve fits is covered in Chapter 11 where various data analysis techniques are presented. Figure 6.37 shows where to look to find the waveform graph and the

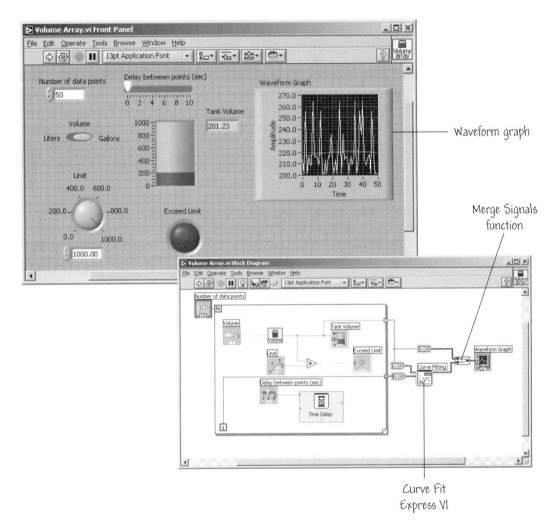

FIGURE 6.36
Using the Build Array function to organize volume data for graphing purposes.

Curve Fit Express VI. It is not important at this point to know the details about waveform graphs or polynomial curve fits—that information will come later!

When you have finished editing and debugging of the VI, save it in the **Users Stuff** folder as **Volume Array.vi**.

Experiment with your new VI. Set the variable **Number of data points** to 10 and run the VI with no delay between data points. The VI will display the volume measurements and curve fit the data with a smooth line. Run the VI several times and observe the results—the data points should be easily fitted with the smooth

Select **Analysis**

Curve Fit
Express VI

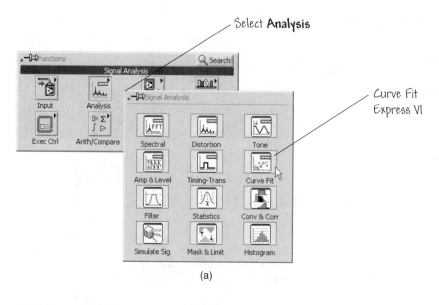

(a)

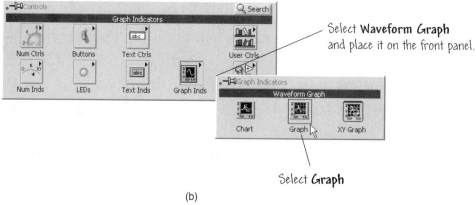

Select **Waveform Graph**
and place it on the front panel.

Select **Graph**

(b)

FIGURE 6.37
Finding waveform graphs and polynomial curve-fitting functions.

curve fit. Experiment with the VI by switching between **Liters** and **Gallons** and increasing the number of data points.

6.9   RELAXED READING: LABVIEW MONITORS MEDIEVAL BELL TOWER

A long-term investigation is being carried out by the Politecnico di Milano on the Torrazzo cathedral bell tower. The bell tower, shown in Figure 6.38, is the symbol of Cremona, Italy, the city of Stradivarius. The date of construction can

FIGURE 6.38
Torrazzo bell tower.

be traced back to the 13th century. At a height of 112 m (367 ft), the Torrazzo is the tallest masonry bell tower in Europe. Signs of crack damage and surface deterioration appearing on the tower walls motivated an on-site investigation to understand the tower dynamic and static behavior mainly under the action of wind and thermal gradients. A PC-based system has been set up for continuous measurement, on-line analysis, and monitoring of slow (such as walls and air temperature and crack openings) and fast (such as wind velocity and direction) signal vibrations.

The challenge was to construct a system for long-term investigations of the static and dynamic behavior of the Torrazzo medieval bell tower. The solution was a hybrid system with high-frequency measurements by means of a DAQ board for vibrations, and low-frequency measurement with a Fieldpoint network for crack monitoring. The LabVIEW-based system is remote-monitored via modem.

Static measurements are mainly related to openings in cracks in the tower. Crack openings depend predominantly on thermal actions during the day (sun radiation) and during the year (seasonal). The air and wall temperatures are measured using 40 RTDs and the crack openings are measured with 16 LVDTs. The transducers and sensor are placed in various sections of the tower. The main purpose of static measurements is to understand whether the crack openings/closings follow a natural cycle during the year or continually grow larger as time passes.

The dynamic measurements of the complex tower vibrations and their correlation with wind and bell actions provide both direct and diagnostic information. Direct information can be helpful in understanding the risk of structural damage under the action of the wind or bell movements. The direct information can be used for diagnostic purposes: if one notices changes, for example, in the natural frequencies of the tower, then an alarm signal would be sounded that something has changed in the tower's structure, and appropriate action can be taken to avert further problems.

The dynamic signals are measured by nine servoaccelerometers around the tower. Instantaneous speed and direction of the wind are measured by means of four cup anemometers and four wind vanes installed close to the top (85 m) on the four sides of the tower. The dynamics of crack openings are also acquired from LVDTs. The lowest transducer is placed near the base of the tower (about 15 meters), the highest at about 100 meters.

The system development posed many challenges. One of the main challenges was the measurement and data acquisition system, which had strict requirements:

- The signal and supply cables had to be as short as possible, for economic reasons and for measurement reliability.

- A good mass reference had to be given to all signals to avoid ground loops.

- The system had to be flexible and scalable, to allow for component failure or new sensors being added during the measurement campaign.

- Sufficient data had to be acquired to develop a correct and detailed knowledge of the tower behavior without resorting to redundancy and extensive memory requirements.

- The system had to be controlled remotely by modem.

When conducting an unattended long dynamic measurement campaign, it is important to be able to handle signals over a wide range (when signals are high), and a good resolution (when signals are low). The characteristics of the dynamic signals are not completely known *a priori*. To accommodate the range of possible signals, a 16-bit NI PCI-6031 E DAQ board was chosen for dynamic data acquisition.

It is also important to avoid acquiring *too* much data. The LabVIEW software was tasked to acquire data on a ten-minute basis (continuous scan), and store the data to disk only if certain criteria are satisfied. Otherwise, store only statistical

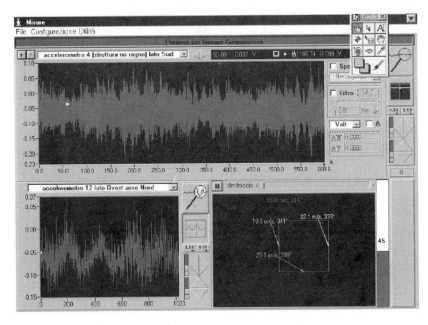

FIGURE 6.39
Screen capture of data acquisition program. (Top-left display: stored signal of an accelerometer, bottom left: real-time signal, bottom right: real-time wind measurement)

data for all channels (i.e., mean and RMS). For example, the user can decide to store just ten files for wind coming from direction 0–10° and velocity 5–10 m/s. When these conditions occur, the program stores waveform data until ten files are on the disk. Afterwards only the statistical data are saved. This is repeated in all directions and for different wind velocities. In this way, in a few months' time, enough data will be collected to cover a 360° circle with wind up to 27 m/s, and allow a nearly complete dynamic analysis.

In the case of the static measurement campaign, the sampling rate, and, therefore, the amount of data are not a great problem. On the other hand, the high channel count and the distribution of sensors along the tower suggested the installation of a FieldPoint network connected to a PC in a room located at 35 m. The network is made of two FP1001 network modules in two locations at 25 and 75 meters, which communicate with the PC by means of a PCI 485/2 board. The two network modules control analog input modules for the LVDT and RTD signals. This solution allows signals to have short cable paths to the data acquisition and conditioning points and to carry only two serial cables (from the top and the bottom of the tower) instead of hundreds from each sensor to a DAQ board. A program written with LabVIEW controls the FieldPoint network and collects data onto a dedicated PC.

The Torrazzo measurement campaign is being conducted in a hostile and noisy environment with logistic problems and with many transducers of different kinds. The acquisition system is required to be reliable and easy to use, perfectly tailored to the necessities, and at the same time, as cheap as possible. LabVIEW provided the software development environment that made a practical solution possible.

For further information, contact:

Giovanni Moschioni
Politecnico di Milano
Facoltà di Ingegneria di Lecco
Milano, Italy

6.10   SUMMARY

The main topics of the chapter were arrays and clusters. Arrays are a variable-sized collection (or grouping) of data elements that are all the same type, such as a group of floating-point numbers or a group of strings. LabVIEW offers many functions to help you manipulate arrays. In this chapter we discussed the following array functions:

- Array Size
- Initialize Array
- Build Array
- Array Subset
- Index Array

A cluster is a fixed-sized collection of data elements of mixed types, such as a group containing floating-point numbers and strings. As with arrays, Lab-VIEW offers built-in functions to help you manipulate clusters. In this chapter we discussed the following cluster functions:

- Bundle
- Unbundle

The important concept of polymorphism was introduced. Polymorphism is the ability of a function to adjust to input data of different types, dimensions, or representations. For example, you can add a scalar to an array or add two arrays of differing lengths together.

KEY TERMS

Array: Ordered and indexed list of elements of the same type.

Array shell: Front panel object that contains the elements of an array. It consists or an index display, a data object window, and an optional label.

Array Size: Array function that returns the number of elements in the input array.

Array Subset: Array function that returns a portion of an array.

Auto-indexing: Capability of loop structures to assemble and disassemble arrays at their borders.

Build Array: Array function that concatenates multiple arrays or adds extra elements to an array.

Bundle: A cluster function that assembles all the individual input components into a single cluster.

Cluster: A set of ordered, unindexed data elements of any data type, including numeric, Boolean, string, array, or cluster.

Index Array: Array function accessing elements of arrays.

Initialize Array: Array function that creates an n-dimensional array with elements containing the values that you specify.

Polymorphism: The ability of a function to adjust to input data of different dimension or representation.

Unbundle: A cluster function that splits a cluster into each of its individual components.

E6.1   Open While Loop Indexing.vi found in the Exercises&Problems\Chapter 6 folder in the Learning directory and click on the **Broken Run** button to access the error list. Highlight the error and press the **Show Error** button. This error is a result of an array wired to a terminal that is a scalar value. Recall that While Loops disable indexing on tunnels by default. To eliminate the error, right-click on the loop tunnel and choose **Enable Indexing**. Run the VI. After a few moments, press the **Stop** button on the front panel. An array will appear in the numeric indicator titled Temperature Measurements.

E6.2   Open Cluster Order.vi found in the Exercises&Problems\Chapter 6 folder in the Learning directory and click on the **Broken Run** button to access the error list. Highlight the error and press the **Show Error** button. This error is a result of the cluster elements not being in the same order as the elements that are bundled. To eliminate the error, right-click on the cluster border on the front panel and select **Reorder Controls in Cluster**. Change the order so **Sinc Data Points** is the first element, **Limit** is the second element, and **Over Limit?** is the third element. Now you can run the VI.

E6.3   Open Unbundle By Name.vi found in the Exercises&Problems\Chapter 6 folder in the Learning directory and view the block diagram. This VI is the same as the Cluster Unbundle Demo.vi discussed earlier in this chapter. However, in this VI the Unbundle By Name function is used instead of the Unbundle function. The advantage to using the Unbundle by Name is that you can easily identify the cluster values that are unbundled. You can use Unbundle By Name only if every object in the cluster has an owned label associated with it. Using the LabVIEW Help as a source of information, write a short description of the cluster function Unbundle By Name.

E6.4   Open Reversed Array.vi found in the Exercises&Problems\Chapter 6 folder in the Learning directory and view the block diagram. Add code to this VI that will display the data that is in the Array indicator in reversed order in the Reversed Array indicator.

E6.5   Open Rolling Dice.vi found in the Exercises&Problems\Chapter 6 folder in the Learning directory and view the block diagram. This program uses the random number generator to simulate rolling a six-sided die. The program should keep track of the number of times each value is rolled by incrementing the appropriate value in a 1D array. You will need to wire the Index Array, Increment, and Replay Array Subset nodes appropriately to run this VI.

P6.1 Complete the crossword puzzle.

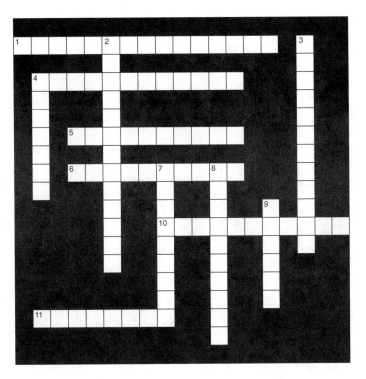

Across

1. Array function that creates an n-dimensional array with elements containing the value that you specify.
4. Front panel object that contains the elements of a cluster.
5. Array function that concatenates multiple arrays or adds extra elements to an array.
6. Array function accessing elements of arrays.
10. Array function that returns a portion of an array.
11. A cluster function that splits a cluster into each of its individual components.

Down

2. Capability of loop structures to assemble and disassemble arrays at their borders.
3. The ability to adjust to input data of different dimension or representation.
4. A set of ordered, unindexed data elements of any data type.
7. Array function that returns the number of elements in the input array.
8. Ordered and indexed list of elements of the same type.
8. Front panel object that contains the elements of an array.
9. A function that assembles all the individual input components into a single cluster.

P6.2   Create a VI that calculates the dot (inner) product of two n-dimensional vectors. Double check your math by comparing your calculation using arrays and math functions with the results of the Dot Product.vi that can be found in the Functions palette.

Review: If the two vectors are denoted by v_1 and v_2, where

$$v_1 = \begin{bmatrix} v_1(0) \\ v_1(1) \\ \vdots \\ v_1(n) \end{bmatrix} \quad \text{and} \quad v_2 = \begin{bmatrix} v_2(0) \\ v_2(1) \\ \vdots \\ v_2(n) \end{bmatrix},$$

then the dot product is given by

$$v_1 \cdot v_2 = v_1(0)v_2(0) + v_1(1)v_2(1) + \cdots + v_1(n)v_2(n).$$

P6.3   Create a VI that calculates the cross product of two 3-dimensional vectors.

Review: If the two vectors are denoted by v_1 and v_2, where

$$v_1 = \begin{bmatrix} v_1(0) \\ v_1(1) \\ v_1(2) \end{bmatrix} \quad \text{and} \quad v_2 = \begin{bmatrix} v_2(0) \\ v_2(1) \\ v_2(2) \end{bmatrix},$$

then the cross product is given by

$$v_1 \times v_2 = \begin{bmatrix} v_1(1)v_2(2) - v_1(2)v_2(1) \\ v_1(2)v_2(0) - v_1(0)v_2(2) \\ v_1(0)v_2(1) - v_1(1)v_2(0) \end{bmatrix}.$$

P6.4   Create a VI that performs matrix multiplication for two input matrices **A** and **B**. The matrix **A** is an $n \times m$ matrix, and the matrix **B** is an $m \times p$ matrix. The resulting matrix **C** is an $n \times p$ matrix, where **C** = **AB**. Double check your math by comparing your calculation using arrays and math functions with the results of the AxB.vi.

P6.5   Create a VI that reads 20 temperature measurements using the Demo Temp.vi found in the Exercises&Problems\Chapter 6 folder in the Learning directory. With each temperature measurement, bundle the time (including seconds) and date of the measurement. Include a Time Delay Express VI to slow the loop so it executes four times per second. Run the program and review the time stamps in the output array to be sure, there are four samples per second.

P6.6 Build a VI that generates and plots 500 random numbers on a waveform graph indicator. Compute the average of the random numbers and display the result on the front panel. Use the **Statistics Express VI** to compute of the average of the random numbers.

P6.7 Create a VI that computes and plots the second order polynomial $y = Ax^2 + Bx + C$. The VI should use controls on the front panel to input the coefficients A, B, and C, and also should use front panel controls to enter the number of points N to evaluate the polynomial over the interval x_0 to x_{N-1}. Plot y versus x on a waveform graph indicator.

P6.8 Create a VI with a cluster of six buttons labeled Option 1 through Option 6. When executing, the VI should wait for the user to press one of the buttons. When a button is pressed, use the Display Message To User Express VI to indicate which option was selected. Repeat this process until the user presses the **Stop** button. Be sure to include the Time Delay Express VI to give the user time to press the buttons.

Hint 1: This program will require a simplified version of the state machine architecture along with a user menu. A typical state machine in LabVIEW consists of a While Loop, a Case Structure, and a Shift Register. Each state of the state machine is a separate case in the Case Structure. You place VIs and other code that the state should execute within the appropriate case. A Shift Register stores the state to be executed upon the next iteration of the loop. This is required in a typical state machine because there are times when the results of the state you are currently in control the state you will go to next. In the program described above, no two states are dependent upon one another, so the shift register is optional.

Hint 2: You can use latched Boolean buttons in a cluster to build a menu for a state machine application. The Cluster to Array function converts the Boolean cluster to a Boolean array, where each button in the cluster represents an element in the array. The Search 1D Array function searches the 1D array of Boolean values created by the Cluster to Array function for a value of TRUE. A TRUE value for any element in the array indicates that the user clicked a button in the cluster. The Search 1D Array function returns the index of the first TRUE value it finds in the array and passes that index value to the selector terminal of a Case structure. If no button is pressed, Search 1D Array returns an index value of -1 and the -1 case executes, which does nothing. The While Loop repeatedly checks the state of the Boolean cluster control until you click the **Stop** button.

CHAPTER 7

Charts and Graphs

Graphs and charts are used to display data in a graphical form. Three types of charts are discussed—scope, strip, and sweep charts—and two types of graphs—waveform and XY graphs. Charts and graphs are different! Data is presented on charts by appending new data as it becomes available to the existing plot, much in the same manner as on a strip chart you might find in a laboratory. On the other hand, graphs are used to display pregenerated arrays of data in a more traditional fashion, such as the typical x-y graph. In this chapter, you will learn about charts and graphs, the type of data required by each, and several ways to utilize graphics. Since data presentation is an important component of communication, it is necessary to be able to customize the appearance of charts and graphs. The subject of customizing charts and graphs using the palette and the legend (and other tools) is discussed in the chapter.

GOALS

1. Learn about charts and graphs and recognize their similarities and differences.
2. Understand the three modes of a chart: strip, scope, and sweep.
3. Learn to customize the appearance of charts and graphs.

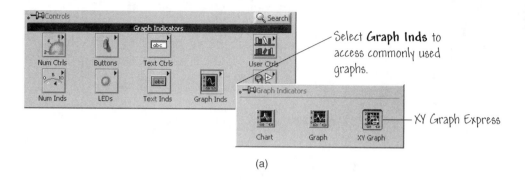

Select **Graph Inds** to access commonly used graphs.

XY Graph Express

(a)

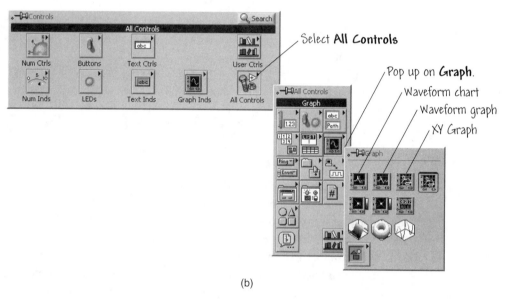

Select **All Controls**

Pop up on **Graph**.

Waveform chart

Waveform graph

XY Graph

(b)

FIGURE 7.1
The graphs and chart are located in the **Graph** subpalette of the **Controls** palette.

7.1 WAVEFORM CHARTS

A **waveform chart** is a special kind of indicator—located in the **Graph Indicators** palette of the **Controls** palette, as illustrated in Figure 7.1. The waveform chart can also be found on the **Controls»All Controls»Graph** palette.

There is only one type of waveform chart, but it has three different update modes for interactive data display—**strip chart**, **scope chart**, and **sweep chart**, as shown in Figure 7.2. You select the update mode that you want to use by popping up on the waveform chart and choosing one of the options from the Advanced»Update Mode menu. This process is illustrated in Figure 7.3. This can also be selected from the **Appearance** menu of the **Properties** dialog box. The strip chart, scope chart, and sweep chart all handle the incoming data a little

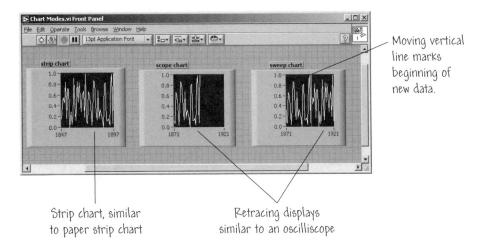

Moving vertical line marks beginning of new data.

Strip chart, similar to paper strip chart

Retracing displays similar to an oscilliscope

FIGURE 7.2
The waveform chart has three update modes.

differently. The strip chart has a display that scrolls so that as each new data point arrives, the entire plot moves to the left—the oldest data point falls off the chart, and the latest data point is added at the rightmost part of the plot. This action is very similar to a paper strip chart commonly found in laboratories.

The scope and sweep charts more closely resemble the action of an oscilloscope. When the number of data points is sufficient so that the plot reaches the right border of the plotting area of the scope chart, the entire plot is erased, and

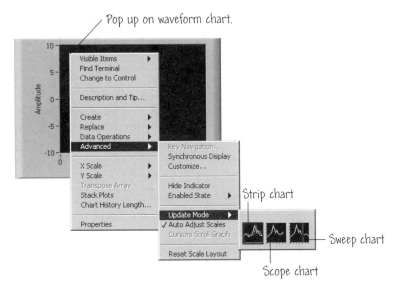

FIGURE 7.3
Changing waveform chart modes: strip chart, scope chart, and sweep chart.

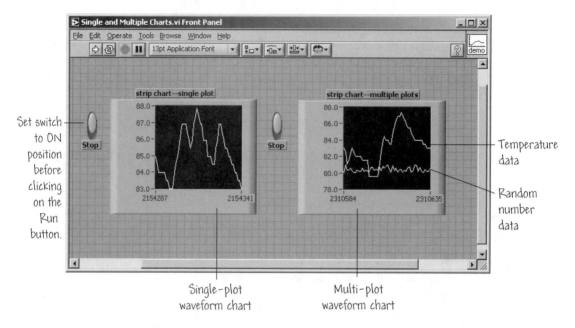

Set switch to ON position before clicking on the Run button.

Single-plot waveform chart

Multi-plot waveform chart

Temperature data

Random number data

FIGURE 7.4
Examples of single- and multiple-plot waveform charts.

the plotting begins again from the left side. The sweep chart acts much like the scope chart except that, rather than erasing the plot when the data reaches the right border, a moving vertical line marks the beginning of new data and moves across the display as new data is added. The scope chart and the sweep chart run faster than the strip chart.

You can run the waveform chart demonstration shown in Figure 7.2 by opening **Chart Modes.vi** *located in* **Chapter 7** *of the* **Learning** *directory. Run the VI in* **Run Continuously** *mode and examine the three different waveform charts: the strip chart, the scope chart, and the sweep chart. Select* **Highlight Execution** *if you want to observe the data as it plots in slow motion.*

Waveform charts may display single or multiple traces. An example of a multiple plot waveform chart is shown in Figure 7.4.

To generate a single-plot waveform chart, you wire a scalar output directly to the waveform chart. The data type displayed in the waveform chart will match the input. In the example shown in Figure 7.5, a new temperature value is plotted on the chart each time the While Loop iterates.

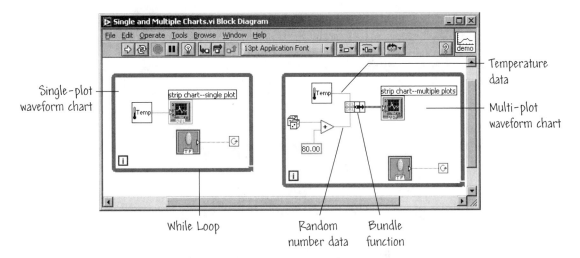

FIGURE 7.5
Block diagrams associated with single- and multiple-plot waveform charts.

As illustrated in Figure 7.4, waveform charts can display multiple plots. To generate a multi-plot waveform chart, you bundle the data together using the Bundle function or use the Merge Signal function located on the **Signal Manupulation** palette on the **Functions** palette. (See Chapter 6 for a review). In the example in Figure 7.5, the Bundle function groups the output of the two different data sources—Digital Thermometer.vi and Random Number (0–1) function—that acquire temperature and random number data for plotting on the waveform chart. The random number is biased by the constant 80 so that the plots are easily displayed on the same y-axis scale. Remember that Digital Thermometer.vi simulates acquiring temperature data presented in degrees Fahrenheit in the general range of 80°. You can add more plots to the waveform chart by increasing the number of Bundle function input terminals—that is, by resizing the Bundle function.

You can run the single- and multi-plot waveform chart demonstration shown in Figure 7.4 by opening Single and Multiple Charts.vi *located in* Chapter 7 *of the* Learning *directory. Run the VI and examine the single- and multi-plot waveform chart. You can stop the plotting of each chart using the vertical switch button. After stopping the plotting with the vertical switch, remember to reset the switch to the up position before the next execution of the VI.*

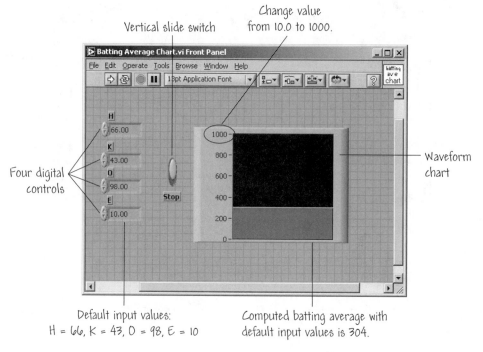

Change value
from 10.0 to 1000.

Vertical slide switch

Four digital
controls

Waveform
chart

Default input values:
H = 66, K = 43, O = 98, E = 10

Computed batting average with
default input values is 304.

FIGURE 7.6
The front panel for a waveform chart to plot a baseball batting average.

Practice with Waveform Charts

In this exercise you will construct a VI to compute and display a baseball batting average on a waveform strip chart. This VI utilizes the **Batting Average subVI.vi** you developed in Chapter 4. Figure 7.6 shows the front panel that you can use as a guide as you construct your VI.

The front panel has six items:

- Four numeric controls
- One vertical slide switch
- One waveform strip chart

Open a new VI front panel and place the six objects listed above in the window. Label the four numeric controls **H, K, O**, and **E**. Add a free label under the vertical slide switch indicating the switch position for stopping the execution. Place a waveform chart in the front panel window and locate the front panel objects in approximately the same locations as seen in Figure 7.6.

Since a baseball batting average always has a value between 0 and 1000 (with 1000 representing a perfect batting average), we need to rescale the waveform chart *y*-axis to insure the scaling is always 0 to 1000. Using the **Labeling** tool,

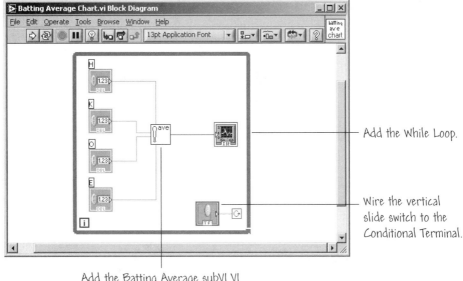

Add the While Loop.

Wire the vertical slide switch to the Conditional Terminal.

Add the Batting Average subVI VI and wire the terminals.

FIGURE 7.7
The block diagram for a waveform chart to plot a baseball batting average.

double-click on the 10.0 on the waveform chart y-axis scale, type 1000, and click on the **Enter** button. This changes the maximum value of the y-axis to 1000. The minimum value is 0 by default, so no additional adjustment is required.

Input default values for the number of hits (**H**), strike-outs (**K**), put-outs (**O**), and fielding errors (**E**). Enter the following data: **H** $= 66$, **K** $= 43$, **O** $= 98$, and **E** $= 10$. From the pull-down menu **Operate**, select **Make Current Values Default** so that it is not necessary to input the default values each time the VI is opened again.

Figure 7.7 shows the block diagram that you can use as a guide for building your VI. The numeric control terminals and waveform chart and vertical switch terminals will automatically appear on the block diagram after you have placed their associated objects on the front panel.

The block diagram has two additional items:

- While Loop
- Batting Average subVI.vi

Switch to the block diagram window of your VI. First enclose the four numeric controls, waveform chart, and vertical slide switch terminals inside a While Loop. The only other object that is needed is the subVI to compute the batting average. The Batting Average subVI.vi is located in Chapter 4 in the Learning

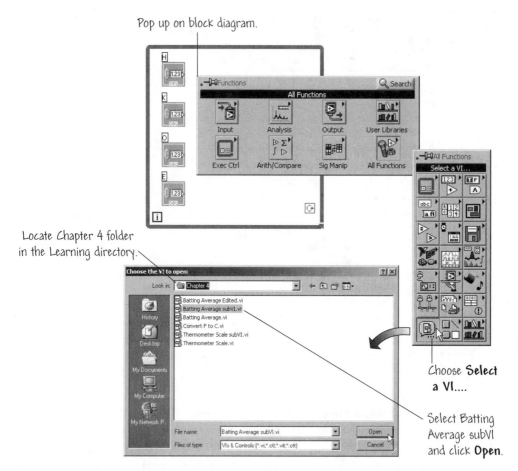

Pop up on block diagram.

Locate Chapter 4 folder in the Learning directory.

Choose **Select a VI....**

Select Batting Average subVI and click **Open**.

FIGURE 7.8
Placing the Batting Average subVI.vi on the block diagram.

directory. You can access the subVI by popping up on the block diagram and from the **Functions** palette choosing **Select a VI...** as illustrated in Figure 7.8. You will need to navigate through the file structure to reach the **Chapter 4** directory. Once there, you select the desired subVI—this is demonstrated in Figure 7.8.

Once all the required objects are on the block diagram, wire the diagram as shown in Figure 7.7. As an optional step, you can remove the x-axis scale. Pop up on the waveform chart on the front panel and under the **Visible Items** category choose **X Scale** to deselect the X scale. In this example, the value of the batting average is being computed and plotted each iteration of the While Loop—the X scale does not have any physical meaning, other than as a count of the total loop iterations.

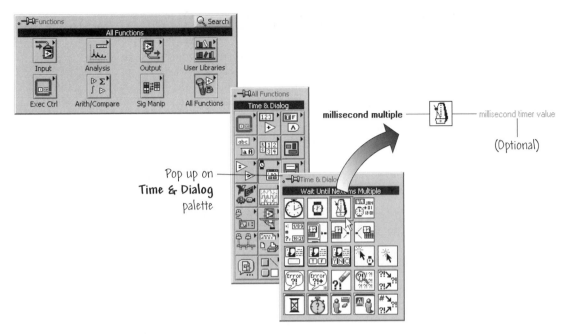

FIGURE 7.9
The Wait Until Next ms Multiple function.

Once everything is ready for execution, run the VI. With the default values you should find that the computed batting average is 304. Using the **Operating** tool, increase the number of hits or the number of strike-outs and observe how it affects the batting average. As you vary the various input parameters (that is, the number of hits, strike-outs, put-outs, and errors) you should see a line across the waveform chart move up or down, depending on which inputs you are varying (hits make the batting average increase, while increasing any of the other three parameters will reduce the batting average!). Save your VI in the Users Stuff directory and name it Batting Average Chart.vi.

You can find a completed VI for the waveform chart demonstration shown in Figure 7.6 in Chapter 7 *of the* Learning *directory. The VI is called* Batting Average Chart.vi.   ◆

In the previous example, the While Loop executed as quickly as the computer system would allow. In many situations you may want to acquire and plot data at specified intervals. This can be accomplished by controlling the loop timing using the Wait Until Next ms Multiple function. This timing control function, located in the **Time & Dialog** subpalette, as shown in Figure 7.9, waits until the

Iterate
For Loop
ten times.

Waveform chart

Place the function
Wait Until Next
ms Multiple inside
the For Loop.

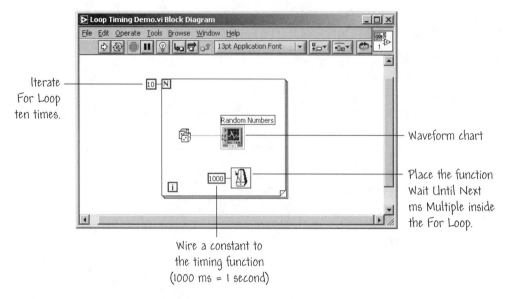

Wire a constant to
the timing function
(1000 ms = 1 second)

FIGURE 7.10
Controlling the loop timing and rate at which data is plotted on a waveform chart.

millisecond timer is a multiple of the specified input value before returning the (optional) millisecond timer value (the output of the function need not be wired!). Therefore, you can control the loop iteration to execute only once every specified number of milliseconds. As with the Wait (ms) function, the timer resolution is system dependent and may be less accurate than 1 millisecond.

The example shown in Figure 7.10 shows how to control the timing of a For Loop to execute once every second (that is, once per 1000 ms). Placing the Wait Until Next ms Multiple function within the For Loop, each iteration will take approximately 1 second, and the random number will be plotted at 1-second intervals. Can you predict how much time it will take to complete the execution of the For Loop shown in Figure 7.10? It should take around 10 seconds. You can test it out by opening and running the VI shown in Figure 7.10. It is called Loop Timing Demo.vi and can be found in Chapter 7 of the Learning directory.

**Practice
with
Timing**

Open the VI called Batting Average Chart.vi that you constructed in the previous example (it should be located in the Users Stuff directory). Add the capability to control the loop timing by introducing a Wait Until Next ms Multiple function within the While Loop, as shown in Figure 7.11.

While Loop synchronized
to 25 ms

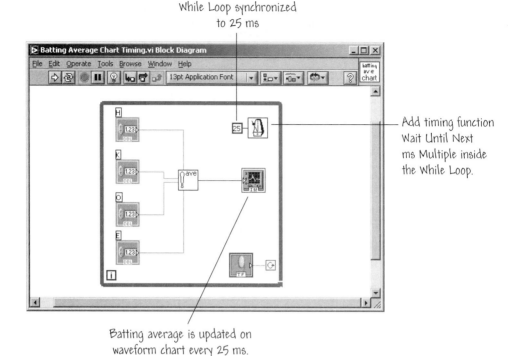

Add timing function
Wait Until Next
ms Multiple inside
the While Loop.

Batting average is updated on
waveform chart every 25 ms.

FIGURE 7.11
Controlling the rate on the waveform chart displaying a computed baseball batting average.

Set the loop timing for 25 ms. When the VI is now executing the batting average is computed and the waveform chart is updated at 25 millisecond intervals. The effect is that you should be able to watch the computed batting average trends change more clearly than when the While Loop was executing as fast as possible. Vary the loop-timing parameter and observe the effect on the waveform chart.

Save your updated VI in the Users Stuff directory and name it Batting Average Chart Timing.vi.

You can find a completed VI for the waveform chart timing demonstration in Chapter 7 *of the* Learning *directory. The VI is called* Batting Average Chart Timing.vi.   ◆

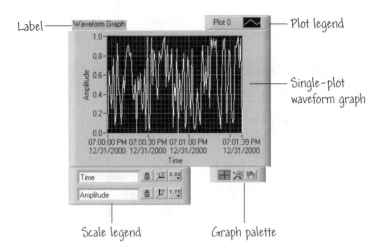

FIGURE 7.12
Example of a waveform graph.

7.2 WAVEFORM GRAPHS

A **waveform graph** is an indicator that displays one or more data arrays. This is equivalent to the familiar 2D plot with horizontal and vertical axes. There are two types of graphs: waveform graphs and XY graphs. In this section the focus is on waveform graphs—XY graphs are the subject of the next section.

Waveform graphs and XY graphs are functionally different, but they look the same on the front panel. Both waveform graphs and XY graphs plot existing arrays of data all at once, unlike waveform charts, which plot new data as they become available. An example of a waveform graph is shown in Figure 7.12. But the waveform graph plots only single-valued functions with uniformly spaced points and is ideal for plotting arrays of data in which the points are evenly distributed. Conversely, XY graphs are general-purpose Cartesian graphs suitable for plotting data available at irregular intervals or plotting two dependent variables against each other. The waveform graphs and XY graphs accept different types of input data!

Waveform graphs are located on the **Graph Indicator** subpalette of the **Controls** palette (see Figure 7.1). For basic, single-plot graphs, an array of Y values (along the vertical axis) can pass directly to a waveform graph. This method assumes the initial X value (along the horizontal axis) is $X_0 = 0$ and the $\Delta X = 1$. The value of ΔX determines the X marker spacing. When passing only the Y values to the waveform graph, the graph icon will appear as an array indicator, as seen in Figure 7.13(a).

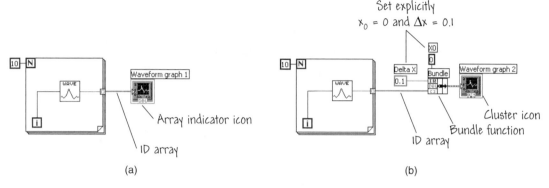

FIGURE 7.13
Single-plot waveform graphs.

If you want to start plotting other than $X_0 = 0$, or if your data points are spaced differently than $\Delta X = 1$, you can wire a cluster consisting of the initial value X_0, ΔX, and a data array to the waveform graph. The graph terminal will then appear as a cluster indicator, as seen in Figure 7.13(b).

If you want to plot more than one curve on a single graph, you can pass a 2D array of data to create a multiple-plot waveform graph. Two methods for wiring multiple-plot waveform graphs are illustrated in Figure 7.14. The graph icon assumes the data type to which it is wired.

Graphs always plot the *rows* of a two-dimensional array. The two-dimensional array in Figure 7.14 has two rows with 10 columns per row—a 2×10 array. If your data is given in columns, you must transpose the array before graphing. Once you have wired your 2D array of data to the graph terminal, go to the front panel, pop up on the graph, and select **Transpose Array**.

The waveform, single-plot graph example shown in Figure 7.13 is called Waveform Graphs demo.vi. *The waveform multiple-plot graph example shown in Figure 7.14 is called* Multiple graphs demo.vi. *Both VIs can be found in* Chapter 7 *of the* Learning *directory.*

Practice with Waveform Graphs

Open a new VI and build the front panel shown in Figure 7.15. Place three objects on the front panel:

- Numeric control labeled No. of points
- Numeric control labeled Rate of growth, r
- Waveform graph labeled Population

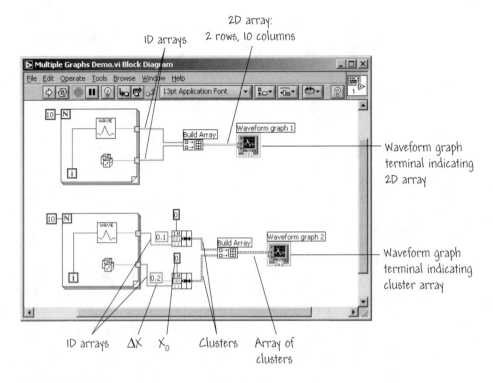

FIGURE 7.14
Two ways to wire multiple-plot waveform graphs.

We would like the waveform graph y-axis scale to be driven by the data values, rather than remaining at some predetermined fixed values. To modify the y-axis limits of the waveform graph, pop up on the waveform graph and change the **Y Scale** to **Autoscale Y**, as illustrated in Figure 7.16.

The block diagram that you need to construct is shown in Figure 7.15. You will need to add the following objects:

- For Loop with the loop counter wired to the digital control **No. of points**

- Shift registers on the For Loop to transfer data from one iteration to the next

- A Formula Node containing the equation

$$y = rx(1 - x),$$

where the source of the parameter r is the digital control **Rate of growth**, **r**.

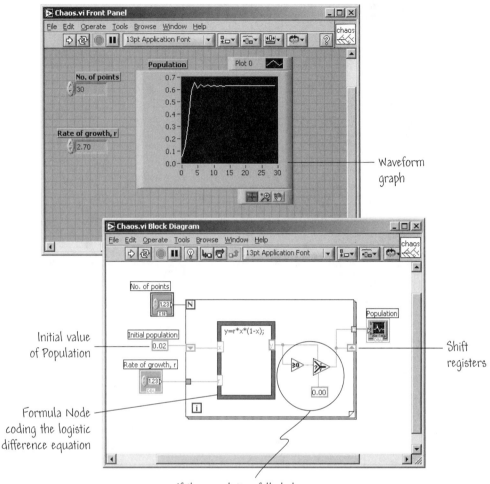

Waveform
graph

Initial value
of Population

Shift
registers

Formula Node
coding the logistic
difference equation

If the population falls below zero,
set it to exactly zero—the population is extinct.

FIGURE 7.15
Front panel and block diagram arrangement to experiment with chaos.

The equation in the Formula Node is known as the *logistic difference equa-tion*, and is given more formally as

$$x_{k+1} = rx_k(1 - x_k)$$

where $k = 0, 1, 2, \cdots$, and x_0 is a given value. In the block diagram (see Fig-ure 7.15), the initial value is wired as $x_0 = 0.02$. The logistic difference equation has been used as a model to study population growth patterns. The model has

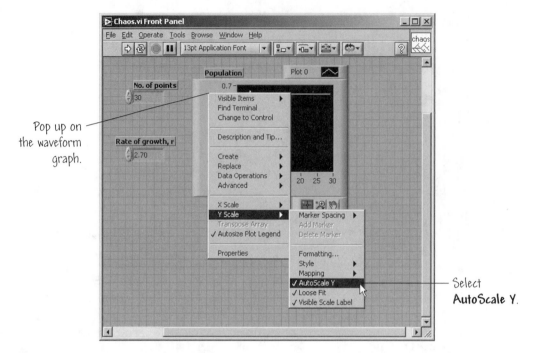

FIGURE 7.16
Changing the waveform graph *y*-axis limits to allow autoscaling.

been scaled so that the values of the population vary between 0 and 1, where 0 represents extinction and 1 represents the maximum conceivable population.

Wire the block diagram as shown in Figure 7.15 and prepare the VI for execution. Set the variable **No. of points** initially to 30 and the variable **Rate of growth**, **r** to 2.7. Execute the VI by clicking on the **Run** button. The value of the population is shown graphically in the waveform graph. You should observe the population reach a steady-state value around 0.63.

Experiment with the VI by changing the values of the **Rate of growth**, **r**. For values of $1 < r < 3$, the population will reach a steady-state value. You will find that as $r \rightarrow 3$, the population begins to oscillate, and in fact, the "steady-state" value oscillates between two values. When the parameter $r > 4$, the behavior appears erratic. As the parameter increases, the behavior becomes chaotic.[1] Verify that when the parameter $r = 4.1$, the population exceeds the maximum conceivable value (that is, it exceeds 1.0 on the waveform graph) and subsequently becomes extinct! Save your updated VI in the Users Stuff directory and name it Chaos.vi.

---

1. For more on chaos, see *Chaos: Making a New Science*, by James Gleick, Penguin Books (New York, 1987).

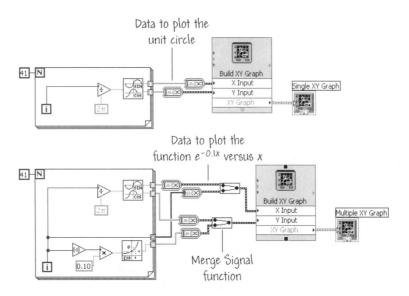

FIGURE 7.17
Examples of the XY graphs: single and multiple plots.

  The working version of the chaos example shown in Figure 7.15 is called **Chaos.vi** *and is located in* **Chapter 7** *of the* **Learning** *directory.* ◆

7.3   XY GRAPHS

As previously mentioned, waveform graphs are ideal for plotting evenly sampled waveforms. But what if your data is available at irregular intervals, or what if you want to plot two dependent variables against each other (e.g., x versus y)? The **XY graph** is well-suited for use in situations where you want specify points using their (x, y) coordinates. The XY graph is a general-purpose Cartesian graph that can also plot multivalued functions—such as circles and ellipsoids. The Express XY Graph VI is located on the **Graph Indicator** subpalette of the **All Controls** palette.

A single-plot XY graph and its corresponding block diagram are shown in Figure 7.17. The Merge Signal function is used to combine the X and Y arrays into a single output for use with the XY Graph. The Express XY Graph VI makes creating XY graphs simple. The Express XY Graph VI is located on the **Graph Indicators** palette on the **Controls** palette, as shown in Figure 7.1. When placed on the Front Panel, the Express XY Graph VI places a corresponding terminal on the Block Diagram and an Express VI for quick configuration of your X and Y inputs. Wire the data arrays into the X Input and Y Input and they will automatically be configured for the graph, as shown in Figure 7.17.

*The XY graph example shown in Figure 7.17 is called **XY Graphs demo.vi** and is located in **Chapter 7** of the **Learning** directory. The single XY graph produces a unit circle, and the multiple XY graph produces a unit circle and a plot of $e^{-0.1x}$ for $0 \leq x \leq 41$.*

Practice with XY Graphs

Suppose that you need to borrow $1000 from a bank and are given the option of choosing between simple interest and compound interest. The annual interest rate is 10% and, in the case of compound interest, the interest accrues annually. In either case, you pay the entire loan off in one lump sum at the end of the loan period. Construct the VI shown in Figure 7.18 to compute and graph the amount due at the end of each loan period, where the loan period varies from $N = 1$ year to $N = 20$ years in increments of 1 year.

Create a multiple-plot XY graph showing the final payments due for both the simple interest and the compound interest situations.

Define the following variables: F = final payment due, P = amount borrowed, i = interest rate, N = number of interest periods. The relevant formulas are:

- Simple interest

$$F = P(1 + iN)$$

- Compound interest

$$F = P(1 + i)^N$$

Open a new front panel and place the following three items:

- Numeric control labeled **P**
- Numeric control labeled **i**
- XY Graph

The XY graph in Figure 7.18 has been edited somewhat. Pop up on the first plot in the legend and select **Common Plots** and choose the scatter plot (top row, center). Then pop up in the same place again and select a **Point Style**. Repeat this process for the second plot listed in the legend.

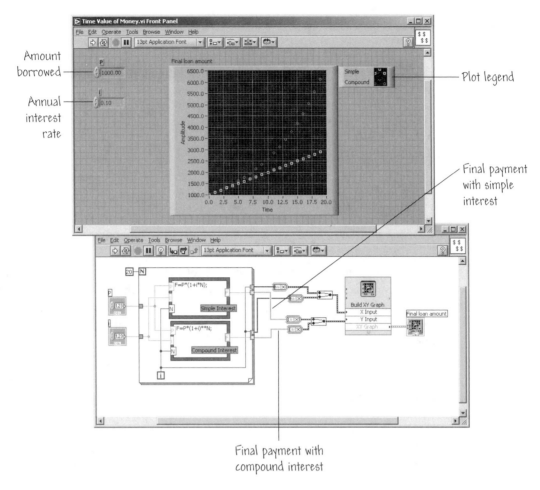

Amount borrowed

Annual interest rate

Plot legend

Final payment with simple interest

Final payment with compound interest

FIGURE 7.18
A VI to compute and plot the time value of money.

One solution to coding the VI is shown in Figure 7.18. In this block diagram, the formulas are coded using Fomula Nodes. Can you think of other ways to code the same equations?

Once your VI is working, you can use it to experiment with different values of annual interest and different initial loan amounts. Based on your investigations, would you prefer to have simple interest or compound interest?

The XY graph example shown in Figure 7.18 is called Time Value of Money.vi *and is located in* Chapter 7 *of the* Learning *directory.*   ◆

7.4   CUSTOMIZING CHARTS AND GRAPHS

Charts and graphs have editing features that allow you to customize your plots. This section covers how to configure some of the more important customization features. In particular, the following items are discussed:

- Autoscaling the chart and graph x- and y-axes

- Using the Plot Legend

- Using the Graph Palette

- Using the Scale Legend

- Chart customizing features, including the scrollbar and the digital display

By default, charts and graphs have the **plot legend** showing when first placed on the front panel. Using the **Positioning** tool, you can move the scale and plot legends and the graph palette anywhere relative to the chart or graph. Figure 7.19 shows some of the more important components of the chart and graph customization objects.

7.4.1   Axes Scaling

The x- and y-axes of both charts and graphs can be set to automatically adjust their scales to reflect the range of the plot data. The **autoscaling** feature is enabled or disabled using the **AutoScale X** and **AutoScale Y** options from the **X Scale** or **Y Scale** submenus of the short cut menu, as shown in Figure 7.20.

The use of autoscaling may cause the chart or graph to update more slowly.

The X and Y scales can be varied manually using the **Operating** or **Labeling** tool—just in case you don't want to use the autoscale feature. For instance, if the graph x-axis end marker has the value 10, you can use the **Operating** tool to change that value to 1 (or whatever other value you want!). The graph x-axis marker spacing will then change to reflect the new maximum value. In typical situations, the scales are set to fit the exact range of the data. You can use the **Loose Fit** option (see Figure 7.20) if you want to round the scale to multiple values of the scale increment—in other words, with a loose fit the scale marker numbers are rounded to a multiple of the increment used for the scale. For example, if the scale markers increment by 5, then with a loose fit, the minimum and maximum scale values are set to a multiple of 5.

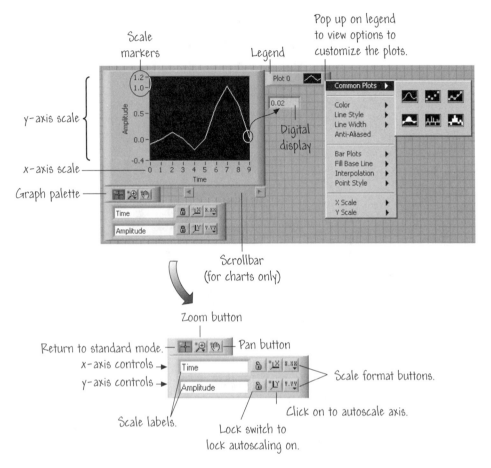

FIGURE 7.19
The chart and graph customization objects.

On both the **X Scale** or **Y Scale** submenus, you will find the **Formatting**...
option. Choosing the option opens up a dialog box, as shown in Figure 7.21, that
allows you to configure many components of the chart or graph. This dialog can
also be opened by popping up and selecting **Properties** to change the properties
of the graph. You can modify the *x*- and *y*-axis characteristics individually. The
following choices are available:

- The **Appearance** tab is used to specify which elements of the object are
visible. This provides a way to choose to show the graph palette, plot legend,
scroll bar, scale legend, and cursor legend. You can also select these items
on the **Visible Items** pull-down menu by popping up on the graph.

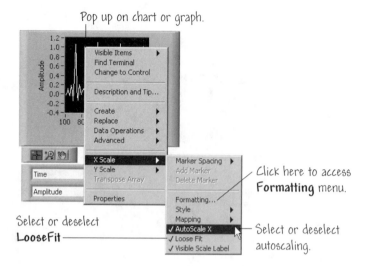

FIGURE 7.20
Using the **Autoscaling** feature of charts and graphs.

- Within the **Format and Precision** area, you change the format and precision of numeric objects. You can choose the number of digits of precision for the scale marker numbers, as well as the notation. You can choose among floating point, scientific, SI notation, hexadecimal, octal, binary, absolute time, and relative time. The choices depend on the format choice.

- The **Plots** tab is used to configure the appearance of plots on a graph or chart. You can select the line type (e.g., solid versus dashed), point markers and scale titles.

- Under **Scales** tab you format scales and grids on graphs and charts. For example, you can set the axis origin $X0$ (or $Y0$) and the axis marker increment ΔX (or ΔY). The grid style and colors provide control over the gridlines: no gridlines, gridlines only at major tick mark locations, or gridlines at both major and minor tick marks. This option also allows you to change the color of the grid lines. Here you can also specify the scale as either a linear or a logarithmic scale. The **Scales** menu lets you select major and minor tick marks for the scale, or none at all. A major tick mark corresponds to a scale label, while a minor tick mark denotes an interior point between labels. You have to click on the **Scale Style** icon to view the choices. With this menu, you can make the axes markers invisible.

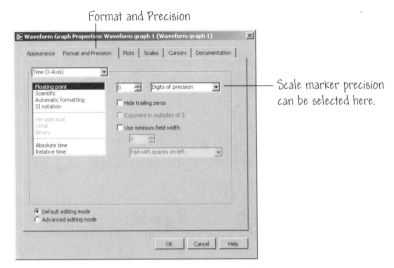

Precision Formatting

(a)

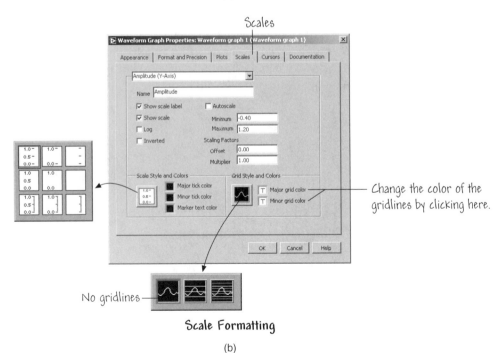

Scale Formatting

(b)

FIGURE 7.21
Scale formatting.

- The **Cursors** tab is used to add cursors to a graph or chart and to configure the appearance of the cursors.

- Use the **Documentation** tab to describe the purpose of the object and to give users instructions for using the object.

7.4.2 The Legend

The legend provides a way to customize the plots on your charts and graphs. This is where you choose the data point style, the plot style, the line style, width and color, and other characteristics of the appearance of the chart or graph. For example, on multiple-plot graphs you may want one curve to be solid line and the other curve to be a dashed line or one curve to be red and the other blue. An example of a legend is shown in Figure 7.22. When you move the chart or graph around on the front panel, the legend moves with it. You can change the position of the legend relative to the graph by dragging the legend with the **Positioning** tool to the desired location.

The legend can be visible or hidden—and you use the **Visible Items** submenu of the chart or graph short cut menu (see Figure 7.22) to choose. After you customize the plot characteristics, the plot retains those settings regardless of whether the legend is visible or not. If your chart or graph receives more plots than are defined in the legend, they will have the default characteristics.

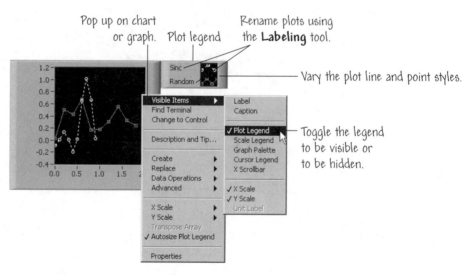

FIGURE 7.22
The legend for charts and graphs.

When the legend first appears, it is sized to accomodate only a single plot (named Plot 0, by default). If you have a multiple-plot chart or graph, you will need to show more plot labels by dragging down a corner of the legend with the **Resizing** cursor to accomodate the total number of curves. Each curve in a multiple-plot chart or graph is labeled as Plot 0, Plot 1, Plot 2, and so forth. You can modify the default label by assigning a name to each plot in the legend with the **Labeling** tool. Choosing names that reflect the physical value of the data depicted in the chart makes good sense—for instance, if the curve represents the velocity of an automobile you might use the label Automobile velocity in km/hr. The legend can be resized on the left to give the plot labels more room or on the right to give the plot samples more room. This is useful when you assign long names to the various curves on a multiple-plot chart or graph. You can pop up on each plot in the legend and change the plot style, line, color, and point styles of the plot. The short cut menu is shown in Figure 7.19.

7.4.3 The Graph Palette and Scale Legend

The graph palette and scale legend allows you access to various aspects of the chart and graph appearances while the VI is executing. You can autoscale the axes and control the display format of the axes scales. To aid in the analysis of the data presented on the charts and graphs, buttons allow you to pan and zoom in and out. The graph palette and scale legend shown in Figure 7.23 can be displayed using the **Visible Items** submenu of the chart or graph short cut menu.

On the left of the scale legend are three buttons that control the autoscaling of the axes. Clicking on the autoscaling button (either the x- or y-axis buttons) will cause the chart or graph to autoscale the axis. When you press the lock button, the chart or graph will autoscale either of the axes scales continuously.

The x-axis and y-axis labels can also be modified by entering the desired labels (using the **Labeling** tool) into the text area provided on the left side of the scale legend. The upper text is for the x-axis.

The scale format buttons (on the right of the scale legend) provide run-time control over the format and precision of the x and y scale markers, respectively. As shown in Figure 7.23, you can control the **Format** of the axes markers, the **Precision** of the marker numbers (from 0 to 6 digits after the decimal point), and the **Mapping Mode**. Remember that in the edit mode, you can control the format, precision, and mapping mode through short cut menus on the chart or graph. The chart or graph scale legend gives you the added feature of being able to control the format and precision of the x and y scale markers while the VI is executing.

The three buttons on the graph palette allow you to control the operational mode of the chart or graph. You are in the standard mode under normal circumstances where the button with the cross-hair (left side of the palette) is selected.

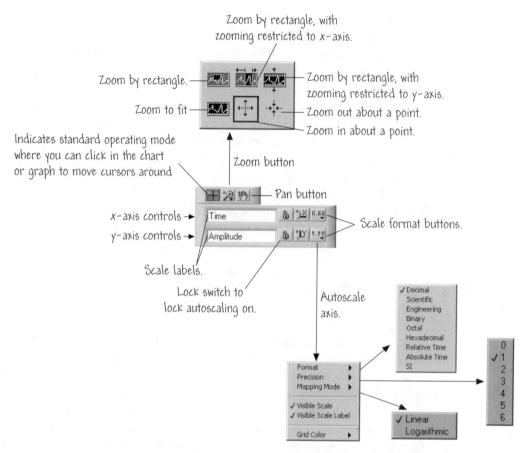

FIGURE 7.23
The chart and graph palette.

This indicates that you can click in the chart or graph area to move cursors around. Other operational modes include panning and zooming. When you press the pan button (depicted in Figure 7.23), you can scroll through the visible data by clicking and dragging sections of the graph with the pan cursor. Selecting the zoom button gives you access to a short cut menu that lets you choose from several methods of zooming.

As depicted in Figure 7.23, you have several zoom options:

▪ **Zoom by Rectangle**. In this mode, you use the cursor to draw a rectangle around the area you want to zoom in and when you release the mouse button, the axes will rescale to zoom in on the desired area.

- **Zoom by Rectangle**—with zooming restricted to X data. This is similar in function to the **Zoom by Rectangle** mode, but the Y scale remains unchanged.

- **Zoom by Rectangle**—with zooming restricted to Y data. This is similar in function to the **Zoom by Rectangle** mode, but the X scale remains unchanged.

- **Zoom to Fit**. This autoscales all x- and y-scales on a graph or chart.

- **Zoom In about a Point**. With this option you hold down the mouse on a specific point on the chart or graph to continuously zoom in until you release the mouse button.

- **Zoom Out about a Point**. With this option you hold down the mouse on a specific point and the graph will continuously zoom out until you release the mouse button.

For the zoom in and zoom out about a point modes, <shift>-clicking will zoom in the other direction.

7.4.4   Special Chart-Customizing Features

Charts have the same customizing features as graphs—with two additional options: **Scrollbars** and **Digital Displays**. The **Visible Items** submenu (of the chart short cut menu) is used to show or hide the optional digital display(s) and a scrollbar.

Charts have scrollbars that can be used to display older data that has scrolled off the chart. The chart scrollbar is depicted in Figure 7.19. The scrollbar can be made visible by popping up on the chart and selecting the scrollbar from the **Visible Items** submenu, as illustrated in Figure 7.24.

There is one digital display per plot that displays the latest value being plotted on the chart. You can place a digital display on the chart by selecting the **Digital Display** from the **Visible Items** submenu, as illustrated in Figure 7.24. The digital display can be moved around relative to the chart using the **Positioning** tool, but by default it will initially be placed on the legend. The last value passed to the chart from the block diagram is the latest value for the chart and the value that is displayed on the digital display. If you want to view past values contained on the chart data you can navigate through the older data using the scrollbar. There is a limit to the amount of data the chart can hold in the buffer. By default, a chart can store up to 1024 data points. When the chart reaches the limit, the oldest point(s) are discarded to make room for new data. If you want to change the size of the

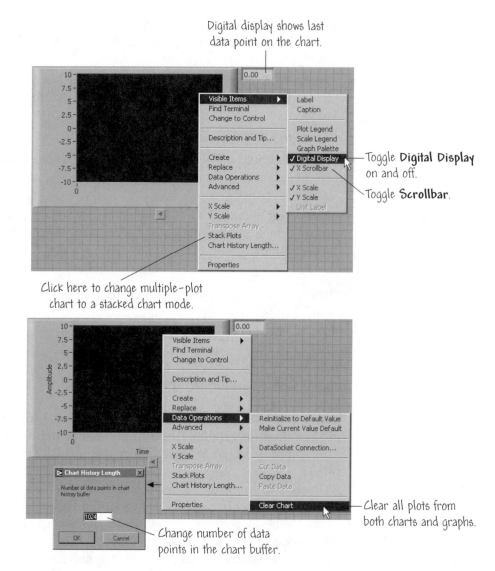

FIGURE 7.24
The scrollbar and the digital display for charts.

chart data buffer, select **Chart History Length**. . . from the short cut menu and specify a new value of up to 100,000 points.

When you are not running the VI, you can clear the chart by popping up on the chart and selecting **Clear Chart** from the **Data Operations** menu, as shown in Figure 7.24. However, if you are running the VI (you are in the run mode), then

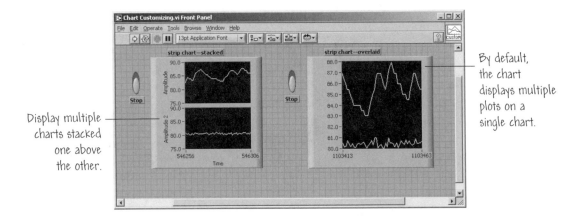

Display multiple charts stacked one above the other.

By default, the chart displays multiple plots on a single chart.

FIGURE 7.25
Stacked and overlaid charts.

Clear Chart becomes a short cut menu option and it will automatically appear in the chart short cut menu after clicking on the **Run** button. This allows you to clear the chart while the VI is executing.

If you have a multiple-plot chart, you can choose between overlaid plot or stacked plot. When you display all plots on the same set of scales, you have an overlaid plot. When you give each plot its own scale, you have a stacked plot. You can select **Stack Plots** or **Overlay Plots** from the chart short cut menu (see Figure 7.24) to toggle between stacked and overlaid modes. Figure 7.25 illustrates the difference between stacked and overlaid plots.

The multiple-plot chart example shown in Figure 7.25 is called Chart Cus-tomizing.vi *and is located in* Chapter 7 *of the* Learning *directory. The two charts on the block diagram display the same type of data, except that the chart on the left uses the stacked chart, and the chart on the right uses the overlaid chart.*

7.4.5   Special Graph-Customizing Features: Cursor Legend

To add cursors to a graph, you first bring the **Cursor Legend** to the front panel graph by popping up on the graph and selecting Visible Items≫Cursor Legend from the short cut menu, and clicking anywhere in a cursor legend row to activate a cursor, as illustrated in Figure 7.26. To add multiple cursors to the graph, the cursor legend displays the exact value of that point. The cursor legend is shown in Figure 7.27.

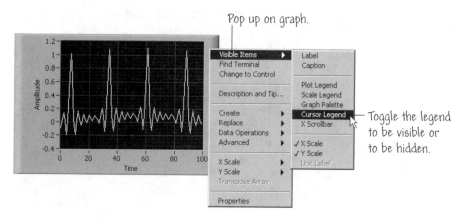

FIGURE 7.26
Placing the **Cursor Legend** on the front panel.

You can place and label cursors on all graphs. To lock a cursor onto a plot, select the lock icon associated with that icon on the **Cursor Legend**. A graph can have any number of cursors and multiple cursors can be moved at the same time. The **Operating** tool is used to click any of the following buttons on the right of the cursor legend to customize your cursor:

- **Cursor Movement Selector**: Click this button to move the cursor using the cursor mover (see Figure 7.27). If you click the Cursor Movement Selector for more than one cursor, the cursors move in parallel.

- **Formatting Ring**: Customize the appearance of each cursor.

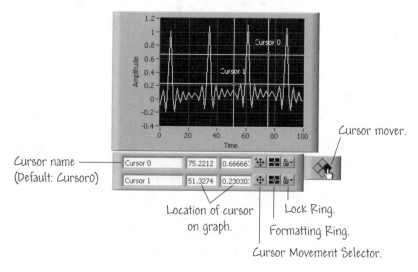

FIGURE 7.27
The **Cursor Legend** is associated with graphs.

- **Lock Ring**: Customize the action of each cursor or associate a particular cursor with a particular plot.

On the **Formatting Ring** you can choose to customize the appearance of your cursor. The following options are available:

- **Color**: Lets you choose the color of your cursor from the Color Picker.
- **Cursor Style**: Provides various cursor styles.
- **Point Style**: Provides various point styles for the intersection of your cursor.
- **Line Style**: Provides various solid and dotted line styles.
- **Line Width**: Provides various line widths.
- **Show Name**: Displays the name of the cursor on the graph. You can use the **Positioning** tool to move the name in relation to the cursor.
- **Bring to Center**: Centers the cursor on the graph without changing the X and Y scales.
- **Go to Cursor**: Changes the X and Y scales to show the cursor at the center of the graph.

The **Lock Ring** provides three options to customize the action of your cursor:

- **Free**: Click the cursor on the graph or enter values in the X and Y coordinates in the cursor legend to move the cursor.
- **Snap to Point**: The cursor moves to the closest plot point. The cursor can switch to another plot in this mode.
- **Lock to Plot**: The cursor locks to a particular plot. The cursor cannot switch to another plot in this mode. If you have more than one plot, LabVIEW lists the plots at the bottom of the Lock Ring menu. Click the plot you want to associate with each cursor.

7.4.6   Using Context Help

For students new to LabVIEW, it can often be confusing when you try to wire data to charts and graphs. Do you use a **Build Array** function, a **Bundle** function, or both? What order do the input terminals use? You can find valuable information in the Context Help window. For example, if you select **Show Context Help** from the **Help** menu and put your cursor over a Waveform Graph terminal in the diagram, you will see the information shown in Figure 7.28. The Context Help window shows you what data types to wire to the Waveform Graph, how to specify

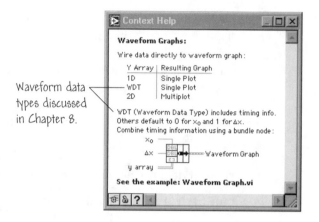

Waveform data types discussed in Chapter 8.

FIGURE 7.28
Using Context Help to determine which data types to wire to the Waveform Graph.

point spacing with the **Bundle** function, and which example to use when you want to see the different ways you can use a Waveform Graph. The Context Help window also shows similar information for XY Graphs and Waveform Charts.

BUILDING BLOCK

7.5   BUILDING BLOCKS: MEASURING VOLUME

The work on the VI called Volume Array.vi created in Chapter 6 will be used as the basis for a new VI that displays the volume data on a chart. The goal is to make the necessary modifications to the existing Volume Array.vi so you can plot the data on a strip chart. You will also control the timing of the loop using the Time Delay function. Use Figure 7.29 as a guide for your VI construction.

The **Limit** input found on the front panel of the VI existed on Multiple Volume Points.vi, which you constructed in the building blocks exercise of Chapter 5. You may want to revisit that during your work on this exercise. When you have finished constructing the VI, save it in the Users Stuff folder as Volume Chart.vi.

Run the VI and observe the operation of the strip chart. The parameters that you can play with are controlled by the Limit knob, the Volume Units horizontal switch, and the Delay between points (sec) horizontal slide. You can easily change the chart mode to scope or sweep mode (even while the VI is running) by popping up on the chart and selecting the desired chart mode in the **Update Mode** menu.

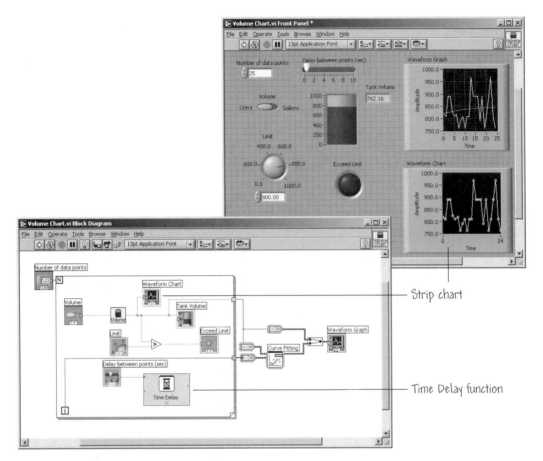

FIGURE 7.29
The Volume Chart.vi block diagram and front panel.

7.6   RELAXED READING: SPACECRAFT ATTITUDE DETERMINATION ON THE WEB

Real-time spacecraft attitude determination involves estimation of the orientation of the spacecraft in space using the sensor data as it is being telemetered to the ground from the spacecraft. Attitude determination is very important in the early phase of a space mission because it is essential that the spacecraft be oriented in the desired way and that all attitude sensors be operational. During this phase, the spacecraft analysts monitor the real-time attitude determination process occurring in the mission control center.

In the 1990s, a prototype for a real-time spacecraft attitude determination system (RTADS) existed in LabVIEW. The LabVIEW real-time attitude determination system was investigated as potential flight dynamics functionality. The current version, built specific to X-ray Timing Explorer (XTE) spacecraft, is capable of handling magnetometer, sun and star tracker data and of displaying the estimated real-time attitude data and its comparison with the on-board estimated and telemetered attitude data in text, as well as real-time graphical plots.

With the support of the Computer Sciences Corporation (CSC) Leading Edge Forum (LEF), researchers began working in the area of real-time data rendering over the Internet. The existing RTADS was Web-enabled using the National Instruments Internet Developers Toolkit. The feasibility of interactive monitoring by spacecraft analysts via third-party browsers was firmly established.

The Web-enabled prototype, called RAW (Real-time Attitude on the Web), shown in Figure 7.30, in its current form reads spacecraft telemetry data from a file on hard disk on the workstation on which it resides. The major functional steps of the RTADS prototype include extraction of relevant sensor data from among magnetometer, digital sun sensor, and two star trackers, conversion to engineering units, and generation of "observation vectors" and transformation to body coordinates. If the star tracker is a selected sensor, star identification is carried out using *a priori* attitude. Reference vectors are generated, and in the final step, the attitude determination process is performed and the results are displayed on a LabVIEW front panel as text and as graphical plots.

The Web interface is accomplished using the G Web server and its built-in functions in conjunction with html documents. The key VI provided by NI was the VI-CTRL.vi designed to control custom VIs. This VI was very helpful in minimizing the CGI programming effort involved in Web-enabling. Most of the browser-based input controls were accomplished using this VI in conjunction with image maps. The Web document also embeds video help, audio help, and email buttons activated using JavaScript.

Some of the features of the LabVIEW real-time attitude determination system on the Web include:

- **Restricted access**: Access to the front panels and chosen sub-directories are restricted using username and passwords, which were configured without any programming effort using a Windows-like interface. IP address-based restrictions are also incorporated in the same way.

- **Browser-based inputs**: Several user-selectable browser-based inputs are provided, such as mode of TLM/Ephemeris data, type of estimation method, sensor selection and standard deviation, mode of *a priori* attitude, *a priori* attitude, and plot size. Additionally, browser-based controls are provided for stopping, closing the front panel, and restarting the program.

- **Visual and e-mail alarms**: Visual and e-mail alarms are provided, which are activated when the error between estimated attitude angles and the telemetered on-board attitude angles exceed a pre-set threshold.

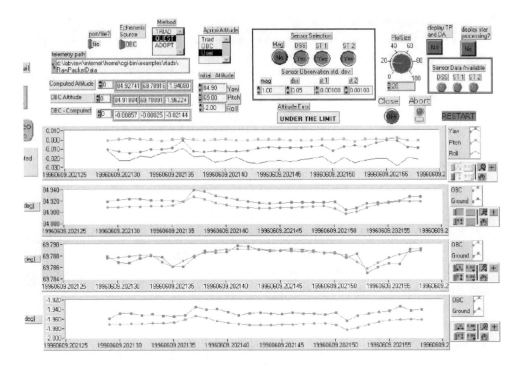

FIGURE 7.30
Front panel of Real-time Attitude on the Web being run.

- **Multiple component types**: The system has both LabVIEW and MATLAB components. Existing MATLAB components were easily incorporated using MATLAB script nodes without having to re-code.

- **On-line visual and audio help files**: Audio and video help files to aid browser-input interface embedded into the html document and activated using JavaScript.

To sum up, some of the advantages of this approach are:

- Offers off-site monitoring

- Requires minimal programming effort

- Legacy modules in other programming languages can be incorporated and Web enabled

- Can be served from workstations on which LabVIEW is not installed if a stand-alone executable is built and shipped.

Some of the scenarios where such systems can be put to use are for spacecraft monitoring and control for low-cost constellation missions that cannot afford spacecraft-specific control centers and for remote access by payload scientists at universities and international institutions.

For further information, contact:

Sheela Belur
bvsheela@yahoo.com
Computer Sciences Corporation

7.7   SUMMARY

You can display plots of data using charts or graphs. Three types of charts were discussed—scope, strip, and sweep—and two types of graphs—waveform and XY. Both charts and graphs can display multiple plots at a time. Charts append new data to old data by plotting one point at time. Graphs display a full block of data. The waveform graph plots only single-valued functions with points that are evenly distributed. Conversely, the XY graph is a general-purpose Cartesian graph that lets you plot unevenly spaced, multivalued functions. You can customize the appearance of your charts and graphs using the legend and the palette. The format and precision of the X and Y scale markers and zoom in and out are controlled with the palette even while the VI is executing.

KEY TERMS

Autoscaling: The ability of graphs and charts to adjust automatically to the range of plotted values.

Legend: An object owned by a chart or graph that displays the names and styles of the plots.

Plot: A graphical representation of data shown on a chart or graph.

Palette: An object owned by a chart or graph from which you can change the scaling and format options while the VI is running.

Scope chart: A chart modeled after the operation of an oscilloscope.

Strip chart: A chart modeled after a paper strip chart recorder, which scrolls as it plots data.

Sweep chart: Similar to a scope chart, except that a line sweeps across the display to separate new data from old data.

Waveform chart: Displays one point at a time on one or more plots.

Waveform graph: Plots single-valued functions or array(s) of values with points evenly distributed along the x-axis.

XY graph: A general-purpose, Cartesian graph used to plot multivalued functions, with points evenly sampled or not.

EXERCISES

E7.1 Open Multiple Graphs Demo.vi found in the Exercises&Problems\Chapter 7 folder in the Learning directory. Right-click on each graph and select **Properties**. In the **Plots** tab, change the line style, point style, and color of each plot. Go to the **Cursors** tab and select **Add**. The **Allow dragging** checkbox should be checked. Below the checkbox, look for the pull-down menus. Select **Lock to plot** from the first menu and **Plot0** from the second menu. Add a second cursor for Plot1. Select **OK** to exit the Properties window. Practice moving the cursors to different values on the graph.

E7.2 Create a new VI where temperature data, created with the Digital Thermometer.vi in the Activity directory, is displayed on a graph. At the beginning of each execution of the VI, clear the graph.

E7.3 Create a VI that plots an ellipse

$$r^2 = \frac{A^2 B^2}{A^2 \sin^2 \phi + B^2 \cos^2 \phi}$$

where r, A, and B are input parameters and $0 \le \phi \le 2\pi$.

E7.4 Create a VI that graphs the function $\sin x$ where $x = 0 \ldots n\pi$ and the integral $y = \int_0^{n\pi} \sin x\, dx$. The value of n should be an input on the front panel.

E7.5 Open the NI Example Finder and go to the **Search** tab. Type in "charts." Open Charts.vi. Run the VI and examine the different types of update modes, specifically the differences between the strip chart, scope chart, and sweep chart. When you are done examining the VI, stop and close the VI. Do not save any changes.

E7.6 Open the NI Example Finder and go to the **Browse** tab. Select browse according to **Task**. Open Building User Interfaces≫Displaying Data≫Graphs and Charts≫3D Graph Properties_Torus.vi.. Run the VI and change the **Plot Style**. Use your mouse to click on the 3D graph and rotate the position of the graph. Experiment with changing the **Transparency** setting. Stop the VI. Switch to the block diagram. The 3D Parametric Surface Graph is a special type of LabVIEW graph that uses ActiveX to generate the 3D image. This graph and other types of 3D graphs are available in the **Controls≫Graph** palette.

PROBLEMS

P7.1 Create a new VI to plot a circle using an XY Graph.

P7.2   Create a new VI where temperature data, created with the **Digital Thermome-ter.vi** in the **Activity** directory, is displayed on a strip chart. Compute and display the running average of the temperature data.

P7.3   Open a blank VI and place a Simulate Signal Express VI on the block diagram. Configure the Express VI to generate a 50-Hz sine wave. Click **OK** to exit the Express VI configuration page. Right-click on the **Sine** terminal and select Create≫Graph Indicator. Run the VI.

P7.4   Complete the crossword puzzle.

Across

1. Plots single-valued functions or array(s) of values with points evenly distributed along the x-axis.
3. An object owned by a chart or graph that displays the names and styles of the plots.
4. Similar to a scope chart, except that a line sweeps across the display to separate new data from old data.
5. A graphical representation of data shown on a chart or graph.
7. The ability of graphs and charts to adjust automatically to the range of plotted values.
8. A general-purpose, Cartesian graph used to plot multivalued functions, with points evenly sampled or not.

Down

1. Displays one point at a time on one or more plots.
2. An object owned by a chart or graph from which you can change the scaling and format options while the VI is running.
4. A chart modeled after the operation of an oscilloscope.
6. A chart modeled after a paper strip chart recorder, which scrolls as it plots data.

CHAPTER 8

Data Acquisition

The subject of this chapter is data acquisition (DAQ). Focusing on the use of the DAQ Assistant, the basic notions associated with analog and digital I/O are presented. Some common terms and concepts associated with data acquisition are also discussed, including the components of a DAQ system, signal conditioning, and the types of signals encountered.

GOALS

1. Review some basic notions of signals and signal acquisition.
2. Introduce the organization of the DAQ VIs.
3. Understand the basics of analog and digital input and output using DAQ Assistant.

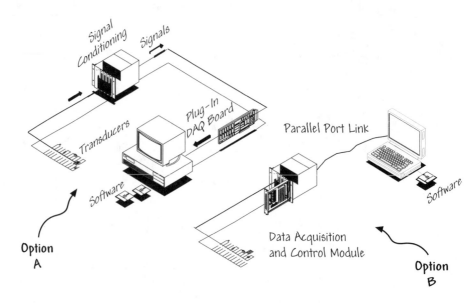

FIGURE 8.1
Two options for a DAQ system.

8.1 COMPONENTS OF A DAQ SYSTEM

In this chapter, we introduce some of the basic notions associated with **data acquisition**. The focus is on the LabVIEW VIs that can be used in a DAQ system—analog and digital signal **input/output** VIs. The subject of data acquisition cannot be adequately covered in one chapter, although the fundamental ideas can be introduced and discussed in enough detail to generate enthusiasm for pursuing other sources of information.[1] The most effective way to learn about data acquisition is by doing it. Reading about it is not enough.

At its most basic level, the task of a DAQ system is the measurement or generation of physical signals. Two options for constructing a DAQ system are illustrated in Figure 8.1. Some students think that having a plug-in DAQ board properly installed in a personal computer is equivalent to having a complete DAQ system. This is not the case at all. In fact, the plug-in DAQ board is only one component of the system. A DAQ system generally has (in addition to the plug-in DAQ board) software for acquiring and manipulating the raw data, analyzing sensors and transducers, signal conditioning, and a suite of and displaying (and storing) the data.

---

1. A good source of information for the beginner is the *Data Acquisition Basics Manual*, available from National Instruments, Inc. Part Number 320997E-0. For the advanced student, see Chapters 14 to 17 of *LabVIEW Graphical Programming, 3rd ed.,* by Gary W. Johnson and Richard Jennings (McGraw-Hill, New York, 2001).

In Option A (see Figure 8.1), the plug-in DAQ board resides in the computer. This computer can be a tower, desktop model, or a laptop with PCMCIA slots. In Option B, the DAQ board is external to the computer. With the latter approach you can build DAQ systems using computers without available plug-in slots (such as some laptops), and the computer and DAQ module communicate through various buses, such as the parallel port. Option B systems are usually more practical for remote data acquisition and control applications where you want to bring the DAQ system into the field.

Before a computer-based system can measure a physical signal, a sensor (or transducer) must convert the physical signal into an electrical signal (such as voltage or current). Generally, you cannot connect signals directly to a plug-in DAQ board in a computer, as you can with most stand-alone instruments. In many cases, the measured physical signal is very low-voltage and susceptible to noise. In these situations, the measured signal may need to be amplified and filtered before conversion to a digital format for use in the computer. A signal-conditioning accessory conditions measured signals before the plug-in DAQ board converts them to digital information. More will be said on signal conditioning later in the chapter. Software controls the DAQ system—acquiring the raw data, analyzing the data, and presenting the results.

8.2   TYPES OF SIGNALS

The concepts of signals and systems arise in a wide variety of fields, including science, engineering, and economics. In an effort to develop an analytic framework for studying certain natural phenomena, the notion of an input-output representation has arisen, and is illustrated in Figure 8.2. In the input-output representation, the input signals are operated on by the system to produce the output signals. Signals are physical quantities that are functions of an independent variable (such as time) and contain information about a natural phenomena. For example, the input signals might be degraded and weak video signals received from a spacecraft approaching a distant asteroid. The system is the DAQ system, which includes

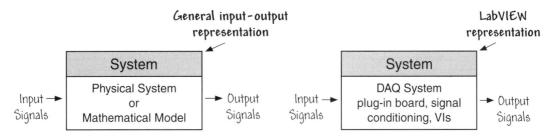

FIGURE 8.2
Modeling physical phenomena with input-output representations.

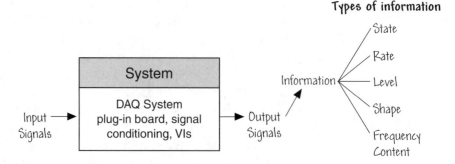

FIGURE 8.3
Measuring and analyzing signals provides information: state, rate, level, shape, and frequency content.

the hardware and software to acquire, process and enhance the spacecraft signals (e.g., the spacecraft images of the asteroid can be enhanced to show features and colors more clearly), and the output signals would be the enhanced asteroid images. In general, physical signals are converted to electrical signals (such as voltage or current) before use by the signal-conditioning and DAQ hardware. This conversion is accomplished by some type of transducer. The list of common transducers includes video cameras, thermocouples, strain gauges, and thermistors. Once the physical signals are measured and converted to electrical signals, the information contained in them may be extracted and analyzed.

Five common classes of information that can be extracted from signals are illustrated in Figure 8.3. The signals evolve either continuously or only at discrete points in time and are known as continuous- or discrete-time signals, respectively. Room temperature is an example of a continuous-time signal (although we may discretize the temperature signal by recording the temperature only at specific sampling points). The end-of-day closing Dow-Jones stock index is an example of a discrete signal, taking on a new value once per working day.

For purposes of discussing data acquisition, we will use the following signal classifications:

- For digital signals, we have two types:
  - on-off
  - pulse train
- For analog signals, we have three types:
  - DC
  - time-domain
  - frequency-domain

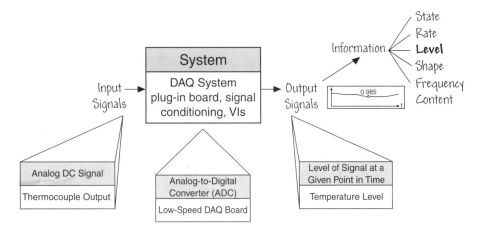

FIGURE 8.7
Analog DC signals are static or slowly varying analog signals.

pressure, strain-gauge output, and fluid level, as illustrated in Figure 8.8. The information contained in the signals is displayed on meters, gauges, strip charts, and numerical readouts.

The DAQ system will return a single value indicating the magnitude of the signal at the requested time. For the measurement to be an accurate representation of the signal, the DAQ system must possess adequate accuracy and resolution capability and adequate sampling rates (usually slow).

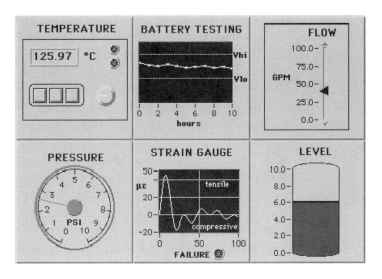

FIGURE 8.8
Common examples of DC signals include temperature, battery voltage, flow rate, pressure, strain-gauge output, and fluid level.

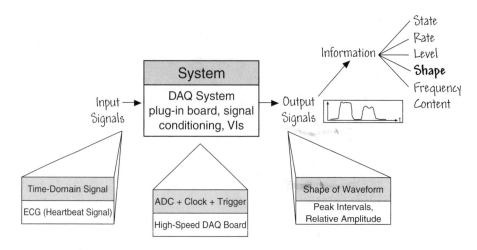

FIGURE 8.9
Analog time-domain signals convey information in the signal level and in the way this level varies with time.

8.2.3   Analog Time-Domain Signals

Analog time-domain signals differ from other signals in that they convey useful information in signal level *and* the way this level varies with time. The information associated with a time-domain signal (also referred to as a waveform) includes such information as time to peak, peak magnitude, time to settle, slope, and shapes of peaks. An analog time-domain signal is illustrated in Figure 8.9.

You must take a precisely timed sequence of individual amplitude measurements to measure the shape of a time-domain signal. DAQ systems that are used to measure time-domain signals usually have an **analog-to-digital conversion** (ADC) component, a sample clock, and a trigger.

The physical signal must be sampled and measured at a rate that adequately represents the shape of the signal. Therefore, when acquiring analog time-domain signals, DAQ systems need a high-bandwidth ADC to sample the signal at high rates. The signal must also be sampled accurately without significant loss of precision.

Generally the signal measurements need to start at a specified time to guarantee that an interesting segment of the signal is acquired. The sample clock accurately times the occurrence of each analog-to-digital (A/D) conversion. In many situations, triggering is necessary to initiate the measurement process at a precise time. The trigger starts the measurement at the proper time based on some external condition specified by the user. There are an unlimited number of different time-domain signals, some of which are shown in Figure 8.10.

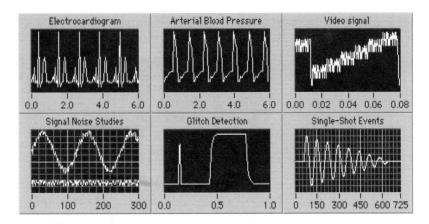

FIGURE 8.10
Six different time-domain signals.

8.2.4   Analog Frequency-Domain Signals

Analog frequency-domain signals are similar to time-domain signals in that they also convey information in the way the signals vary with time. However, the information extracted from a frequency-domain signal is based on the signal frequency content, as opposed to the shape of the signal. An analog frequency-domain signal is shown in Figure 8.11.

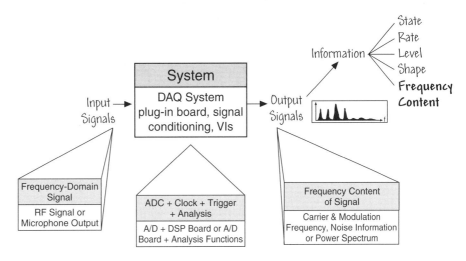

FIGURE 8.11
The information extracted from a frequency-domain signal is based on the signal frequency content.

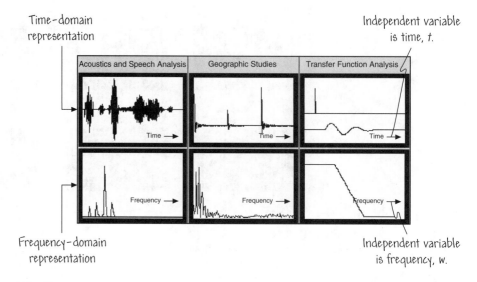

FIGURE 8.12
Three examples of frequency-domain signals: speech and acoustics analysis, geophysical signals, and transfer function frequency response.

As with DAQ systems that measure time-domain signals, a system used to measure a frequency-domain signal must also include an ADC, a sample clock, and a trigger to accurately capture the waveform. Additionally, the DAQ system must include the necessary analysis capabilities to extract frequency information from the signal. You can perform this type of digital signal processing (DSP) using application software or special DSP hardware designed to analyze the signal quickly and efficiently.

In short, a DAQ system that acquires analog frequency-domain signals should possess a high-bandwidth ADC capability to sample signals at high rates and an accurate clock to take measurements at precise intervals. Also, it is common to use triggers to initiate the measurement process precisely at prespecified times. Finally, a complete DAQ system should provide a library of analysis functions, including a function to convert time-domain information to frequency-domain information. Chapter 11 discusses some of the LabVIEW analysis functions, but time and space limitations make it impossible to cover all those available to VI developers.

Figure 8.12 shows some examples of frequency-domain signals. Each example in the figure includes a graph of the originally measured signal as it varies with respect to time, as well as a graph of the signal frequency spectrum. While you can analyze any signal in the frequency domain, certain signals and application areas lend themselves especially to this type of analysis. Among these areas are speech and acoustics analysis, geophysical signals, vibration, and studies of

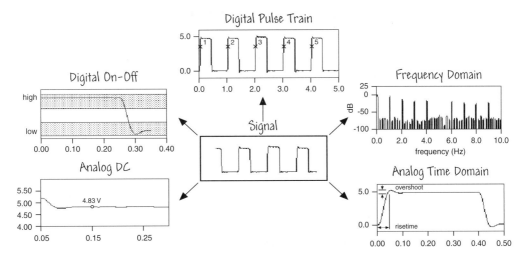

FIGURE 8.13
A series of voltage pulses can provide information for all five signal classes.

system transfer functions.

8.2.5 One Signal—Five Measurement Perspectives

The five classifications of signals presented in the previous discussions are not mutually exclusive. A particular signal may convey more than one type of information, and you can classify a signal as more than one type and measure it in more than one way. In fact, you can use simpler measurement techniques with the digital on-off, pulse train, and DC signals, because they are just simpler cases of the analog time-domain signals.

The measurement technique you choose depends on the information you want to extract from the signal. In many cases you can measure the same signal with different types of systems, ranging from a simple digital input board to a sophisticated frequency analysis system. Figure 8.13 demonstrates how a series of voltage pulses can provide information for all five signal classes.

8.3 COMMON TRANSDUCERS AND SIGNAL CONDITIONING

When measuring a physical phenomena, a transducer must convert this phenomena (such as temperature or force) into a measurable electrical signal (such as voltage or current). Table 8.1 lists some common transducers used to convert physical phenomena into measurable quantities.

TABLE 8.1 Phenomena and Transducers

Phenomenon	Transducer
Temperature	Thermocouples
	Resistance temperature detectors (RTDs)
	Thermistors
	Integrated circuit sensor
Light	Vacuum tube photosensors
	Photoconductive cells
Sound	Microphone
Force and pressure	Strain gauges
	Piezoelectric transducers
	Load cells
Position (displacement)	Potentiometers
	Linear voltage differential transformer (LVDT)
	Optical encoder
Fluid flow	Head meters
	Rotational flowmeters
	Ultrasonic flowmeters
pH	pH electrodes

All transducers output an electrical signal that is often not well suited for a direct measurement by the DAQ system. For example, the output voltage of most thermocouples is very small and susceptible to noise, and often needs to be amplified and filtered before measuring. Figure 8.14 shows some common types of transducer and signal pairs and the required signal conditioning. A highly expandable signal conditioning system—dubbed Signal Conditioning eXtensions for Instrumentation (**SCXI**) made by National Instruments, Inc.—conditions low-level signals in a noisy environment within an external chassis located near the sensors. The close proximity improves the signal-to-noise ratio of the signals reaching the DAQ boards. Of course, LabVIEW works with signal conditioning systems other than SCXI.

Some **signal conditioning** (such as linearization and scaling) can be performed in the software. LabVIEW provides several VIs in the **Data Acquisition**

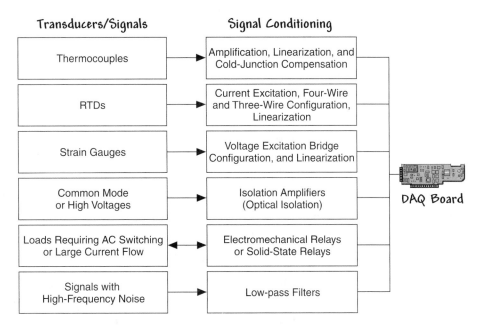

FIGURE 8.14
Common types of transducers/signals and the required signal conditioning.

palette for such purposes. The remainder of this section is devoted to short descriptions of some of the basic ideas involved in signal conditioning. Further information on signal conditioning and SCXI hardware (i.e., setup procedures for SCXI hardware, hardware operating modes, and programming considerations for SCXI) can be found in other LabVIEW documentation (check the NI website).

Some common types of signal conditioning follow:

- **Transducer excitation**: Certain transducers (such as strain gauges) require external voltages or currents to excite their own circuitry in a process known as transducer excitation. The process is similar to a television needing power to receive and decode video and audio signals. The necessary excitation in a DAQ system can be provided by the plug-in DAQ boards and the signal-conditioning peripherals, and sometimes by external instruments.

- **Linearization**: Common transducers (such as strain gauges, thermistors, RTDs, and thermocouples) generate voltages that are nonlinear with respect to the phenomena they represent. The NI-DAQ driver and LV SW can perform **software linearization** to scale a transducer voltage to the correct units of strain or temperature.

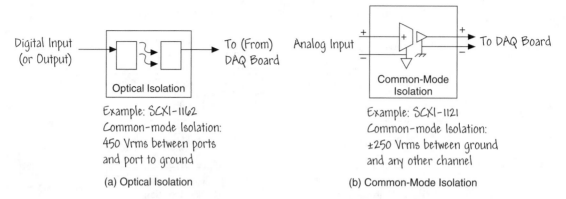

FIGURE 8.15
Two common methods for isolating signals.

- **Isolation**: Another common use for signal conditioning is to isolate the transducer signals from the computer. For example, when the signal being monitored contains large voltage spikes that could damage a computer or harm a person, you should not connect the signal directly to a DAQ board without some type of isolation. Figure 8.15 shows two common methods for isolating signals.

- **Filtering**: Another form of signal conditioning is the filtering of unwanted signals from the desired signal. A common filter reduces 60 Hz AC powerline noise present in many signals. Other well-known types of filters include low-pass, high-pass, and notch filters. Some DAQ boards and signal-conditioning devices have built-in filters.

- **Amplification**: This is the most common type of signal conditioning. Amplification maximizes the use of the available voltage range to increase the accuracy of the digitized signal and to increase the signal-to-noise ratio (SNR). Low-level signals should be amplified at the DAQ board or at an external signal-conditioning peripheral positioned near the source of the signal, as shown in Figure 8.16. One reason to amplify low-level signals close to the signal source, instead of at the DAQ board, is to increase the signal-to-noise ratio. Consider the case where you amplify the signal only on the DAQ board. Then the DAQ board will also measure and digitize any noise that enters the lead wires along the path as the signal travels from the source to the DAQ board. On the other hand, the ratio of signal voltage to noise voltage that enters the lead wires is larger if you amplify the signal close to the signal source. Table 8.2 shows how the SNR changes with the location of amplification.

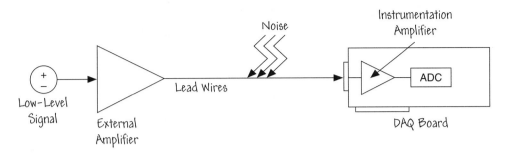

FIGURE 8.16
Low-level signals should be amplified at the DAQ board or at an external signal-conditioning peripheral positioned near the source of the signal.

 You can minimize the effects of external noise on the measured signal by using shielded or twisted-pair cables and by minimizing the cable length. Keeping cables away from AC power cables and computer monitors will also help minimize 50/60 Hz noise.

TABLE 8.2   Effects of Amplification on Signal-to-Noise Ratio (S.C. → Signal-Conditioning Peripheral)

	Signal Voltage	S.C. Amplification	Noise in Lead Wires	DAQ Board Amplification	Digitized Voltage	SNR
Amplify only at DAQ board	0.01 V	None	0.001 V	×100	1.1 V	10
Amplify at S.C. and DAQ board	0.01 V	×10	0.001 V	×10	1.01 V	100
Amplify only at S.C.	0.01V	×100	0.001 V	None	1.001 V	1000

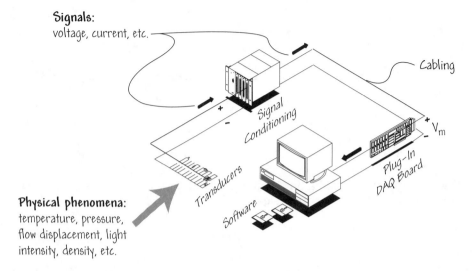

Signals:
voltage, current, etc.

Cabling

Signal
Conditioning

Transducers

Software

Plug-In
DAQ Board

$+$ V_m
$-$

Physical phenomena:
temperature, pressure,
flow displacement, light
intensity, density, etc.

FIGURE 8.17
A DAQ system highlighting the signals and the cabling.

8.4   SIGNAL GROUNDING AND MEASUREMENTS

Up to this point, we have discussed three components of the DAQ system: signals, transducers, and signal conditioning. Now you might be tempted to think that all that remains is to wire the signal source to the DAQ board and begin acquiring data. However, a few important items must be considered:

- The nature of the signal source (grounded or floating)

- The grounding configuration of the amplifier on the signal conditioning hardware or DAQ board

- The cabling scheme to connect all the components together

A DAQ system is depicted in Figure 8.17 highlighting the signals and the cabling.

8.4.1   Signal Source Reference Configuration

Signal sources come in two forms: referenced and nonreferenced. Referenced sources are usually called **grounded signals**, and nonreferenced sources are called **floating signals**. A schematic of a grounded signal source is shown in Figure 8.18(a).

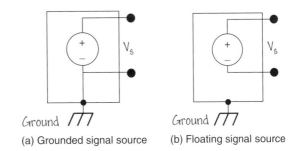

(a) Grounded signal source   (b) Floating signal source

FIGURE 8.18
Grounded signal sources have voltage signals that are referenced to a system ground. Floating signal sources contain a signal that is not connected to an absolute reference.

Grounded signal sources have voltage signals that are referenced to a system ground, such as earth or a building ground. Devices that plug into a building ground through wall outlets, such as signal generators and power supplies, are the most common examples of grounded signal sources. Grounded signal sources share a common ground with the DAQ board.

Floating signal sources contain a signal (e.g., a voltage) that is not connected to an absolute reference, such as earth or a building ground. Some common examples of floating signals are batteries, battery-powered sources, thermocouples, transformers, isolation amplifiers, and any instrument that explicitly floats its output signal. As illustrated in Figure 8.18(b), neither terminal of the floating source is connected to the electrical outlet ground.

8.4.2   Measurement System

A schematic of a measurement system is depicted in Figure 8.19. A measurement system can be placed in one of three categories:

- Differential
- Referenced single-ended (RSE)
- Nonreferenced single-ended (NRSE)

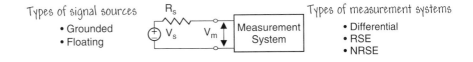

FIGURE 8.19
Types of signal sources and measurement systems.

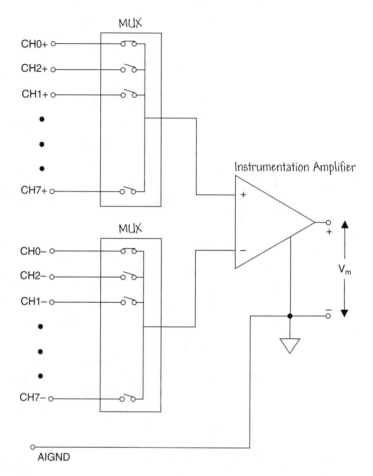

FIGURE 8.20
An 8-channel differential measurement system.

In a **differential** measurement system, you do not need to connect either input to a fixed reference, such as earth or a building ground. DAQ devices with instrumentation amplifiers can be configured as differential measurement systems. Figure 8.20 depicts the 8-channel differential measurement system used in the MIO series devices. A **channel** is a pin or wire lead where analog or digital signals enter or leave a DAQ device. For this device, the analog input ground pin labeled AIGND is the measurement system ground. The analog multiplexers (labeled MUX in the figure) increase the number of available measurement channels while still using a single instrumentation amplifier.

An ideal differential measurement system, shown in Figure 8.21, reads only the potential difference between its two terminals—the $(+)$ and $(-)$ inputs. It completely rejects any voltage present at the instrumentation amplifier inputs with

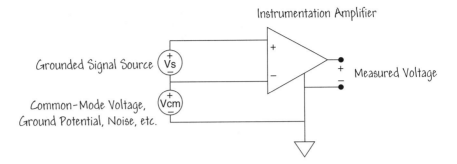

FIGURE 8.21
An ideal differential measurement system completely rejects common-mode voltage.

respect to the amplifier ground. In other words, an ideal differential measurement system completely rejects the common-mode voltage.

A **referenced single-ended** (RSE) measurement system measures a signal with respect to building ground and is sometimes called a grounded measurement system. Figure 8.22 depicts a 16-channel version of an RSE measurement system.

DAQ devices often use a **nonreferenced single-ended** (NRSE) measurement system, which is a variation of the RSE measurement system. In an NRSE measurement system, all measurements are made with respect to a common reference, because all of the input signals are already grounded. Figure 8.23 depicts an NRSE measurement system. AISENSE is the common reference for taking measurements and all signals in the system share this common reference. AIGND is the system ground.

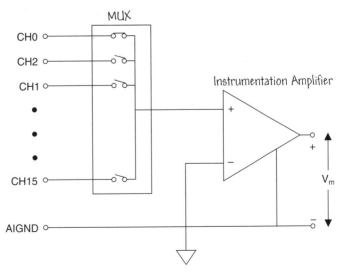

FIGURE 8.22
A 16-channel RSE measurement system.

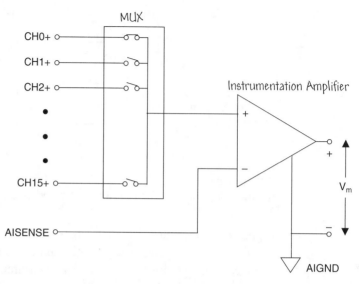

FIGURE 8.23
An NRSE measurement system.

A grounded signal source is best when using a differential or NRSE measurement system. An RSE measurement system can be used with a grounded signal source if the signal levels are high and the cabling has a low impedance. The measured signal voltage will be degraded with the RSE system, but the degradation is usually small relative to the signal voltage. Floating signal sources can be measured with differential, RSE, and NRSE measurement systems. In general, a differential measurement system is preferable because it rejects the ground loop-induced errors and reduces the noise picked up in the environment. On the other hand, the single-ended configuration allows for twice as many measurement channels and is acceptable when the magnitude of the induced errors is smaller than the required accuracy of the data. You can use single-ended measurement systems when all input signals meet the following criteria:

1. High-level signals (normally, greater than 1 V).

2. Short or properly shielded cabling traveling through a noise-free environment (normally less than 15 ft).

3. All signals can share a common reference signal at the source.

4. A summary of analog input connections is given in Figure 8.24.

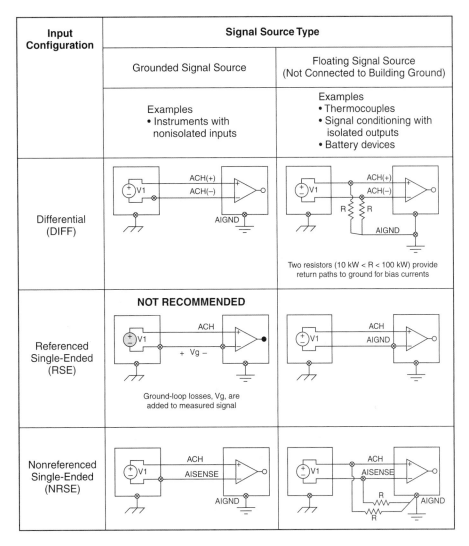

FIGURE 8.24
Summary of analog input connections.

8.5   ANALOG I/O CONSIDERATIONS

When preparing to configure a DAQ board you need to consider the quality of the analog-to-digital conversion. There are many questions that arise when making analog signal measurements with your DAQ system. For example, what are the signal magnitude limits? How fast does the signal vary with time? The latter

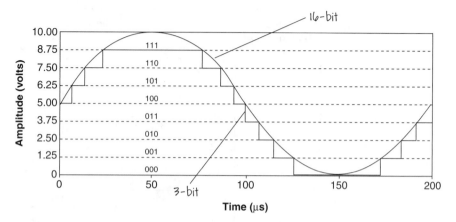

FIGURE 8.25
16-bit versus 3-bit resolution (5 kHz sine wave).

question is important because the sample rate determines how often the A/D conversions take place. The four parameters of concern are

- Resolution
- Range
- Signal limit settings
- Sampling rate

Depending on the type of DAQ board you have, these four parameters will be set either in the hardware (using dip switches and jumpers on the board) or set using software (with the **Measurement and Automation Explorer** or in LabVIEW as discussed later in this chapter).

The number of bits used to represent an analog signal determines the **ADC resolution** of the analog-to-digital conversion. You can compare the resolution on a DAQ device to the number of divisions on a ruler. For a fixed ruler length, the more divisions you have on the ruler, the more precise the measurements that you can make. A ruler marked off in millimeters can be read more accurately than a ruler marked off in centimeters. Similarly, the higher the ADC resolution, the higher the number of divisions of the ADC range and, therefore, the more accurately the analog signal can be represented. For example, a 3-bit ADC divides the signal range into 2^3 divisions, with each division represented by a binary or digital code between 000 and 111. The ADC then translates measurements of the analog signal to one of the digital divisions. Figure 8.25 shows a sine wave digital image obtained by a 3-bit ADC. Clearly, the digital signal does not represent the

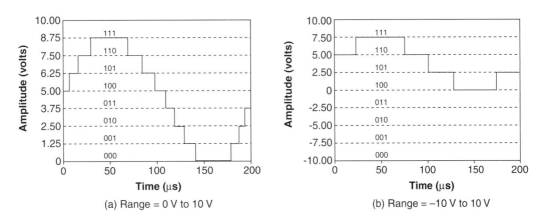

FIGURE 8.26
The 3-bit ADC in (a) has eight digital divisions in the range from 0 to 10 volts—if you select a range of −10.00 to 10.00 volts as in (b), the same ADC now separates a 20-volt range into eight divisions.

original signal adequately because there are too few divisions to represent the varying voltages of the analog signal. By increasing the resolution to 16 bits, however, the ADC's number of divisions increases from 2^3 to 2^{16}. The ADC can now provide an adequate representation of the analog signal.

The **range** refers to the minimum and maximum analog signal levels that the ADC can handle. You should attempt to match the range to that of the analog input signal to take best advantage of the available resolution. Fortunately, many DAQ devices feature selectable ranges. Consider the example shown in Figure 8.26(a), where the 3-bit ADC has eight digital divisions in the range from 0 to 10 volts. If you select a range of −10.00 to +10.00 volts, as shown in Figure 8.26(b), the same ADC now separates a 20-volt range into eight divisions. The smallest detectable voltage increases from 1.25 to 2.50 volts, and you now have a much less accurate representation of the signal.

The **limit settings** are the maximum and minimum values of the signal you are measuring. The closer the limit setting is to the incoming analog signal maximum and minimum, the more digital divisions will be available to the ADC to represent the signal. Using a 3-bit ADC and a range setting of 0.00 to 10.00 volts, we see in Figure 8.27 the effects of a limit setting between 0 and 5 volts and 0 and 10 volts. With a limit setting of 0 to 10 volts, the ADC uses only four of the eight divisions in the conversion; with a limit setting of 0 to 5 volts, the ADC now has access to all eight digital divisions. This makes the digital representation of the signal more accurate.

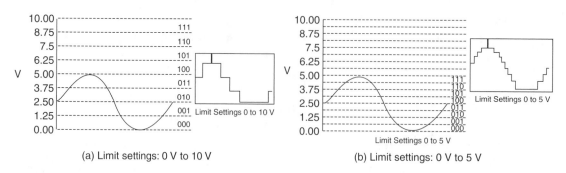

(a) Limit settings: 0 V to 10 V   (b) Limit settings: 0 V to 5 V

FIGURE 8.27
Precise limit setting allows the ADC to use more digital divisions to represent the signal.

The resolution and range of a DAQ board and the limit settings determine the smallest detectable change in the input voltage. This change in voltage represents 1 **least significant bit** (LSB) of the digital value and is often called the **code width**. The smallest code width is calculated with the following formula:

$$V_{cw} = \frac{range}{2^{resolution}},$$

where the resolution is given in bits. For example, a 12-bit DAQ board with a 0 to 10-V range detects a 2.4-mV change. This is calculated as follows:

$$V_{cw} = \frac{range}{2^{resolution}} = \frac{10}{2^{12}} = 2.4 \text{ mV},$$

while the same board with a -10 to 10-V range detects only a change of 4.8 mV:

$$V_{cw} = \frac{range}{2^{resolution}} = \frac{20}{2^{12}} = 4.8 \text{ mV}.$$

A high-resolution ADC provides a smaller code width for a given range. For example, consider the two preceding examples, except that the resolution is now 16-bit. A 16-bit DAQ board with a 0 to 10-V range detects a 0.15-mV change:

$$V_{cw} = \frac{range}{2^{resolution}} = \frac{10}{2^{16}} = 0.15 \text{ mV},$$

while the same board with a -10 to 10-V range detects only a change of 0.3 mV:

$$V_{cw} = \frac{range}{2^{resolution}} = \frac{20}{2^{16}} = 0.3 \text{ mV}.$$

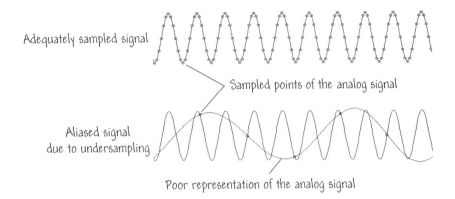

Adequately sampled signal

Sampled points of the analog signal

Aliased signal
due to undersampling

Poor representation of the analog signal

FIGURE 8.28
Sampling too slowly may result in a poor representation of the analog signal.

You also need to determine whether your signal is unipolar or bipolar, as this affects the code width. **Unipolar** signals range from 0 V to a positive value (e.g., 0 V to 10 V). **Bipolar** signals range from a negative value to a positive value (e.g., −5 V to 5 V). A smaller code width is obtained by specifying the range to be unipolar, if indeed the signal is unipolar; and conversely, specifying the range as bipolar if the signal is bipolar. If the maximum and minimum variation of the analog signal is smaller than the value of the range, you will need to set the limit settings to more accurately reflect the analog signal variation.

Some confusion may exist about selecting the limit settings rather than selecting the gain. *When you set the signal limit settings, you are effectively selecting the gain for that signal. Limit settings automatically magnify the magnitude of the signal to create more precise analog-to-digital conversions.*

The **sampling rate** is the rate at which the DAQ board samples an incoming analog signal. Figure 8.28 shows an adequately sampled signal as well as the effects of undersampling. The sampling rate determines how often an analog-to-digital (A/D) conversion takes place. Computing the proper sampling rate requires knowledge of the maximum frequency of the incoming signal and the accuracy required of the digital representation of the analog signal. It also requires some knowledge of the noise affecting the incoming signal and the capabilities of your hardware. A fast sampling rate acquires more points in a given time, allowing, in general, a better representation of the original signal than a slow sampling rate would allow. In fact, sampling too slowly may result in a misrepresentation of the incoming analog signal.

The effect of undersampling is that the signal appears as if it has a different frequency than it truly does. This misrepresentation is called an *alias*. To prevent undersampling, you must sample, at a minimum, twice the rate of the maximum frequency component of the incoming signal. One way to deal with aliasing is to use low-pass filters that attenuate any frequency components in the incoming signal above the Nyquist frequency (defined to be one-half the sampling frequency). These filters are known as *antialiasing filters*.

Signal-Limit Selection

The kind of calculations required to select the signal limits for a DAQ application are illustrated in this example. Keep in mind that the objective is to minimize the code width while making sure that the entire signal fits within the allowable device range.

Assume your transducer output is a sine wave with an amplitude of ± 30 mV and an offset of 10 mV. Your board has limit settings of 0 to $+10$ V, ± 10 V, and ± 5 V. What limit setting would you select for maximum precision if you use a DAQ board with 12-bit resolution?

Table 8.3 shows how the code width of a hypothetical 12-bit DAQ device varies with range and limit settings. The values in Table 8.3 depend on the hardware, and you should consult your DAQ hardware documentation to determine the available device voltage range and limit settings for your device.

Step 1: Select a range. Since the signal input is bipolar (the sine wave contains both positive and negative voltages), we must choose one of the bipolar ADC ranges. For our board, there are two, so to start we choose ± 5 V, for a range of 10 V.

Step 2: Determine the code width. Using the formula for determining the code width, we compute

$$V_{cw} = \frac{range}{2^{resolution}} = \frac{10}{2^{12}} = 2.4 \text{ mV}.$$

Step 3: Choose a limit setting. Since our signal has a maximum magnitude of $+30$ mV, we must choose input settings that allow us to read ± 30 mV. Referring to Table 8.3, we find that (for a range of -5 to 5 V) a limit setting of -50 mV to 50 mV will cover the signal variation of ± 30 mV. Choosing this limit setting yields a code width (or precision) of 24.4 μV.

TABLE 8.3   Code Width for Various Ranges and Limit Settings

Voltage Range	Limit Settings	Code Width
0 to 10 V	0 to 10 V	2.44 mV
	0 to 5 V	1.22 mV
	0 to 2.5 V	610 μV
	0 to 1.25 V	305 μV
	0 to 1 V	244 μV
	0 to 0.1 V	24.4 μV
	0 to 20 mV	4.88 μV
−5 to 5 V	−5 to 5 V	2.44 mV
	−2.5 to 2.5 V	1.22 mV
	−1.25 to 1.25 V	610 μV
	−0.625 to 0.625 V	305 μV
	−0.5 to 0.5 V	244 μV
	−50 to 50 mV	24.4 μV
	−105 to 10 mV	4.88 μV
−10 to 10 V	−10 to 10 V	4.88 mV
	−5 to 5 V	2.44 mV
	−2.5 to 2.5 V	1.22 mV
	−1.25 to 1.25 V	610 μV
	−1 to 1 V	488 μV
	−0.1 to 0.1 V	48.8 μV
	−20 to 20 mV	9.76 μV

Step 4:   Repeat steps 1 through 3 for range of ±10 V. We must also try the other range of ±10 V to see if it yields a smaller code width. Repeating steps 1 through 3 gives us a limit setting of −0.1 V to 0.1 V and a code width of 48.8 μV. This does not improve our accuracy. Thus, we choose the settings as follows:

- Range = −5 to 5 V
- Limit setting = −50 mV to 50 mV   ◆

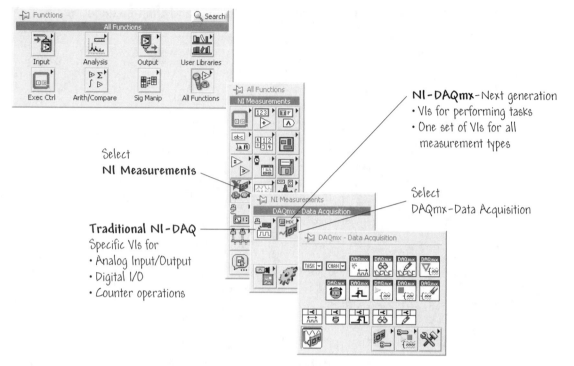

FIGURE 8.29
The **DAQmx Data Acquisition** functions palette.

8.6   DAQ VI ORGANIZATION

The LabVIEW DAQ VIs are organized into two palettes, one for Traditional NI-DAQ and another for NI-DAQmx, as illustrated in Figure 8.29. These are two different device drivers for the National Instruments DAQ hardware line. NI-DAQmx is the next-generation DAQ driver that incorporates significant improvements over Traditional NI-DAQ in both performance and ease of use. For students learning about data acquisition for the first time, the special advantage of NI-DAQmx over previous versions of NI-DAQ is the inclusion of the DAQ Assistant for configuring channels and measurement tasks. The task of writing VIs to acquire data is significantly simplified with the DAQ Assistant, as we shall learn later in this chapter. For these reasons, the remainder of this chapter will cover NI-DAQmx.

The NI-DAQmx VIs are a special type of VI called Polymorphic VIs. The result is a core set of VIs that can adapt to different DAQ functionality, such as analog input, analog output, digital I/O, etc. You access the palette by clicking on Functions≫All Functions≫NI Measurements≫DAQmx, shown in Figure 8.29. In this book, we will focus on the functionality of these three DAQ operations: **Analog Input**, **Analog Output**, and **Digital I/O**.

8.7   DAQ HARDWARE CONFIGURATION

LabVIEW provides utilities designed to help you define which signals are connected to which channels on your data acquisition board and to easily group those into combined tasks. In previous years, significant amounts of time were spent defining the signal types, connections, and transducer equations—and all this before beginning development and programming of the actual DAQ system! For example, if you are using thermocouples, you must perform cold-junction compensation (CJC) calculations and apply appropriate scaling factors to convert raw measured voltages into actual temperature readings. This process is now one of entering the necessary information in dialog boxes, or through wizards, to define an input signal, the type of transducer being used, any scaling factors required, CJC values, and the conversion factors, as well as the timing and triggering information you desire. You can then reference the channel name or task (that you assign) for the input signal(s) and the conversion from voltage to physical units is performed automatically (and transparently). You can save different configuration files for different settings or systems. Not only can you assign **channel** and **task names**, sensor types, engineering units, and scaling information, as well as timing and triggering information, to each channel using the LabVIEW utility, but you can also define the physical quantities you are measuring on each DAQ hardware channel. Once the software has been properly configured, the hardware will be configured correctly to make the measurement for each channel in terms of the physical quantity.

8.7.1   Windows

LabVIEW for Windows installs a configuration utility for establishing all board and channel configuration parameters. This utility is known as the **Measurement & Automation Explorer**, or MAX for short. MAX is also helpful in troubleshooting and self-calibration. The MAX utility reads the information the Device Manager records in the Windows registry and assigns a logical device number to each DAQ board. You use the device number to refer to the board in LabVIEW. Figure 8.30 shows the relationship between the DAQ board and MAX. The Windows Configuration Manager keeps track of all the hardware installed in your system, including National Instruments DAQ boards.

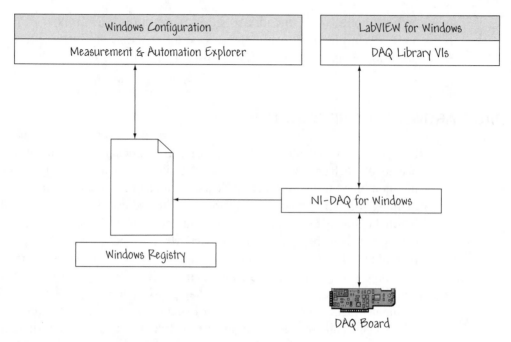

FIGURE 8.30
Windows configuration management.

If you have a Plug & Play (PnP) board, the Windows Configuration Manager automatically detects and configures the board. If you have a non-PnP board (known as a Legacy device) you must configure the board manually using the **Add New Hardware** *option under the Control Panel.*

You can check the Windows Configuration by accessing the Device Manager on your computer. You will find **Data Acquisition Devices**, which lists all DAQ boards installed in your computer as shown in Figure 8.31. Highlight a DAQ board and select **Properties** or double-click on the board, and you see a dialog window with tabbed pages. **General** displays overall information regarding the board. You use **Resources** to specify the system resources to the board such as interrupt levels, DMA, and base address for software configurable boards. **Driver** specifies the driver version and location for the DAQ board.

LabVIEW for Windows DAQ VIs access the National Instruments standard NI-DAQ for Windows 32-bit dynamic link library (DLL). The LabVIEW setup program installs the NI-DAQ DLL in the Windows\System32 directory. NI-DAQ for Windows supports all National Instruments DAQ boards and SCXI.

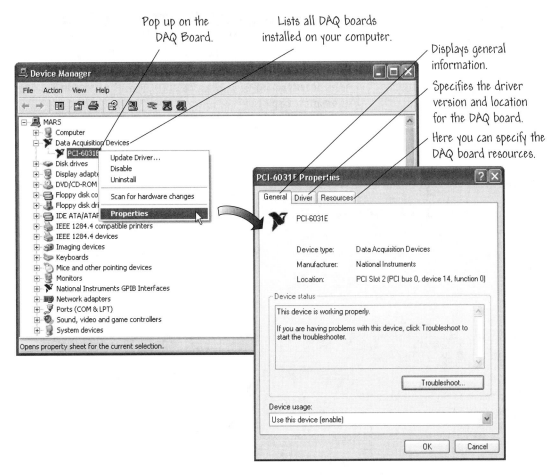

FIGURE 8.31
Checking the Windows Configuration by Accessing the Device Manager.

The nidaq32.dll file, the high-level interface to your board, is loaded into the Windows\System32 directory. The nidaq32.dll file then interfaces with the Windows Registry to obtain the configuration parameters defined by the MAX. You access MAX either by double-clicking on its icon on the desktop or selecting **Measurement & Automation Explorer** from the **Tools** menu in LabVIEW, as illustrated in Figure 8.32.

8.7.2   Macintosh

As of the publishing of this book, there is currently no support for NI-DAQmx on the Mac. This means that there is essentially no DAQ capability for the Mac

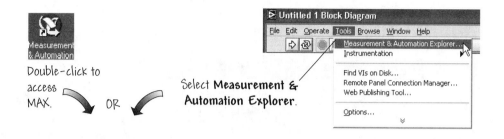

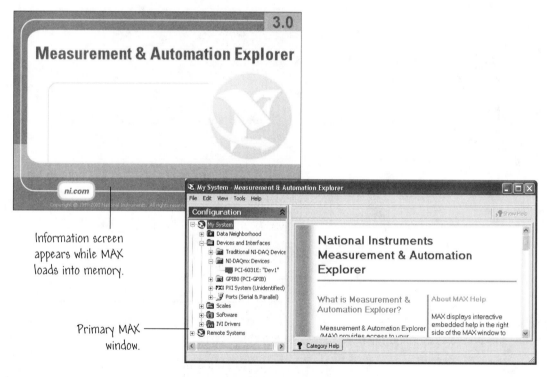

FIGURE 8.32
Accessing the Measurement & Automation Explorer (MAX).

running OS X at this time. For more information on updates to DAQ for the Mac, refer to www.ni.com/mac.

Using the MAX (Windows)

The objective of this exercise is to use MAX to examine the configuration for the DAQ board in your computer and to configure one NI-DAQmx task.

Open **Devices and Interfaces**.

Select **Properties** to obtain more information about the DAQ board.

Click here to hide/show on-line help.

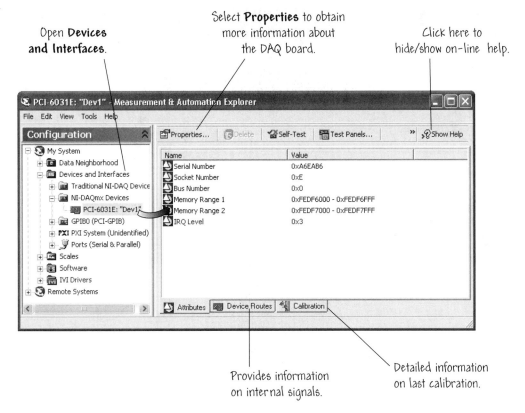

Provides information on internal signals.

Detailed information on last calibration.

FIGURE 8.33
Checking the properties of the DAQ boards.

Start MAX by double-clicking on the desktop icon or by selecting **Measurement & Automation Explorer** from the **Tools** menu in LabVIEW, as illustrated in Figure 8.32.

Depending on your system, MAX may be installed in a different location.

Open the **Devices and Interfaces** section as seen in Figure 8.33. The figure shows what MAX looks like for a PCI-6031E and a PCI-GPIB. The PCI-6031E is an NI-DAQmx device, so open **NI-DAQmx Devices** section as well. The MAX window shows the National Instruments boards and software in your system. Note the device number indicated in the quotes after the DAQ board. The LabVIEW DAQ VIs use the device number to determine which board performs DAQ opera-

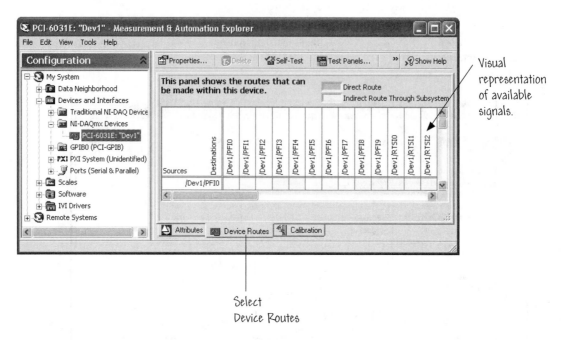

FIGURE 8.34
Accessing information about the DAQ board configuration.

tions. In Figure 8.34, we see that the DAQ board PCI-6031E is "Dev1." You may have a different board installed and some of the options shown may be different.

You can get more information about your DAQ board configuration by examining the properties using the **Attributes**, **Device Routes**, and **Calibration** tabs. With the DAQ board highlighted, MAX displays the attributes of the device, such as the system resources that are being used by the device, the serial number, socket number, and memory range, as shown in Figure 8.33. Clicking the **Device Routes** tab provides detailed information about the internal signals that can be routed to other destinations on the device, as shown in Figure 8.34. This is a powerful resource that gives you a visual representation of the signals that are available to provide timing and synchronization with components that are on the device and other external devices. The **Calibration** tab provides information about the last time the device was calibrated both internally and externally. If you right-click the NI-DAQmx device in the configuration tree and select **Self Calibrate**, the DAQ device will be calibrated using a precision voltage reference source. Once the device has been calibrated, the **Self Calibration** information updates in the **Calibration** tab.

Select
Create New...

Right-click on
**Data
Neighborhood**.

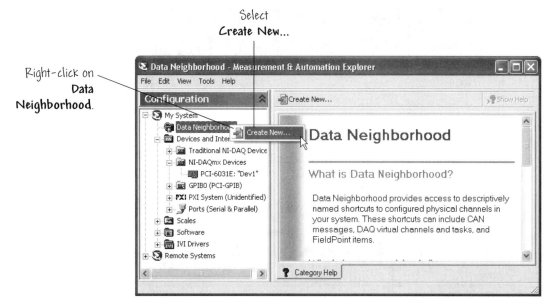

FIGURE 8.35
Configuring a new virtual channel.

*Press the **Show/Hide** button in the top right corner of the MAX window to
hide the online help and show the DAQ board information.*

Next, we want to configure an NI-DAQmx task. An NI-DAQmx task is a
shortcut to configuring a data acquisition task that allows you to save a configuration complete with channel, timing, and triggering information. Right-click
on the **Data Neighborhood** icon and choose **Create New...**, as shown in Figure 8.35. The window shown in Figure 8.36 will appear. In the window, select
NI-DAQmx Task and press the **Next** button, as depicted in Figure 8.36. You will
now configure a channel to take a reading from a temperature sensor (Analog
Input).

After pressing the **Next** button, a window for configuring the input will
appear, and you can select the input as shown in Figure 8.37. We will select
the Analog Input≫Temperature and this will automatically open up a page
on which we select the type of temperature measurement to take. As shown in
Figure 8.37, we select the **Thermocouple**. Once we do so, a new page opens up
to permit us to assign a new local channel to the task. In this case, we select **ai0**
associated with Dev1—our PCI 6031E DAQ board. Once you have completed
the job of creating the local channel, select **Next** to proceed.

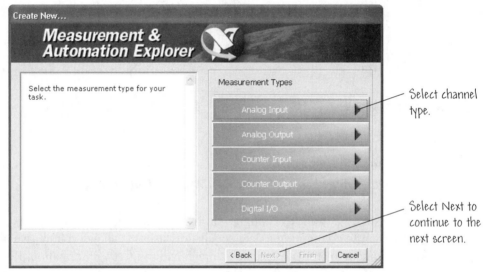

Select NI-DAQmx Task.

Click here to create the new channel configuration.

Select channel type.

Select Next to continue to the next screen.

FIGURE 8.36
Creating a new virtual channel configuration.

The next dialog box that appears allows you to name your task, as shown in Figure 8.38. By default, the name **My Temperature Task** is provided—we will accept this name this time around. You can change the name to something more descriptive depending on your application. Once you have named the NI-DAQmx task, select **Finish**.

*As you proceed through the configuration process, you can always go back to a previous page using the **Back** button.*

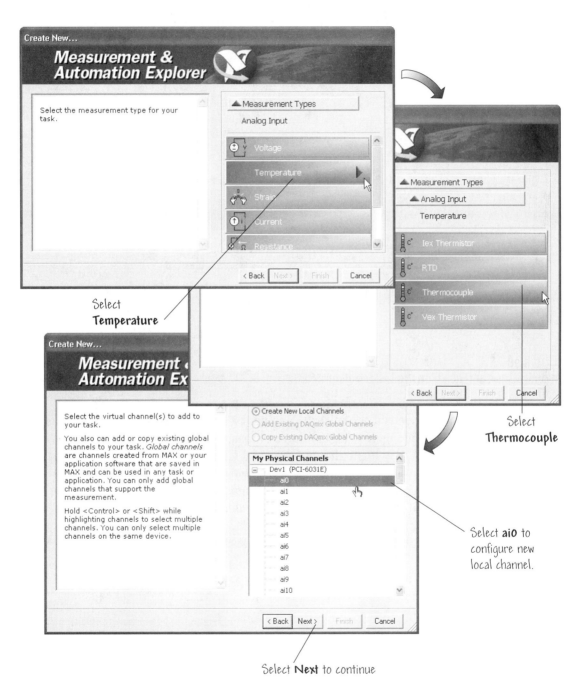

FIGURE 8.37
Selecting the measurement type and creating a new local channel.

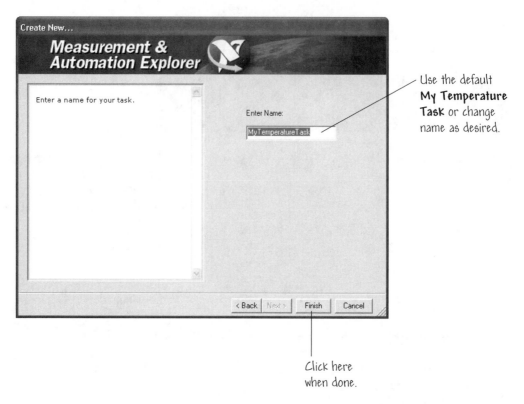

FIGURE 8.38
Naming the task and finishing the task configuration.

When you are finished defining the channel, MAX entries should appear as shown in Figure 8.39. In future sessions, you can edit the configurations through the MAX by highlighting the desired task and right-clicking on **Temperature**. MAX provides important configuration information. For example, as seen in Figure 8.39, the temperature minimum and maximum values are shown to be 0°C and 100°C, respectively. These values can be readily changed by editing the screen directly. Also, if you click on the double arrow ≫ under the **Channel List**, the screen will expand to show the device type and physical channel. You can right-click on **Temperature** to open a dialog box in which the channel assignment can be changed should the need arise in future sessions. When you are finished configuring the NI-DAQmx task, close the MAX window to exit.   ◆

8.7.3 Channels and Tasks

In Section 8.7.2, we created an NI-DAQmx task to acquire temperature readings from a thermocouple using a NI PCI-6031E DAQ board. In the sections

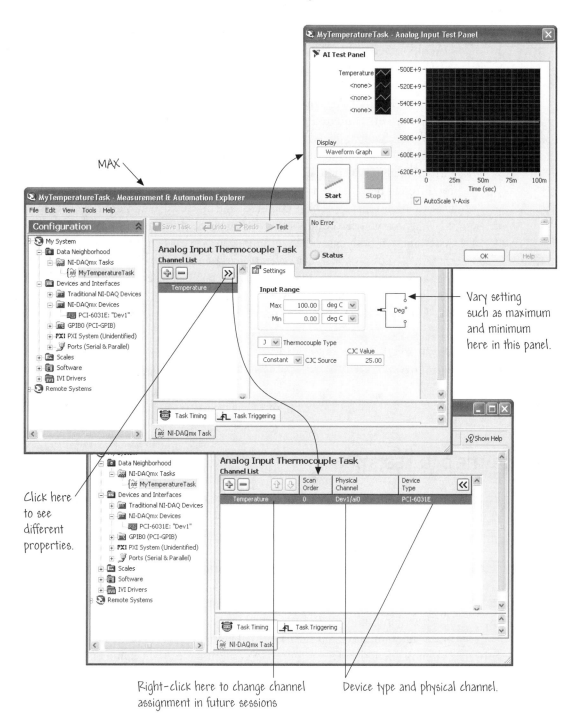

FIGURE 8.39
The MAX final configuration.

following this one, we will use the DAQ Assistant Express VI to create NI-DAQmx other tasks. But what exactly is an NI-DAQmx task? And what are NI-DAQmx channels?

If you are familiar with LabVIEW 6i (or earlier versions), you know that when developing DAQ applications, you configured a set of *virtual channels* using MAX. This configuration was a collection of property settings that included a physical channel, the type of measurement specified in the channel name, and scaling information. In the traditional NI-DAQ (prior to LabVIEW 7 Express), virtual channels were a simple method to remember which channels are used for different measurements. NI-DAQmx *channels* are similar to the virtual channels of traditional NI-DAQ channels and must be configured accordingly. In NI-DAQmx, you can configure virtual channels either in MAX or using the DAQ Assistant—in both cases the configuration process is almost identical.

An NI-DAQmx *task* is a collection of one or more channels, timing, triggering, and other properties that apply to the task itself. For example, a task might represent a measurement you want to perform. In the previous section, we created a task to measure temperature from one channel on a DAQ device.

To complete the exercises using the DAQ Assistant in the next section, you must have installed NI-DAQmx and an NI-DAQmx-supported device.

8.8 USING THE DAQ ASSISTANT

In this section, we introduce the DAQ Assistant Express VI. When you place the DAQ Assistant Express VI on the block diagram, the DAQ Assistant automatically appears. The DAQ Assistant is a graphical interface that we can use to configure measurement tasks and channels (described in Section 8.7).

The DAQ Assistant is located on the **Functions≫Input** and **Functions≫All Functions≫NI Measurements≫DAQmx-Data Acquisition** palettes. To launch the DAQ Assistant, place the DAQ Assistant Express VI on the block diagram, as illustrated in Figure 8.40. When the DAQ Assistant Express VI is placed on the block diagram, the DAQ Assistant will automatically appear. Once the DAQ Assistant is open, the steps required to configure the NI-DAQmx task are basically the same as described in Section 8.7 using MAX.

The general process of constructing a data acquisition VI using the DAQ Assistant Express VI is as follows:

- Open a new VI.
- Place the DAQ Assistant Express VI on the block diagram.

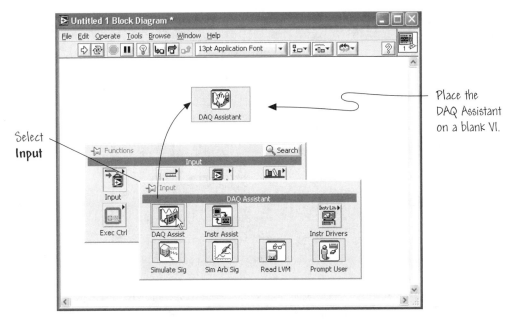

FIGURE 8.40
Locating and placing the DAQ Assistant Express VI on the block diagram.

- The DAQ Assistant appears to configure the measurement task.

- Configure, name, and test the NI-DAQmx task.

- Click the **OK** button to return to the block diagram.

- Edit the front panel and block diagram to complete the VI.

- If desired, generate an NI-DAQmx Task Name control so that the task can be used in other applications.

Practice with the DAQ Assistant

In this example, you will construct an NI-DAQmx task to acquire voltage readings from a measurement device attached to a PCI-6031E DAQ board.

The first step is to open a new VI and place a DAQ Assistant Express VI on the block diagram. The DAQ Assistant will automatically launch to begin the process of creating and configuring the NI-DAQmx task. In the first dialog box, click the **Analog Input** button to display the **Analog Input** options, as illustrated in Figure 8.41. Select **Voltage** to create a new voltage analog input task. The dialog box then displays a list of channels on each DAQ device installed. The number of channels listed depends on the number of channels you have on the DAQ device. In our case, we find a list **ai0**, **ai1**, ... Select **ai0** and click the **Finish** button.

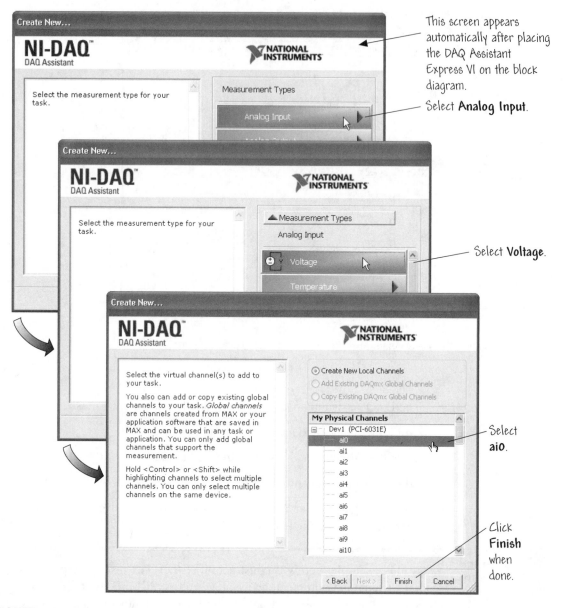

This screen appears automatically after placing the DAQ Assistant Express VI on the block diagram.

Select **Analog Input**.

Select **Voltage**.

Select **ai0**.

Click **Finish** when done.

FIGURE 8.41
Using the DAQ Assistant to configure an analog input channel.

At this point, the DAQ Assistant opens a new window, shown in Figure 8.42, which displays the options for configuring the selected channels. In the **Input Range** section of the **Settings** tab, enter 10 for the **Max** value and enter −10 for the **Min** value. In the **Task Timing** tab, select the **Acquire N Samples** option. Enter a value of 1000 in the **Samples To Read** input box.

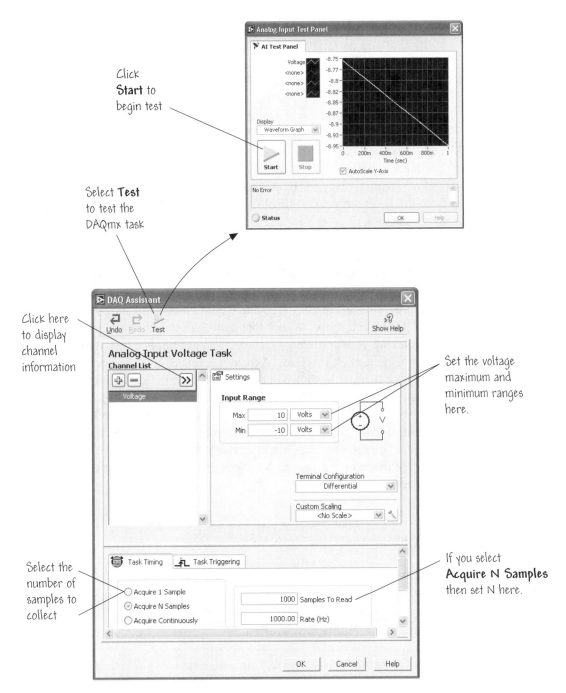

FIGURE 8.42
Configuring the channel settings and testing the DAQmx task.

At this point, the NI-DAQmx task is ready to test. You can test the task to verify that you correctly configured the channel. Click the **Test** button, shown in Figure 8.42, to access the **Analog Input Test Panel** dialog box that will automatically open. Testing the channel is easy—click the **Start** button once or twice to confirm that you are acquiring data. Once you have verified the proper functioning and configuration, click the **OK** button to return to the DAQ Assistant. Once there, click the **OK** button to return to the block diagram. The DAQ Assistant Express VI appears on the block diagram ready for integration into the VI application.

We are now ready to add a graph to the VI to plot the voltage data acquired from the DAQ device. On the block diagram, right-click the **data** output on the DAQ Assistant Express VI and select Create≫Graph Indicator, as illustrated in Figure 8.43. The waveform graph will be added and automatically wired. Switch the view to the front panel. Notice that the waveform graph legend displays the channel name **Voltage**. The DAQmx task is now ready to acquire voltage data and to display the output on the waveform graph.

It is possible to rename the channels configured using the DAQ Assistant. You might, for instance, decide later that a more descriptive name would be helpful in organizing your configured channels. To accomplish this, return to the block diagram, right-click the DAQ Assistant Express VI, and select **Properties** to rename the channel, as shown in Figure 8.44. Right-click **Voltage** in the **Channel List** listbox and select **Rename** to display the **Rename a channel or channels** dialog box. In the **New Name** text box, enter **Extreme Voltage**, and click the **OK** button. You are done—the channel has been renamed. To verify the name change, click the **OK** button and return to the block diagram. Then, switch to the block diagram and run the VI. Notice that **Extreme Voltage** appears in the waveform graph plot legend.

You also can select the name of the channel and press the <F2> key to access the **Rename a channel or channels** *dialog box.*

Save the VI as Read Voltage.vi in the Users Stuff folder in the Learning directory. ◆

8.8.1   DAQmx Task Name Constant

If you configure an NI-DAQmx task using the DAQ Assistant, then the task is a *local task*; hence, it cannot be saved to MAX for use in other applications. If you want to make the task available to other applications, you can use the DAQ Assistant to generate an NI-DAQmx Task Name control so that the task can be saved to MAX and used in other applications. This is readily accomplished by right clicking on the DAQ Assistant Express VI in the block diagram and

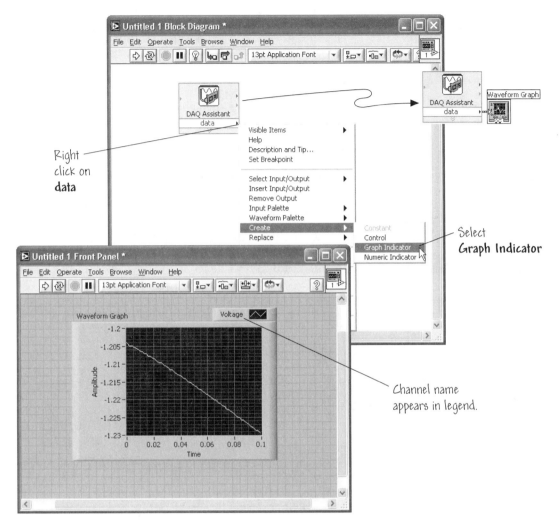

FIGURE 8.43
Adding a waveform graph to the DAQ Assistant Express VI.

selecting **Convert to Task Name Control** in the pull-down menu, as illustrated in Figure 8.45. The DAQmx Task Name Constant is a LabVIEW data type used by the DAQ VIs for communicating to the DAQ boards. After converting the DAQ Assistant Express VI to a DAQmx Task Name Constant, the DAQ Assistant will appear so that you can re-configure the task, if needed. After exiting the DAQ Assistant, the DAQmx task is available in MAX and hence ready for use in other applications. You will probably want to rename the task.

If you want to use one of the DAQmx tasks in MAX in an application, you first place a DAQmx Task Name Constant on the block diagram. You can

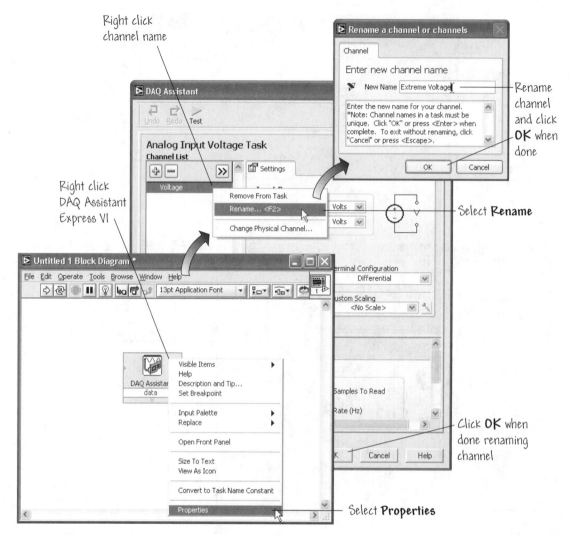

FIGURE 8.44
Renaming a channel.

find the DAQmx Task Name Constant on the **Functions≫All Functions≫NI Measurements≫DAQmx-Data Acquisition** palette, as shown in Figure 8.46. The task names can be entered in two different ways. You can use the **Operating** Constant and choose the desired task name found in the MAX utility, as shown in Figure 8.46. Another way to enter the DAQ task name is to use the **Labeling** tool to enter the task channel number. The task name must refer to one of the tasks configured using the MAX, as discussed in the previous sections.

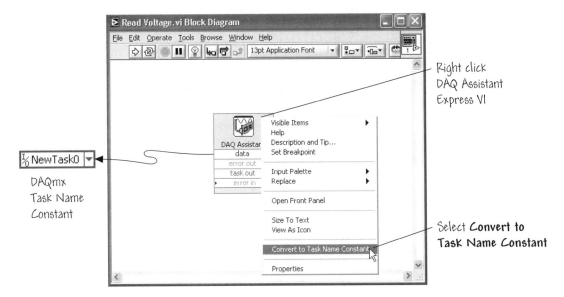

FIGURE 8.45
Create a Task Name Constant to make a DAQmx task available to other applications through MAX.

8.9   ANALOG INPUT

On the first screen of the DAQ Assistant (see Figure 8.41) you choose the measurement type. The available types are analog input, analog output, digital I/O. counter input, and counter output. In this section we focus on analog input. In subsequent sections we will cover analog output and digital I/O.

The analog input is used to perform analog-to-digital (A/D) conversions. In the DAQ Assistant, clicking on **Analog Input** opens the screen that lists the available analog input measurement types: voltage, temperature, strain, current, resistance, frequency, and custom voltage with excitation. In the *Practice with the DAQ Assistant* example in Section 8.8, the voltage measurement type was selected. Each measurement type has its own characteristics, such as resistor values for current measurements or strain gauge parameters for strain measurements.

Once you have selected the virtual channels to add to your task, the DAQ Assistant settings and testing screen opens. The DAQ Assistant is shown in Figure 8.42. In the lower section of the DAQ Assistant are two tabs: Task Timing and Task Triggering. An important step in configuring the DAQmx task is to configure the timing and triggering.

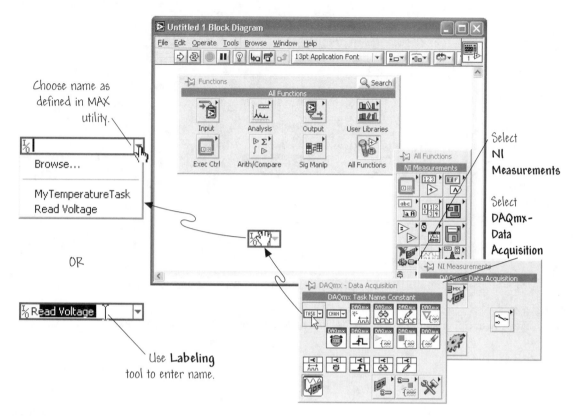

FIGURE 8.46
Entering DAQmx task names.

8.9.1 Task Timing

When performing analog input, the task can be timed to acquire a single sample, *n* samples, or to acquire data continuously. Acquiring a single sample is an *on-demand* operation. When **Acquire 1 Sample** is selected (see Figure 8.42), NI-DAQmx acquires one value from an input channel and immediately returns the value. This operation does not require any buffering or hardware timing. For example, if you periodically monitor the fluid level in a tank, you would acquire single data points. You can connect the transducer that produces a voltage representing the fluid level to a single channel on the measurement device and initiate a single-channel, single-point acquisition when you want to know the fluid level.

It is possible to acquire multiple samples for one or more channels by acquiring single samples in a repetitive manner. This process is, however, inefficient and time consuming. Moreover, you do not have accurate control over the time between each sample. The preferable way to acquire multiple samples is to use hardware timing. This method employs a buffer in computer memory to acquire data more efficiently. For these types of applications, set the sample mode to **Acquire *n* Samples**. With NI-DAQmx, you can acquire multiple samples for a single channel or multiple channels. For example, you can monitor both the fluid level and the temperature in a tank using two transducers connected to two channels on the device.

If you want to view, process, or log a subset of the samples as they are acquired, you need to continually acquire samples. For these types of applications, set the sample mode to **Acquire Continuously**.

8.9.2   Task Triggering

When a device controlled by NI-DAQmx does something, it performs an action, such as producing a sample or starting a waveform acquisition. Every NI-DAQmx action needs a stimulus. When the stimulus occurs, the action is performed. The stimuli to begin the actions are called **triggers**. The start trigger starts the acquisition. The DAQ Assistant provides three start triggers: analog edge, analog window, and digital edge. The reference trigger establishes the reference point in a set of input samples. The DAQ Assistant provides the three reference triggers: analog edge, analog window, and digital edge. Data acquired up to the reference point is known as **pretrigger data**. Data acquired after the reference point is known as **posttrigger data**. These are all examples of hardware triggers. If you do not configure a hardware trigger, the DAQ task will start when run. This is referred to as a software trigger.

Practice with Analog Input

In this example we will construct a VI that acquires an analog signal and displays the output on a meter. For the DAQ device, we are using the DAQ Signal Accessory from National Instruments to produce the temperature measurements. The sensor outputs a voltage proportional to the temperature and is hard-wired to channel 0 of the DAQ device.

If you do have an NI DAQ Accessory, you can still follow this example up to the point of actually acquiring temperature data from the device.

To begin the development, open a blank VI and place a Meter on the front panel, The Meter is located on the **Controls≫Numeric Indicators** palette. Adjust the scales on the meter to read in the range 0.0 to 0.4. Then place a Vertical

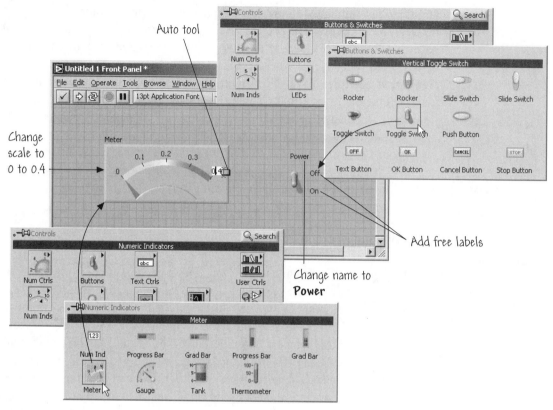

FIGURE 8.47
Configuring the front panel of the temperature acquisition VI.

Toggle Switch on the front panel. The Vertical Toggle Switch is located on the **Controls≫Buttons & Switches** palette. Configure the toggle switch to a default value of FALSE and a mechanical action of **Latch When Pressed**. Create two free labels titled **Off** and **On**. The process of building the front panel is illustrated in Figure 8.47. The needed controls and indicators should be on the front panel before continuing to the block diagram. When the front panel is ready, switch to the block diagram to continue.

On the block diagram, place a While Loop and enlarge the loop enough to be able to add several programming elements within. Then place a DAQ Assistant Express VI (see Section 8.8 for instructions on using the DAQ Assistant). When the DAQ Assistant appears on the screen, proceed to configure a DAQmx task to read an analog input channel and return the voltage. The steps to accomplish this are illustrated in Figure 8.48. In brief, the steps are as follows:

1. Select Analog Input≫Voltage as the measurement type.

2. Select **Dev1≫ai0** as the physical channel.

3. Click the **Finish** button.

4. When the **Analog Input Voltage Task** dialog box appears, configure the **Task Timing** to **Acquire 1 Sample**.

5. Click the **OK** button to close the **Analog Input Voltage Task Configuration** dialog box.

The settings specified for the task are now saved in the DAQ Assistant Express VI. At this point, we need to add several programming elements to the block diagram to complete the VI. First, we want the loop to execute every 100 ms. To accomplish this goal, place a Time Delay function on the block diagram. This function is located on the **Functions≫Execution Control** palette. When the Time Delay function is placed on the block diagram, a dialog box will appear in which the time delay value is set. The default value is a 1 s delay—change the delay to 0.1 s.

It is recommended that you include the capability to stop the program execution in the case where the DAQ Assistant Express VI produces an error. Fortunately, the Express VI outputs include a status indicator that is accessible by the Unbundle by Name function, located on the **Functions≫All Functions≫Cluster** palette. This function is used to access the **status** from the error cluster when wired to the DAQ Assistant Express VI as shown in Figure 8.49.

You will also want to be able to halt the program execution using the toggle switch on the front panel. Actually, you will need to be able to halt program execution if an error occurs *or* if the user clicks the power switch (i.e., the toggle switch). To accomplish this task, place the Or function on the block diagram inside the While Loop. The Or function is located on the **Functions≫Arithmetic & Comparison≫Express Boolean** palette. Wire the block diagram using Figure 8.49 as a guide.

*If you located the While Loop on the **Functions≫Execution Control** palette, the **loop condition** will be automatically wired to a **Stop** button. Delete the **Stop** button and wire the toggle switch instead.*

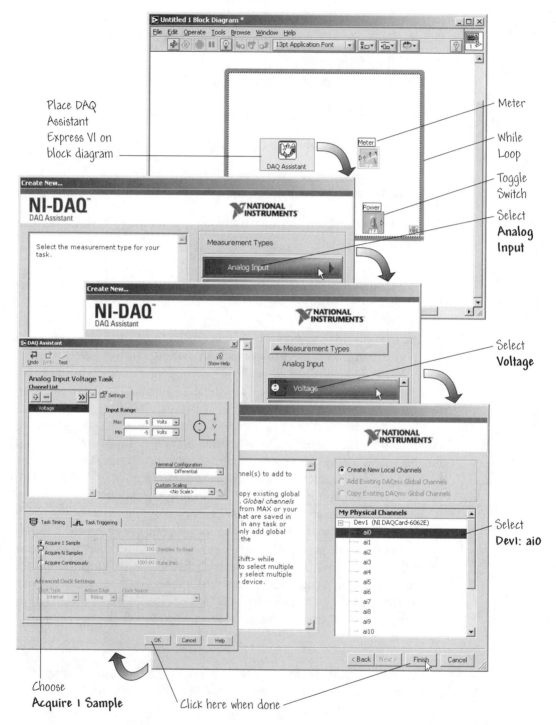

FIGURE 8.48
Configuring the DAQmx task using the DAQ Assistant.

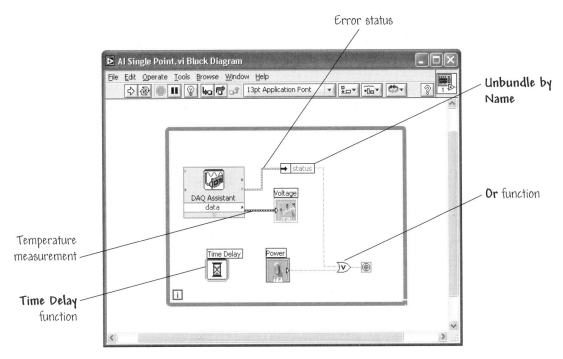

FIGURE 8.49
Acquiring analog input one point at a time.

When you have finished wiring the clock diagram and the **Run** button indicates that the program is ready for execution, save the VI as AI Single Point.vi in the Users Stuff folder in the Learning directory.

Return to the front panel and run the VI. The meter will display the voltage and the temperature sensor outputs. Stop the execution of the VI by clicking the power switch. ◆

8.10 ANALOG OUTPUT

Analog output is used to perform digital-to-analog (D/A) conversions. The first screen of the DAQ Assistant is where you choose the analog output measurement type. Refer to Section 8.8 for an introduction to using the DAQ Assistant. You can also configure a DAQmx analog output task using MAX as described in Section 8.7. We will focus on the use of the DAQ Assistant in this section, but actually the process of configuring the DAQmx task is almost the same using the DAQ Assistant as it is with MAX.

Configuring a DAQmx task to perform analog output consists of the same basic steps as for the analog input (see Section 8.9). In the DAQ Assistant, clicking on **Analog Output** opens the screen that shows the available analog output measurement types: voltage and current.

A compatible device must be installed that can generate a voltage or current task.

Once you have selected the virtual channels to add to your task, the DAQ Assistant settings and testing screen opens, as illustrated in Figure 8.50. In the lower section of the DAQ Assistant are two tabs: Task Timing and Task Triggering. An important step in configuring the DAQmx task is to configure the timing and triggering.

8.10.1 Task Timing

The analog output task can be timed to generate a single sample, *n* samples, or continuous samples. How do you decide when to use single-sample timing? If the signal level is more important than the sample generation rate, then choose **Generate 1 Sample**. You should generate one sample at a time if you need to generate a constant signal in a situation where, for example, you need to generate a known voltage to stimulate a device. You can then use software timing to control the time at which the device generates the signal. Generating a single sample does not require any buffering or hardware timing.

You should use **Generate *n* Samples** if you want to generate a *finite* time-varying signal, such as an AC sine wave. As discussed in Section 8.9 for analog input, one way to generate multiple samples (representing, for example, a time-varying signal) for one or more channels is to generate single samples in a repetitive manner. This is an inefficient and time-consuming way to generate *n* samples. Also, this brute force approach does not provide accurate control over the time between each sample. The best way to proceed is to use hardware timing that uses a buffer in computer memory to generate samples more efficiently. You can use software timing or hardware timing to control when a signal is generated. With software timing, the rate at which the samples are generated is determined by the software and operating system instead of by the measurement device. With hardware timing, a TTL signal, such as a clock on the device, controls the rate of generation. A hardware clock can run much faster and is more accurate than a software loop. As with other functions, you can generate multiple samples for a single channel or multiple channels.

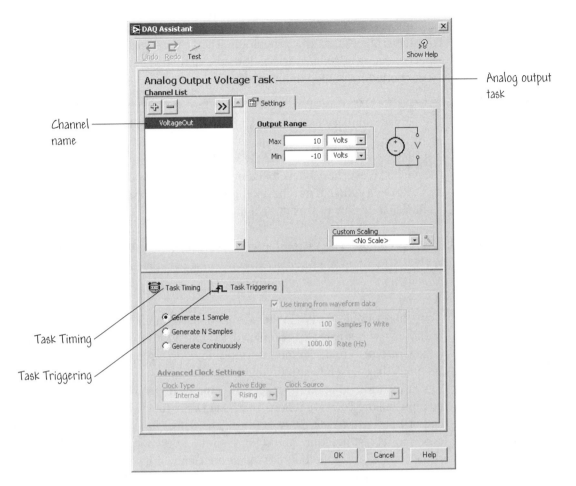

FIGURE 8.50
DAQ Assistant for the analog output DAQmx task.

Some devices do not support hardware timing. Consult the device documentation if you are unsure if the device supports hardware timing.

Continuous sample generation is similar to generating n samples, except that an event must occur to stop the sample generation. If you want to continuously generate signals, such as generating a non-finite signal, set the timing mode to **Generate Continuously**.

8.10.2   Task Triggering

Two very common output NI-DAQmx actions are producing a sample and starting a sample generation. Every NI-DAQmx action requires a stimulus—when the stimulus occurs, the action is performed. Actions are initiated by triggers. The start trigger starts the generation. The DAQ Assistant provides three start triggers: analog edge, analog window, and digital edge. The reference trigger is not supported for analog output tasks.

Practice with Analog Output

In this example, we develop a VI that outputs an analog voltage using a DAQ device. The VI will output the voltage from 0 to 9.5 V in 0.5 V increments. For the DAQ device, we are using the DAQ Signal Accessory from National Instruments. The goal is to use the VI developed in this example to output an analog signal to the DAQ device, and to use the analog input VI developed in Section 8.9 to read the data and display the result on a meter.

*If you have a DAQ Signal Accessory, connect **Analog Out CH0** to **Analog In CH1**. If you do not have a DAQ Signal Accessory, you can still follow this example up to the point of actually generating analog output to the DAQ device.*

In this example, we begin the process of designing a VI to output analog data by starting with a partially completed VI. Open the Voltage Output VI located in the **Chapter 8** folder in the **Learning** directory. You will find the front panel and block diagram shown in Figure 8.51.

On the block diagram there are two interesting functions: the Time Delay function and the Select function. The Time Delay function is configured to cause the For Loop to execute every 500 ms. The Select function checks whether the loop is in its last iteration, and if so, then the DAQ device outputs 0 volts. It is a good idea to reset the output voltage to a known level to prevent damage to connected devices.

On the block diagram, place the DAQ Assistant Express VI, as illustrated in Figure 8.52. When the DAQ Assistant appears, configure the DAQmx task to generate an analog output voltage. The following steps (shown in Figure 8.53) will produce the desired task:

- Select Analog Output≫Voltage on the first DAQ Assistant screen.

- Select **Dev1≫ao0** as the physical channel and click the **Finish** button.

- In the **Analog Output Voltage Task Configuration** dialog box that appears, configure the **Task Timing** to **Generate 1 Sample** and change the output range minimum to 0 and maximum to 10.

- Click the **OK** button to close the **Analog Output Voltage Task Configuration** dialog box. This saves the settings specified for the task in the DAQ Assistant Express VI.

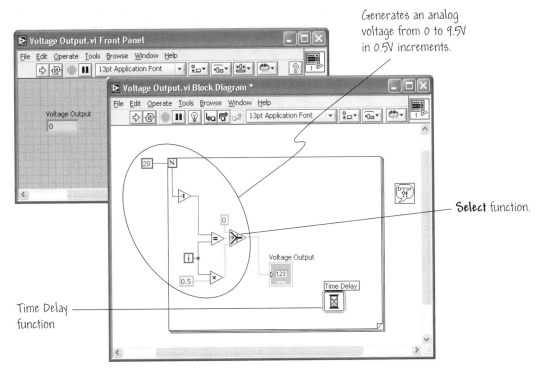

FIGURE 8.51
Generating a voltage output from 0 to 9.5 V in 0.5 V increments.

Once you are back on the block diagram, wire the DAQ Assistant Express VI as shown in Figure 8.54. Save the VI as Voltage Output Done.vi in the Users Stuff folder in the Learning directory.

Open the AI Single Point VI that you completed in Section 8.9. Configure the meter scale minimum to 0.0 and maximum to 10.0. On the block diagram for the AI Single Point VI, double-click the DAQ Assistant Express VI to open the **Analog Input Voltage Task Configuration** dialog box. Right-click **Voltage** in the **Channel List** section and select **Change Physical Channel** as illustrated in Figure 8.55. Select **ai1** for the channel because you wired the DAQ signal accessory to output a voltage on **Analog Out CH0** and acquire the voltage from **Analog In CH1**. Change the voltage range to 0 to 10 and then click the **OK** button to close the dialog box.

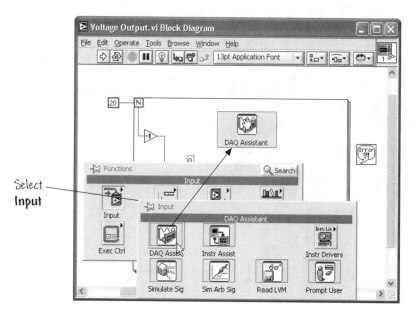

Select
Input

FIGURE 8.52
Placing the DAQ Assistant Express VI on the block diagram.

Run AI Single Point VI. Recall that this VI will run until there is an error or the toggle switch is used to stop the execution of the program. Once AI Single Point VI is executing, run the Voltage Output Done VI. Notice that the Voltage Output Done VI outputs the voltage in 0.5 V increments from 0 to 9.5 V as displayed in the digital indicator in Figure 8.56. At the same time, the meter in AI Single Point VI displays the voltage readings over the same range. When the For Loop executes its last iteration, the VI outputs 0 V to reset the analog output channel, as desired.

What have we accomplished? The Voltage Output Done VI is generating an analog signal output to the DAQ Accessory. This is a digital-to-analog conversion. The AI Single Point VI is acquiring the signal on the DAQ Accessory and displaying the data on the meter. This is an analog-to-digital conversion.   ◆

8.11   DIGITAL I/O

Digital I/O components on DAQ devices consist of parts that generate or accept binary on/off signals that are often used to control processes, generate patterns for

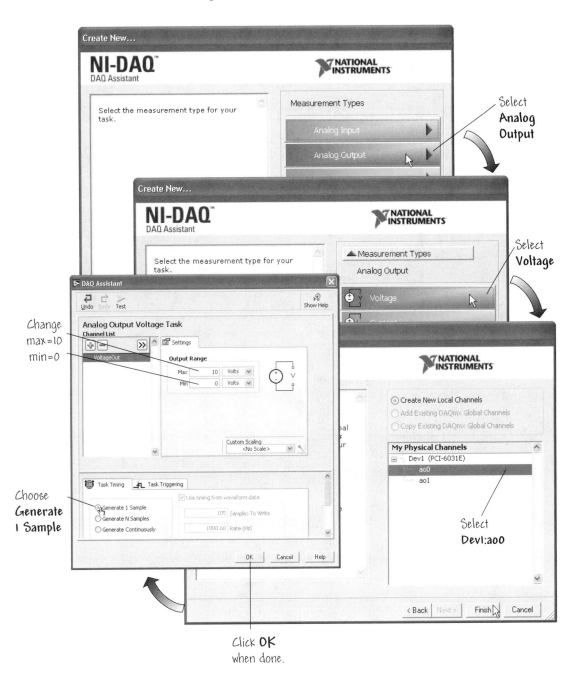

FIGURE 8.53
Configuring the analog output DAQmx task.

Add wires

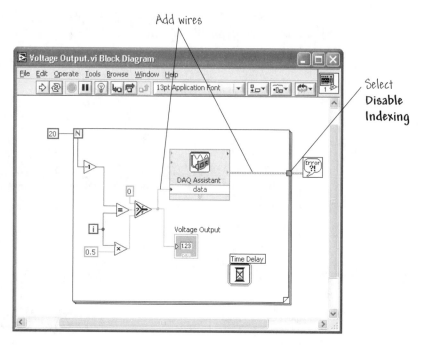

Select
Disable
Indexing

FIGURE 8.54
Wiring the DAQ Assistant Express VI.

testing, and communicate with peripheral equipment. Digital lines are grouped into ports generally consisting of four or eight lines per port. Usually all lines within the same port must all be either input lines or output lines. Since a port contains multiple digital lines, by writing to or reading from a port you can set or retrieve the states of multiple lines simultaneously.

You can use the digital lines in a DAQ device to acquire a digital value based on software timing. On some devices, you can configure the lines individually to either measure or generate digital samples. Each line corresponds to a channel in the task. You can use the digital port(s) in a DAQ device to acquire a digital value from a collection of digital lines, again based on software timing. You can configure the ports individually to either measure or generate digital samples. Each port corresponds to a channel in the task.

**Practice
with
Digital I/O**

Practice with Digital I/O(In this example, we plan to control the digital I/O lines on the DAQ device. As in previous examples in Section 8.9 and 8.10, we are using the DAQ Signal Accessory from National Instruments. Our final VI will turn on the LEDs of port 0 on the DAQ Signal Accessory based on the digital

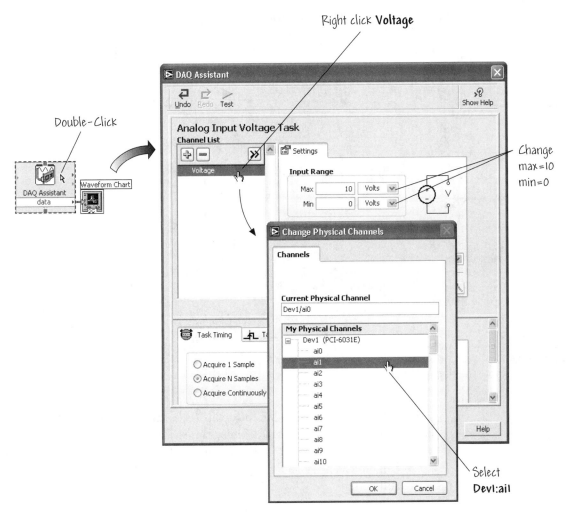

FIGURE 8.55
Reconfiguring the DAQmx task to read from channel ai1.

value set on the front panel. Each LED is wired to a digital line on the DAQ device. The lines are numbered 0, 1, 2, and 3, starting with the LED on the right.

 The LEDs use negative logic. That is, writing a 1 to the LED digital line turns off the LED. Writing a 0 to the LED digital line turns on the LED.

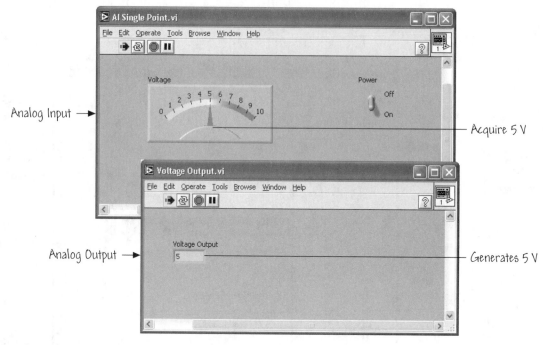

FIGURE 8.56
Generating and acquiring analog signals.

Open the Digital Example VI, located in the **Chapter 8** folder in the **Learning** directory. You will find the block diagram shown in Figure 8.57.

Place the DAQ Assistant Express VI on the block diagram within the While Loop. When the DAQ Assistant appears, configure the DAQmx task for Digital I/O. The following steps (shown in Figure 8.58) will produce the desired task:

- Select Digital I/O≫Port Output.

- Select Dev1≫port0 for the physical channel and click the **Finish** button.

- In the **Digital Output Port Task Configuration** dialog box that appears, select **Invert All Lines In Port** because the LEDs use negative logic.

- Click the **OK** button to close the configuration dialog box.

All of the settings specified for the task are saved internally in the DAQ Assistant Express VI. The Boolean buttons on the front panel are stored in an array to simplify the code. The Array Subset function extracts only the first four elements in the array. The output of the array subset needs to be reversed since element 0 of the array is the most significant bit. The array is then converted to a number with the Boolean Array to Number function, which passes the output to the DAQ Assistant Express VI to write that value to the port.

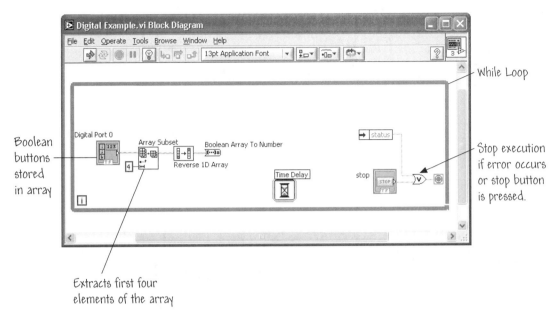

FIGURE 8.57
Digital I/O example block diagram.

Once you are back on the block diagram, wire the DAQ Assistant Express VI as shown in Figure 8.59. When the VI is ready, click the **Run** button on the front panel. Turn the Boolean LEDs on and off on the front panel and observe that the LEDs on the DAQ Signal Accessory turn on and off at the same time and in the same order. You are now performing digital I/O!

Save the VI as Digital Example Done.vi in the Users Stuff folder in the Learning directory. ◆

BUILDING BLOCK

8.12 BUILDING BLOCKS: USING DIGITAL OUTPUT TO MONITOR VOLUME LIMITS

The VI called Volume Chart.vi created in Chapter 7 will be used as the basis for a new VI that uses digital output to monitor the volume limits and provide visual indication that the volume limit has been exceeded.. The goal is to make the necessary modifications to the existing Volume Chart.vi so you can control the LEDs on the DAQ Signal Accessory. Use Figure 8.61 as a guide for your VI construction.

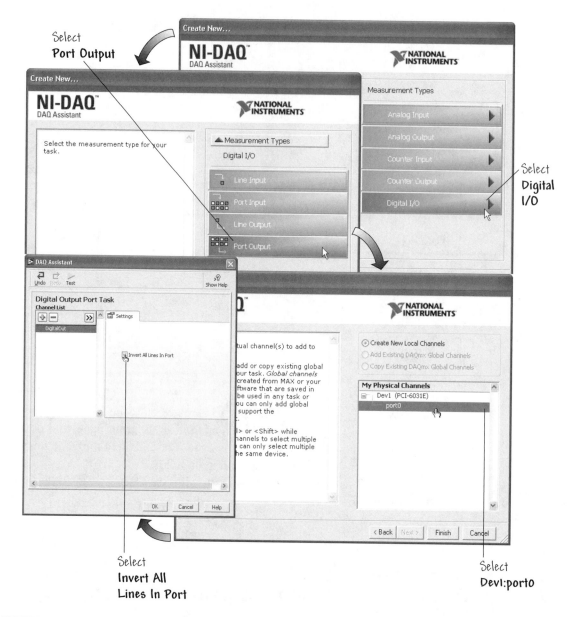

FIGURE 8.58
Configuring the DAQmx task for digital I/O.

In addition to lighting an LED on the LabVIEW Front Panel, we can also program our VI to light a real LED. The following steps describe how to control the four LEDs of the DAQ Signal Accessory, turning all four LEDs on when the volume exceeds your set limit.

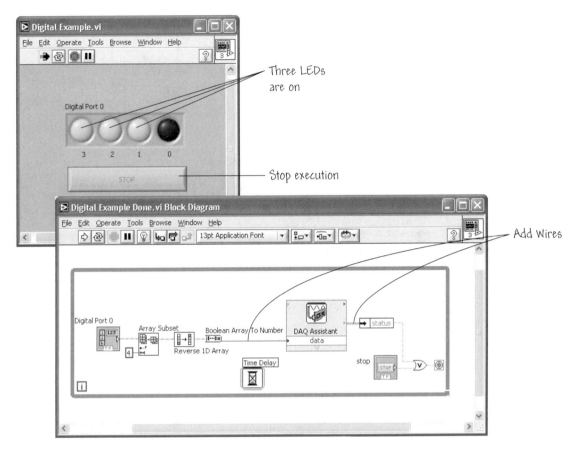

FIGURE 8.59
The Digital Example VI completed.

1. First you need to add the Digital I/O. To do this, place the DAQ Assistant inside the For Loop. Select Digital I/O≫Line Output from the DAQ Assistant dialog box.

2. After you select Line Output, the **DAQ Assistant** will ask you to select the channels you wish to control. In this case, we would like to control digital lines 0–3 of port 0. To select multiple lines, hold the <Shift> key while selecting lines 0–3 (port0/line0 – port0/line3).

3. After selecting the four lines click **Finish**. From the configuration page, select the box that says **Invert Line**. Because the DAQ Signal Accessory LEDs use reverse logic, this allows us to send a True to the **DAQ Assistant** and have the lights turn on. Click on **Test** to test this functionality and click **OK** when you are finished.

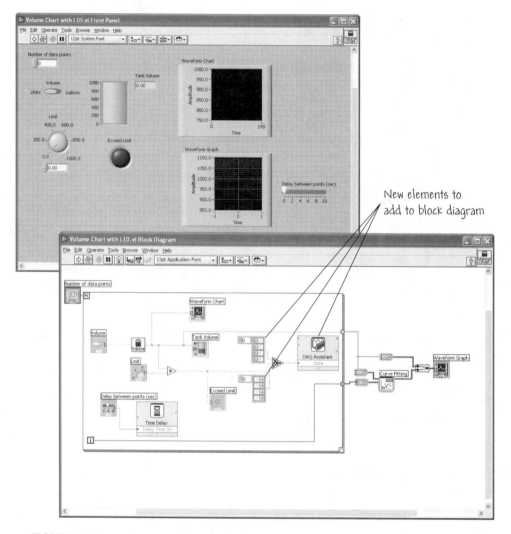

FIGURE 8.60
The front panel and block diagram of the Volume Chart with LED.vi.

4. Now we simply need to add the logic to tell the **DAQ Assistant** what to write out to the four Digital lines. With four lines, we must right an array of booleans where True turns the Boolean on and False will turn it off. Create two different Boolean array constants, each with four elements. Make all of the Booleans True for one array, and all False for the other.

5. To decide which input to send to the DAQ Assistant, place the Select function inside the For Loop and wire the Boolean output of the **Greater?** function to the **s** input of the Select function. When the volume is greater than the limit

FIGURE 8.61
Experimental catalyst reactor.

we want to turn the LEDs on, so wire the Boolean array of True constants to the **t** input of the Select function, and then wire the Boolean array of False constants to the **f** input of the Select function.

When you have finished constructing the VI, save it in the **Users Stuff** *folder as* **Volume Chart with LED.vi.**

Run the VI and observe the operation of the LEDs. By varying the volume limit using the Limit knob on the front panel, you can control how often the limit is exceeded. The LED on the front panel will light when the volume limit is exceeded. In addition, with the changes you have incorporated in the VI, the four LEDs on the DAQ Signal Accessory will also simultaneously light.

8.13   RELAXED READING: FUEL CELL CATALYST RESEARCH

A fuel cell is an electrochemical device that combines hydrogen and oxygen to produce electricity, heat, and water. Fuel cells are combustion-free, hence virtually pollution-free. However, hydrogen is not easily accessible for fuel cell applications, therefore catalyst-based fuel processing systems (FPS) are necessary to create hydrogen from gasoline and natural gas. An experimental catalyst reactor is shown in Figure 8.61.

In collaboration between HydrogenSource, LLC and Bloomy Controls, Inc., a research and engineering effort was undertaken to evaluate the performance of FPS catalysts and related components. The investigations employed experimental chemical reactors that require the control and monitoring of a large number of process parameters, such as temperature, flow, pressure, and gas composition, for periods ranging from a few minutes to several months. Previous manual monitoring methods proved extremely labor intensive, with susceptibility to human error and the inability to provide either the quantity or quality of data required.

Engineers at Bloomy Controls specified, designed, and implemented a series of systems to facilitate the automated operation and data collection to support HydrogenSource's newly constructed FPS laboratory. The research facility required flexible instrumentation and controls because the tests performed in each reactor significantly varied from one experiment to another. The application was especially challenging because the requirements demanded that pairs of reactors share crucial components of expensive analytical instrumentation. The final design possessed the capability to be easily reconfigured for different experiments, the ability to collect large volumes of data while maintaining continuous process control, and the ability to accommodate a wide range of sensors, transducers, and controls.

The dual reactor hardware configuration is representative of a typical layout for controlling a pair of catalyst reactors. Each reactor has its own control PC, and inside each reactor is an E-Series DAQ card interfaced to an associated SCXI chassis. Through these SCXI capabilities, a combination of temperatures, pressures, flow rates, and process alarms are acquired. The relatively large number of parameters, each requiring different methods of signal conditioning, combined with the need to capture signals at fast acquisition speeds for transient measurements, made SCXI the ideal platform for these systems.

An RS-485 interface is employed to control a high-power PID temperature controller, and a series of intelligent mass flow controllers via serial communications commands. A third PC acts as a shared resource server, communicating through RS-485 with a 32-loop PID temperature controller, motion control module, NDIR spectrometer, and motorized sampling valve. A combination of LabVIEW Datalogging and Supervisory Control Module network tags and VI server links are utilized, communicating via the lab's conventional Ethernet connections, so each of the individual reactor PCs acts as a client by sending commands to and receiving data from the server. Each reactor can be controlled independently while simultaneously sharing common lab resources.

FPS catalyst experiments typically include a sequence of timed steps, with process parameters either held at constant values or linearly ramped to desired settings, while data reflecting the system's performance is collected. A sophisticated recipe-driven architecture was created to give experimenters the flexibility to define and run a wide range of test protocols. Using LabVIEW Datalogging and

Supervisory Control Module, two major software components were developed: the Recipe Editor and the Automation Engine.

With the Recipe Editor, the user can set the experimental test attributes, such as the duration of the step in seconds, set points to each controlled process parameter, definitions of high and low alarm values, and recipe sequence controls such as holding and looping. The user can also enter simple descriptive text for documentation.

The front panel of the Automation Engine VI's includes all current values of the parameters in the system, along with descriptive recipe information previously listed. With mode control buttons, users can easily operate the system-start execution of the selected recipe, pause in the current recipe step, immediately advance to next step, or abort execution of the current recipe. In addition, the Automation Engine contains a virtual manual controls panel, providing manual set point overrides on a channel-per channel basis for all controlled values. With this function, users have the option of modifying individual parameter values, independent of the current formula, to fine tune or interact manually with the process.

Several versions of this system have operated successfully for more than a year. Researchers now are conducting complex and lengthy experiments with a minimum of user intervention. It is estimated that the automation systems using National Instruments products have cumulatively saved in excess of 10,000 man hours of manual control, monitoring, and data collection.

For more information, contact:

Robert O. Hamburger, Principal Engineer
Bloomy Controls Inc.
bob.hamburger@bloomy.com

8.14   SUMMARY

DAQ systems consist of the following elements:

- Signals
- Transducers
- Signal-conditioning hardware
- DAQ board or module
- Application software

There are two types of signals:

- Analog—provides level, shape, or frequency content information
- Digital—provides state or rate information

With many transducers it is necessary to provide for signal conditioning. Some types of signal conditioning are

- Amplification
- Transducer excitation
- Linearization
- Isolation
- Filtering

There are two types of signal sources:

- Grounded sources—devices that plug into the building ground
- Floating sources—isolated from the building ground system

There are three types of measurement systems:

- Differential—use this type whenever possible!
- Referenced single-ended—use single-ended types if you require more channels
- Nonreferenced single-ended

Multifunction DAQ boards typically include:

- Analog-to-digital convertors (ADCs)
- Digital-to-analog convertors (DACs)
- Digital I/O ports
- Counter/timer circuits

When configuring the DAQ board, you should consider how the following parameters will affect the quality of the digitized signal:

- Resolution: Increasing resolution increases the precision of the ADC.
- Range: Decreasing range increases precision.
- Limit settings: Changing limit settings to reflect the signal range increases precision.

The Measurement & Automation Explorer is a utility that helps configure the channels on the DAQ board according to the sensors to which they are connected.

The LabVIEW DAQ VIs are organized into palettes corresponding to the type of operation involved—analog input, analog output, counter operations, or digital I/O. Under this palette, the DAQ VIs are organized into six palettes (we covered the first three topics in this chapter):

- **Analog Input**

- **Analog Output**

- **Digital I/O**

- **Counter**

- **Calibration and Configuration**

- **Signal Conditioning**

KEY TERMS

Analog-to-digital converter (ADC): An electronic device (often an integrated circuit) that converts an analog voltage to a digital number.

ADC resolution: The resolution of the ADC measured in bits. An ADC with 16 bits has a higher resolution (and thus a higher degree of accuracy) than a 12-bit ADC.

Bipolar: A signal range that includes both positive and negative values (e.g., −5 V to 5 V).

Channel: Pin or wire lead where analog or digital signals enter or leave a data acquisition device.

Channel name: A unique name given to a channel configuration in the Measurement & Automation Explorer or DAQ Assistant.

Code width: The smallest detectable change in an input voltage of a DAQ device.

Digital-to-analog convertor (DAC): An electronic device (often an integrated circuit) that converts a digital number to an analog voltage or current.

Data acquisition (DAQ): Process of acquiring data from plug-in devices.

Differential measurement system: A method of configuring your device to read signals in which you do not connect inputs to a fixed reference (such as the earth or a building ground).

Floating signal sources: Signal sources with voltage signals that are not connected to an absolute reference or system ground. *Also* called nonreferenced signal sources.

Grounded signal sources: Signal sources with voltage signals that are referenced to system ground, such as the earth or building ground. *Also* called referenced signal sources.

Handshaked digital I/O: A type of digital I/O where a device accepts or transfers data after a signal pulse has been received. *Also* called latched digital I/O.

Immediate digital I/O: A type of digital I/O where the digital line or port is updated immediately or returns the digital value of an input line. *Also* called nonlatched digital I/O.

Input/output (I/O): The transfer of data to or from a computer system involving communication channels and DAQ interfaces.

Limit settings: The settings that you specify as the maximum and minimum voltages on analog input signals.

Linearization: A type of signal conditioning in which the voltage levels from transducers are linearized, so that the voltages can be scaled to measure physical phenomena.

LSB: Least significant bit.

Measurement & Automation Explorer: Provides access to all National Instruments DAQ and GPIB devices.

Non-referenced single-ended (NRSE) measurement system: All measurements are made with respect to a common reference, but the voltage at this reference may vary with respect to the measurement system ground.

Range: The minimum and maximum analog signal levels that the analog-to-digital convertor can digitize.

Referenced single-ended (RSE) measurement system: All measurements are made with respect to a common reference or ground. *Also* called a grounded measurement system.

Sampling rate: The rate at which the DAQ board samples an incoming signal.

SCXI: Signal Conditioning eXtensions for Instrumentation. The National Instruments product line for conditional low-level signals within an external chassis near the sensors so that only high-level signals in a noisy environment are sent to the DAQ board.

Signal conditioning: The manipulation of signals to prepare them for digitizing.

Unipolar: A signal range that is always positive (e.g., 0 V to 10 V).

Update rate: The rate at which voltage values are generated per second.

E8.1 Assume you are sampling a transducer that varies between 80 mV and 120 mV. Your board has a voltage range of 0 to +10 V, ±10 V, and ±5 V. What limit setting would you select for maximum precision if you use a DAQ board with 12-bit resolution? Use the limit setting values in Table 8.3 for your calculations.

E8.2 Open Measurement & Automation Explorer. Locate your DAQ device under Devices and Interfaces ≫ NI-DAQmx Devices. Right-click on the device and select **Test Panels**. Run a test panel to observe the signals connected to your DAQ device. After running a test panel, go to Data Neighborhood and create a DAQmx Global Channel to measure one of your input signals. Test your DAQmx Global Channel.

E8.3 Open the NI Example Finder to Search and enter "benchtop." the Benchtop Function Generator link. To open the **Find Examples**, select Find Examples... in the **Help** pull-down menu. This VI is another good example of how virtual instruments emulate actual instruments. Run this VI.

P8.1 Open a blank VI and place a DAQmx Global Channel control on the front panel. Right-click the control and select **(New Channel) DAQ Assistant**. Configure the DAQ Assistant to acquire a finite number of samples using the internal clock. After configuring and testing your channel with the DAQ Assistant, use the Code Generation feature to generate code for your channel. To do this, right-click on the channel control and select Generate Code≫Configuration and Example. Run the VI.

P8.2 You are acquiring analog data that measures the volume of a tank. You only want to take continual measurements when the machinery is on. Turning on the machinery can act as a trigger, specifically a digital trigger, to start measuring data. Write a VI that acquires analog data only after a digital trigger has occurred.

P8.3 You are controlling a servo motor that requires a +5 V signal to operate. You want the operator of your program to set a Boolean toggle switch to TRUE in order to send a +5 V signal to the motor. When the operator sets the Boolean switch to FALSE, you must stop the motor. Write a VI that will allow the operator to control the servo motor.

Note: You can use either a digital output line or an analog output line.

P8.4   You are monitoring a security system that uses a light sensor to determine when a person has entered a restricted zone. When the sensor is activated, it returns a 0 V TTL signal. Write a VI to monitor the state of the light sensor and turn on a Boolean indicator when the sensor is activated.

Note: You can use either a digital input line or an analog input line.

P8.5   Complete the crossword puzzle.

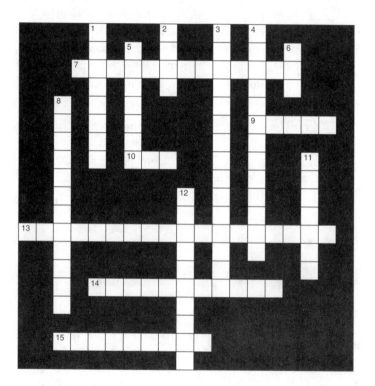

Across

7. The settings specified as max and min voltage on analog input signals.
9. The min and max analog signal levels that the a-to-d converter can digitize.
10. All measurements are made with respect to a common reference or ground.
13. The manipulation of signals to prepare them for digitizing.
14. A unique name given to a channel configuration in the MAX.
15. The smallest detectable change in an input voltage of a DAQ device.

Down

1. A signal range that is always positive.
2. All measurements are made with respect to a common reference, but the voltage may vary with respect to the ground.
3. Process of acquiring data from plug-in devices.
4. A type of signal conditioning in which the voltage levels from transducers are linearized.
5. A signal range that includes both positive and negative values.
6. Least significant bit.
8. The rate at which the DAQ board samples an incoming signal.
11. Pin or wire lead where signals enter or leave a DAQ device.
12. The rate at which voltage are generated.

CHAPTER 9

Strings and File I/O

This chapter introduces strings and file input/output (I/O). You have used strings in a limited fashion throughout the book, but here we discuss them more formally. In instrument control applications, numeric data is commonly passed as character strings, and LabVIEW has many built-in string functions that allow you to manipulate the string data. Writing and reading data from files also utilizes strings. We will discuss how to use high-level file I/O VIs to save data to and retrieve data from a disk file. High-level VIs perform the three basic functions associated with file I/O: opening or creating a file, writing or reading data from the file, and closing the file. We will also discuss the use of key Express VIs for string manipulation and file I/O.

GOALS

1. Practice creating string controls and indicators.

2. Be able to convert a number to a string, and vice versa.

3. Learn to use the file I/O VIs to write and read data to a disk file.

4. Understand how to write data to a file in a format compatible with many common spreadsheet applications.

9.1   STRINGS

A **string** is a sequence of characters—these can be displayable or nondisplayable characters. Strings are often used in **ASCII** text messages. In LabVIEW, strings are also used in instrument control when numeric data is passed as character strings and subsequently converted back to numbers. Another situation requiring the use of strings is in storing numeric data, where numbers are first converted to strings before writing them to a file on disk. In this chapter, we will discuss strings and their various uses in file input/output (I/O).

String controls and indicators are in the **String & Path** subpalette of the **Controls≫All Controls** palette, as illustrated in Figure 9.1. They can also be accessed on the **Text Controls** and **Text Indicators** Express palettes. As discussed in previous chapters (see Chapter 3 for example), you enter and change text inside a string control using the **Operating** tool or the **Labeling** tool. If there is not enough room to fit your text in the default size of the string control, you can enlarge the string controls and indicators by dragging a corner with the **Positioning** tool. If front panel space is limited, you can use a scrollbar to minimize the space that a front panel string control or indicator occupies, as illustrated in Figure 9.2. The Visible Items≫ Scrollbar option is located on the string short cut menu. If there is not enough room to place the scrollbar within the string control or indicator, the scrollbar option will be dimmed. This indicates that you must increase the vertical size if you want a scrollbar.

You can configure string controls and indicators for different types of display, such as passwords, \ codes, and hex. String controls and indicators can display and accept characters that are usually **nondisplayable**—backspaces, carriage returns, tabs, and so on. Choose '/' **Codes Display** from the string's short cut menu to display these characters. In the '/' **Codes Display** mode, non-displayable characters appear as a backslash followed by the appropriate code. A complete list of codes appears in Table 9.1.

To enter a nondisplayable character into a string control, click the appropriate key, like space or tab, or type the backslash character \, followed by the code for the character. As shown in Figure 9.3(a), after you type in the string and click the **Enter** button, any nondisplayable characters appear in backslash code format.

The characters in string controls and indicators are represented internally in ASCII format. You can view the ASCII codes in hex by choosing **Hex Display** from the string's short cut menu, as shown in Figure 9.3(b). You can also choose a password display by enabling the **Password Display** option from the string short cut menu, as shown in Figure 9.3(c). With this option selected, only asterisks appear in the string's front panel display, although on the block diagram the string data reflects the input string. This allows you to set up a security system requiring a password key before proper operation of the VI.

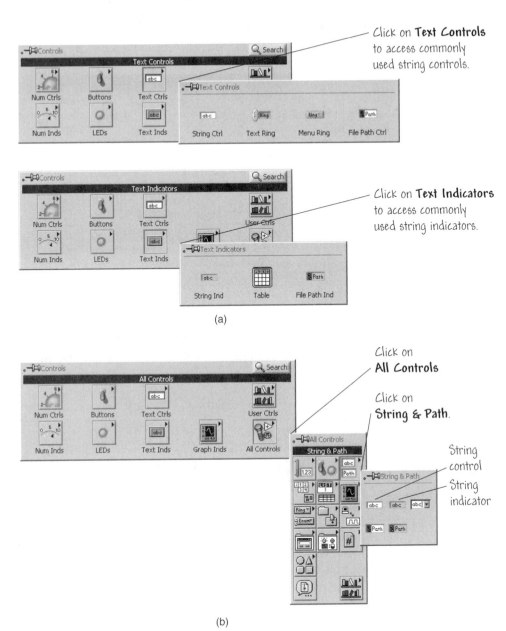

FIGURE 9.1
(a) Accessing string controls and indicators using the Express palettes. (b) Accessing string controls and indicators from the **String & Path** subpalette.

TABLE 9.1   A list of backslash codes

Code	G Interpretation
\00 – \FF	Hex value of an 8-bit character; must be uppercase.
\b	Backspace (ASCII BS, equivalent to \08).
\f	Form feed (ASCII FF, equivalent to \0C).
\n	Linefeed (ASCII LF, equivalent to \0A).
\r	Carriage return (ASCII CR, equivalent to \0D).
\t	Tab (ASCII HT, equivalent to \09).
\s	Space (equivalent to \20).
\\	Backslash (ASCII \, equivalent to \5C).

Practice with Manipulating Strings

In this exercise, you will practice with different ways to manipulate strings using three VIs:

- **Format Into String**: Concatenates and formats numbers and strings into a single output string.

- **Scan from String**: Scans a string and converts valid numeric characters (0 to 9, +, −, e, E, and period) to numbers.

- **Match Pattern**: Searches for an expression in a string, beginning at a specified offset, and if it finds a match, splits the string into three substrings—the substring before the matched substring, the matched substring itself, and the substring that follows the matched substring.

The VI that you will create is intended to interact with a digital multimeter (DMM). Open a new VI and construct a front panel using Figure 9.4 as a guide. Place two string controls, one numerical control, and one string indicator on the front panel. Switch to the block diagram. Select **Format Into String** in the **String** subpalette, as shown in Figure 9.4. In this exercise, this function converts the number you specify to a string and concatenates the inputs to form a command for the DMM. Using the **Positioning** tool, enlarge **Format Into String** so that two inputs appear in the lower left corner. Wire the numeric control **Number** to the first input argument (as shown in Figure 9.4). Wire the string control **Units** to the second input argument; the input type for Format Into String will automatically

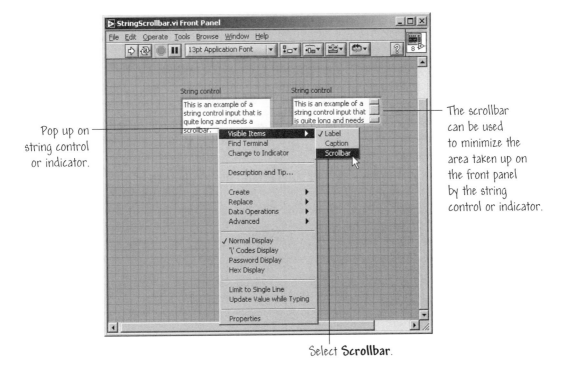

FIGURE 9.2
The scrollbar can be used to minimize the size of string controls and indicators.

change from **DBL** to **abc** upon wiring. Wire the string control **DMM Command** to the **initial string** input at the top left. Finally, wire the output **resulting string** to the string indicator **Command sent to DMM**.

You can create strings according to a format specified using format strings. With format strings you can specify the format of arguments—the field width, base (hex, octal, and so on), and any text that separates the arguments. The format string can be seen in Figure 9.4 (wired at the top of the Format Into String function). A dialog box can be used to obtain the desired format string. On the block diagram, pop up on the Format Into String function and select **Edit Format String**, as illustrated in Figure 9.5. You can also double click on the node to access the dialog box. Notice that the **Current Format Sequence** contains the argument types in the order that you wired them—**Format fractional number** and **Format string**. The **Format fractional number** corresponds to the input value from the **Number** control and **Format string** corresponds to the input value from the **Units** control.

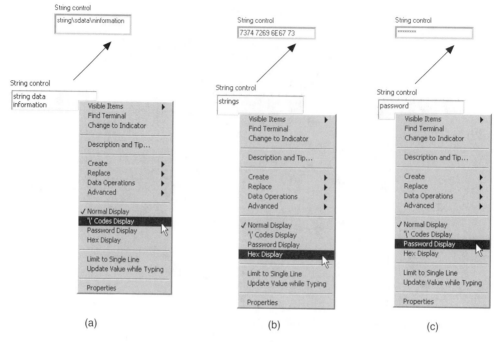

FIGURE 9.3
(a) Displaying characters that are usually nondisplayable. (b) Choosing **Hex Display** to view the ASCII codes in hex. (c) Choosing the **Password Display** option from the pop-up menu.

The only change we want to make is to set the precision of the numeric to 2. To do this, highlight **Format fractional number** in the **Current Format Sequence** list box, click in the **Use Specified Precision** checkbox, and type in the number 2 in the box. When you are finished, press <Enter> (Windows) or <return> (Macintosh) and you should see the **Corresponding Format String** (near the bottom of the dialog box). Press the **Create String** button to automatically insert the correct format string information and wire the format string to the function, as shown in Figure 9.4.

Return to the front panel and type text inside the two string controls and a number inside the digital control, and run the VI. You now have a VI that can concatenate strings and numbers to form a command that can be sent to an external instrument, such as the digital multimeter. Save the VI as String.vi in the Users Stuff directory.

To continue the exercise, we want to add the capability to scan a string and convert any valid numeric characters (0 to 9, +, −, e, E, and period) to numbers. Use Figure 9.6 as a guide and add a string control and numeric indicator to the front panel. Then switch to the block diagram and wire the Scan from String

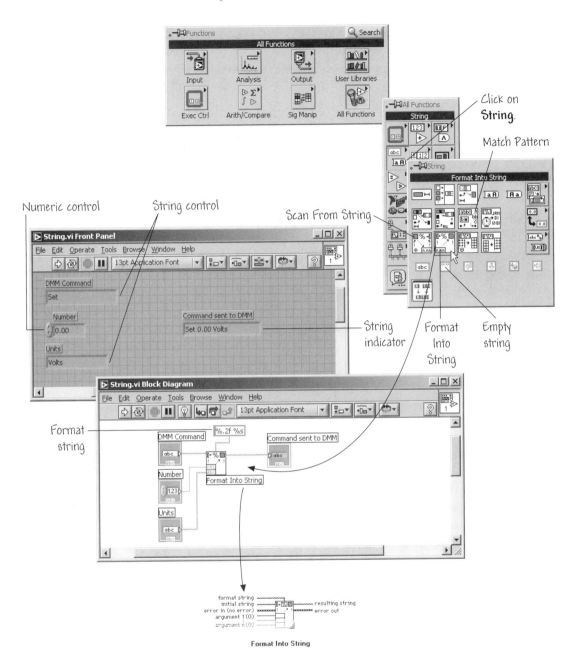

FIGURE 9.4
Constructing a VI using the Format Into String function to build a command to send to the DMM.

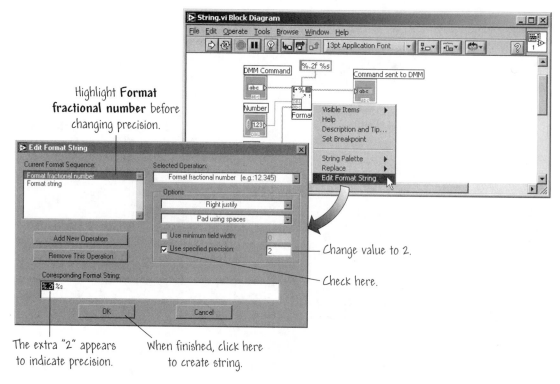

Highlight **Format fractional number** before changing precision.

Change value to 2.

Check here.

The extra "2" appears to indicate precision.

When finished, click here to create string.

FIGURE 9.5
Setting the desired format string using the dialog box.

function, as shown in Figure 9.6. The function itself is located on the **String** palette, as illustrated in Figure 9.4. After the wiring is complete, double click on the function (or pop up on the function and choose **Edit Format String**) to open the dialog box. The current scan sequence should indicate **Scan Number** and the corresponding scan string should show %f. This is fine, so just select **OK** and make sure that the **format string** input (at the top of the **Scan from String** function) is properly wired (see Figure 9.6). To test the VI, type in **5.56 V** in the Present DMM Setting string control and run the VI. You should find that the VI extracts the numeric value 5.56 and displays that value in the numeric indicator DMM Setting - Numeric. Try other voltage values in the input.

Finish with this VI, using Figure 9.7 as a guide. The last addition to the VI uses the Match Pattern function in conjunction with the Scan From String function to detect and extract a series of DMM data points. The Match Pattern function detects the matched substring (in this case, a comma) and outputs the substring before the comma to the **Scan From String** function which scans the

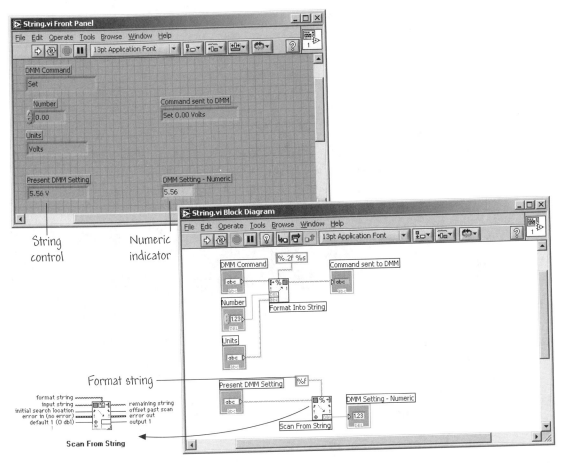

FIGURE 9.6
Constructing a VI using the Scan from String function to extract the present DMM setting from a string.

substring and outputs a detected number. The For Loop iterates until the end of the string is detected—in other words, an empty string is found. You can find the empty string constant on the **Strings** palette, as illustrated in Figure 9.4.

When your VI has been wired properly, enter a few numbers in the string control **DMM Data Points**, such as 3.37, 4.56, 6.89, 5.67. Run the VI and verify that the output array **Data Points - Numeric** contains the four numbers.

A working version of **String.vi** *can be found in* **Learning** \ **Chapter 9**. *Refer to the working version if your VI is not working as you think it should.* ◆

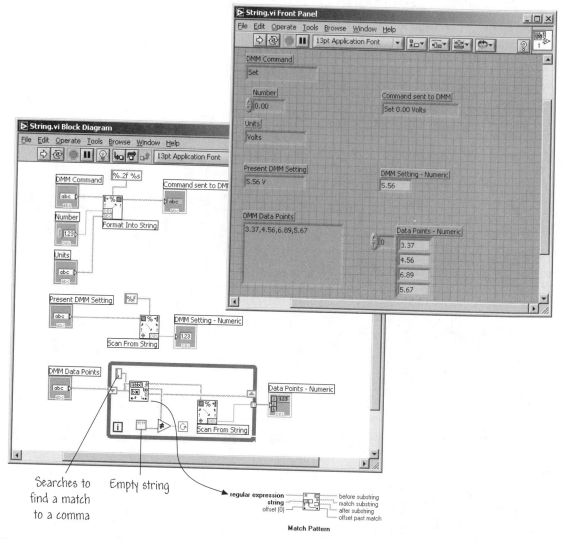

FIGURE 9.7
Constructing a VI using the Match Pattern function to detect and extract a series of DMM data points.

9.1.1   Converting Numeric Values to Strings with the Build Text Express VI

The Build Text Express VI can be used to convert numeric values into strings. The Build Text Express VI is located on the **Functions≫Output** palette, as sh in Figure 9.8. If the input is not a string, this Express VI converts the inp a string based on the configuration of the Express VI.

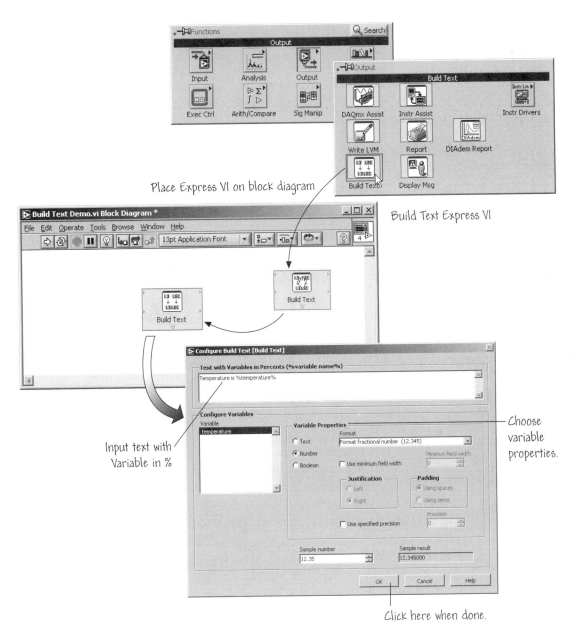

Place Express VI on block diagram

Build Text Express VI

Input text with Variable in %

Choose variable properties.

Click here when done.

FIGURE 9.8
Using the Build Text Express VI.

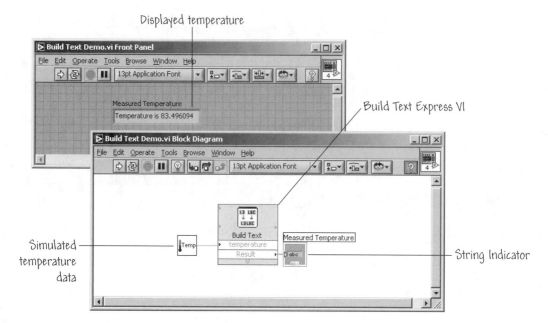

FIGURE 9.9
Converting temperature data to a string.

When you place the Build Text Express VI on the block diagram, the **Configure Build Text** dialog box appears. The dialog box shown in Figure 9.8 shows the Express VI configured to accept one input, **temperature**, and change it to a fractional. The input temperature concatenates on the end of the string Temperature is. A space has been added to the end of the Temperature is string.

A front panel and block diagram illustrating the use of the Build Text Express VI is shown in Figure 9.9. The Digital Thermometer.vi found in the LabVIEW7.0\activity\ directory generates the input temperature data. The output is directed to a string indicator. In the example shown, the displayed temperature is 83.49°.

The VI shown in Figure 9.9 is called Build Text Demo.vi *and is located in* Chapter 9 *of the* Learning *directory.*

9.2 FILE I/O

File input and output (I/O) operations are used to store and retrieve data from files on disk. These operations generally involve three basic steps:

- Open an existing file or create a new file.

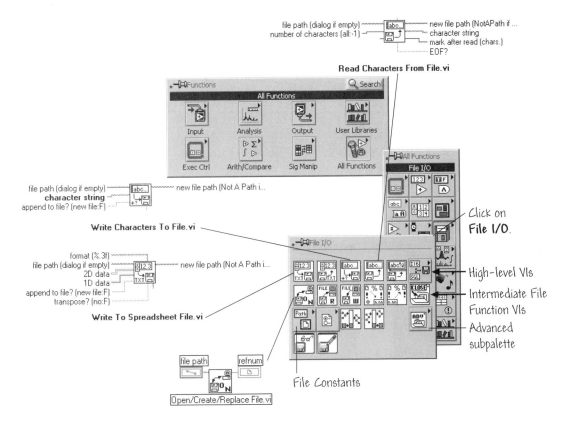

FIGURE 9.10
The **File I/O** palette.

- Write to or read from a file.

- Close a file.

LabVIEW provides file I/O functions that comprise a powerful and flexible set of tools for working with files. Figure 9.10 shows the **File I/O** palette.

You can use the file VIs to write or read the different types of data:

- Strings to text files

- One-dimensional (1D) or two-dimensional (2D) arrays of single-precision numbers to spreadsheet text files.

- 1D or 2D arrays of numbers to binary files.

The file I/O VI library of functions comprises high-level and intermediate File Function VIs. The high-level file VIs call the intermediate File Function VIs to perform complete, easy-to-use file operations. The high-level VIs open or create a file, write or read to it, and close it. If an error occurs, these VIs display a dialog box that describes the problem and gives you the option to halt execution or to continue. The high-level file VIs are located on the top row of the palette (see Figure 9.10) and consist of the following VIs:

- **Write Characters To File**: Writes a character string to a new byte stream file or appends the string to an existing file.

- **Write To Spreadsheet File**: Converts a 2D or 1D array of single-precision (SGL) numbers to a text string and writes the string to a new byte stream file or appends the string to an existing file.

- **Read Characters From File**: Reads a specified number of characters from a byte stream file, beginning at a specified character offset.

- **Read From Spreadsheet File**: Reads a specified number of lines or rows from a numeric text file, beginning at a specified character offset, and converts the data to a 2D, single-precision array of numbers.

- **Read Lines From File**: Reads a specified number of lines from a text file, beginning at a specified character offset.

- **Binary File** VIs: A subpalette that contains VIs that read and write 16-bit (word) integers or single-precision floating-point numbers to binary files.

The intermediate File Function VIs perform one file operation at a time. These VIs perform error detection in addition to their other functions. The most commonly used intermediate File Function VIs are located on the second row of the palette. The remaining functions are located in the **Advanced** subpalette. In addition to reading and writing data, the file I/O functions can move and rename files and directories, return volume information about a file or directory, and return the end-of-file location.

For example, to create a new file or replace an existing file, you can use the intermediate File Function VIs to Open/Create/Replace File, as shown in Figure 9.10. With file I/O, you need to define a path to the file. This can be accomplished by popping up on the left side of the function and selecting **Create Control** to create an input name **file path**. When the file is opened, a reference number is created, and you can read or write to the file. Remember to close an open file before your application completes execution. Many of LabVIEW's high-level I/O VIs automatically open or create new files, read or write data to those files, and close the files upon completion. In the next three subsections we will use three VIs that take care of these three basic operations. In the last subsection, we will discuss briefly an application that uses different VIs to perform the basic file I/O operations individually.

You can store or retrieve data from files in three different formats.

1. **Text (ASCII) Byte Stream**—You should store data in ASCII format when you want to access it from another software package, such as a word processing or spreadsheet program. To store data in this manner, you must convert all data to ASCII strings.

2. **Binary Byte Stream**—These files are the most compact and fastest method of storing data. You must convert the data to binary string format, and you must know exactly what data types you are using to save and retrieve the data to and from files.

3. **Datalog files**—These files are in binary format that only G can access. Datalog files are similar to database files because you can store several different data types into one (log) record of a file. You must use the intermediate File Function VIs when accessing these files.

When dealing with files, you will frequently see the terms **end-of-file**, **refnum**, **not-a-path**, and **not-a-refnum**. The end-of-file (EOF) is the character offset of the end of the file relative to the beginning of the file. Refnum is an identifier that G associates with a file when opened. Not-a-path and not-a-refnum are predefined values that indicates that a path is invalid and that a refnum associated with an open file is invalid, respectively.

9.2.1   Writing Data to a File

In this section we will discuss writing data to a file using the Write Characters To File VI. This VI writes a character string to a new byte stream file or appends the string to an existing file, depending on the VI input parameters. The Write Characters To File VI opens or creates a file, writes the data, and then closes the file.

Writing Data to a File

In this exercise, the objective is to create a VI to append temperature data to a file in ASCII format. This VI uses a For Loop to generate temperature values and store them in a file. During each iteration, the VI converts the temperature data to a string, adds a comma as a delimiting character, and then appends the string to a file.

Open a new front panel and place the objects as shown in Figure 9.11. The front panel contains a digital control and a waveform chart. The chart displays the temperature data. Pop up on the waveform chart and select **Visible Items≫Digital Display** and de-select **Visible Items≫Plot Legend** and **Visible Items≫Graph Palette**. Also make sure that either the waveform chart has y-axis autoscale enabled or that the maximum value of the y-axis is set to at least 90 degrees. The **Number of Points** control specifies how many temperature values to acquire and write to file. Pop up on the **Number of Points** and choose **Representation≫I32**.

Switch to the block diagram and wire the code as shown in Figure 9.11. Add a For Loop and make it large enough to encompass the various components. Place the **Digital Thermometer.vi** on the block diagram—remember that this VI is located in the **Activity** directory, which you can select through **Functions ≫ Select a VI...**, and returns a simulated temperature measurement from a temperature sensor.

Add a shift register to the loop by popping up on the loop border. This shift register contains the path name to the file. It is initially set to Empty Path, which can be found in **Functions≫File I/O≫File Constants**, as shown in Figure 9.11. The Empty Path function initializes the shift register so that the first time you try to write a value to file, the path is empty. When the Write Characters to File VI encounters an empty path for the path name of the file, it will display a dialog box from which you can select a file name. Once that file name has been selected via the dialog box, the shift register will pass the file name on to subsequent iterations of the For Loop.

The Format Into String function converts the temperature measurement (a number) to a string and concatenates the comma that follows it. We need to modify the format string to change the precision to two digits after the decimal point and to add the comma after each data point. The format string can be seen in Figure 9.11 (wired at the top of the Format Into String function). On the block diagram, pop up on the Format Into String function and select **Edit Format String** or double click on the node to access the dialog box. As shown in Figure 9.12(a), the **Current Format Sequence** contains the **Format fractional number**. The first thing to do is to set the precision to 2, as illustrated in Figure 9.12(b). Then select **Add New Operation** and in the **Selected Operation** pull-down menu, choose **Output exact string**. Make sure that the exact string that you input is a comma (as shown in Figure 9.12(c)). When finished, select **Create String** and verify that the string format is "%.2f,"—as desired.

Finish wiring the objects, return to the front panel, and run the VI with the number of points set to 20. A file dialog box prompts you for a file name—select a file name such as **Test**. When you enter a file name, the VI starts writing the temperature values to that file as each point is generated. Save the VI as **Write Temperature to File.vi** in the **Users Stuff** folder. Use any word processing

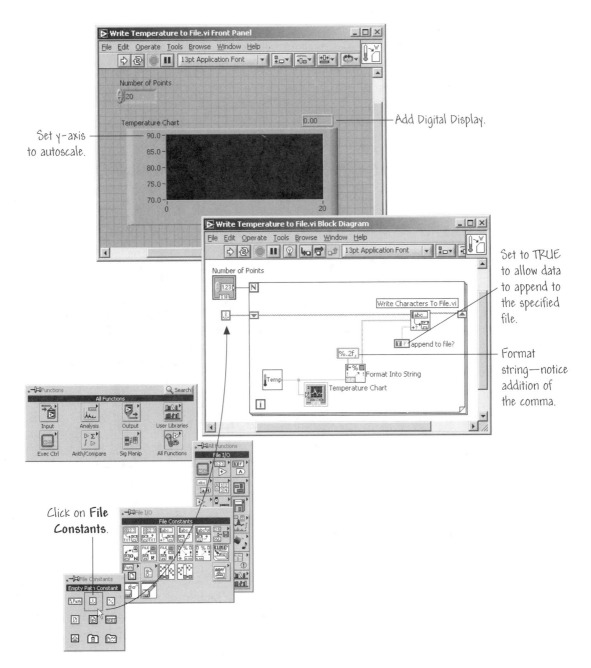

FIGURE 9.11
Using the Write Characters to File VI to write temperature data to a file.

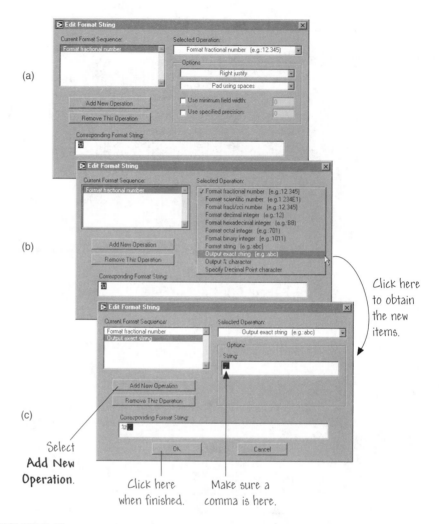

FIGURE 9.12
Setting the desired format string using the dialog box.

software, such as Notepad for Windows or Teach Text for Macintosh, to open that data file and view the contents. You should get a file containing twenty data values (with a precision of two places after the decimal point) separated by commas.

A working version of Write Temperature to File.vi *can be found in the* Chapter 9 *folder in the* Learning *directory.*   ◆

9.2.2   Reading Data from a File

As a natural follow-on to writing data to a file, in this section we discuss reading data from a file using the Read Characters From File VI. This function reads a specified number of characters from a text byte stream file beginning at a specified character offset. The VI opens the file before reading from it and closes it afterwards. In the following example, you will get a chance to construct a VI to read the temperature data that you wrote to a file in the previous exercise.

**Read
Data
from a File**

The goal here is to construct a VI that reads the temperature data file you wrote in the previous example and displays the data on a waveform graph. You must read the data in the same data format in which you saved it. The data was originally saved in ASCII format using string data types, so it must read in as string data.

Open a new front panel and build the front panel shown in Figure 9.13. You will need to place a string indicator and a waveform graph to display the temperature data that is read.

Build the block diagram as shown in Figure 9.13. The Read Characters From File VI reads the data from the file and outputs the information in a string. If no path name is specified, a file dialog prompts you to enter a file name.

The Extract Numbers VI is located in the Examples\General\Strings.llb. This VI takes an ASCII string containing numbers separated by commas, line feeds, or other non-numeric characters and converts them to an array of numbers. It uses the Spreadsheet String To Array function to convert a spreadsheet string (that is, delimiter-separated columns with end-of-line characters between rows) into an array of numbers (by default) or strings. You can access the Extract Numbers VI on Functions≫Select a VI. . ..

Complete the wiring of the block diagram, return to the front panel, and run the VI. When prompted for a file name, select the data file name that contains the temperature data from the previous exercise. You should see the same temperature data values displayed in the graph as you saw in the previous exercise. Save the VI as Read Temperature from File.vi in the Users Stuff and close the VI.

A working version of Read Temperature from File.vi *can be found in the* Chapter 9 *folder in the* Learning *directory.*   ◆

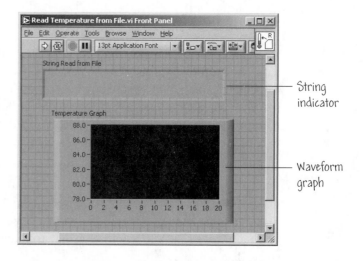

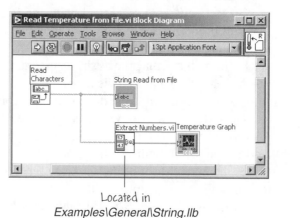

Located in
Examples\General\String.llb

FIGURE 9.13
Using the Read Characters From File VI to read temperature data from a file.

9.2.3 Manipulating Spreadsheet Files

In many instances it is useful to be able to open saved data in a spreadsheet. In most spreadsheets, tabs separate columns and EOL (end-of-line) characters separate rows, as shown in Figure 9.14(a). Opening the file using a spreadsheet program yields the table shown in Figure 9.14(b).

We will concentrate in this section on writing data to a spreadsheet format, with the idea that you will be accessing that data using a spreadsheet application outside of LabVIEW. However, if you want to read spreadsheet data from within

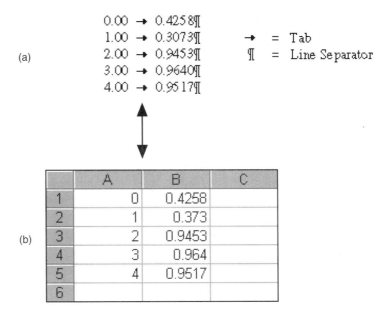

(a)

	→	
→	=	Tab
¶	=	Line Separator

(b)

FIGURE 9.14
A common spreadsheet data format.

LabVIEW, it is possible to read data in text format from a spreadsheet using the Read From Spreadsheet File VI. This VI reads a specified number of lines or rows from a spreadsheet file, beginning at a specified character offset, and converts the data to a 2D, single-precision array of numbers. This is a high-level VI; hence it opens the file beforehand and closes it afterwards.

Write to a Spreadsheet File

The objective of this exercise is to construct a VI that will generate and save data to a new file in ASCII format. This data can then be accessed by a spreadsheet application.

Open a new VI and construct a front panel, as shown in Figure 9.15. This VI generates two data arrays and plots them on a graph. You modify this VI to write the two arrays to a file where each column contains a data array. The front panel contains only one waveform graph.

Open the block diagram and modify the VI by adding the block diagram functions shown in Figure 9.15. The Write to Spreadsheet File VI (for location of this VI see Figure 9.10) converts the 2D or 1D array of single-precision numbers to a text string and writes the string to a new text byte stream file or appends the string to an existing file. If you have not specified a path name, then a file

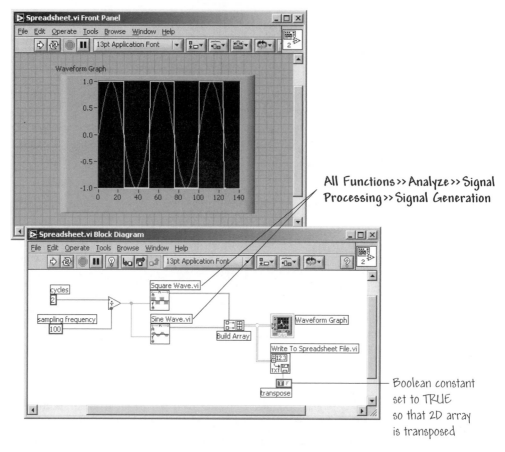

FIGURE 9.15
Using the Write To Spreadsheet File VI to write data in ASCII format to a file.

dialog box appears and prompts you for a file name. You can write either a 1D or 2D array to file—in this exercise we have a 2D array of data, so the 1D input is not wired. With this VI, you can use a spreadsheet delimiter such as tabs or commas in your data. The format string default is %.3f, which creates a string long enough to contain the number with three digits to the right of the decimal point.

This VI is a high-level VI that opens or creates the file before writing to it and closes it afterwards. You can use Write to Spreadsheet File VI to create a text file readable by most spreadsheet applications.

The Boolean constant (see Figure 9.15) connected to the Write To Spreadsheet File VI controls whether or not the 2D array is transposed before writing it to file. To change the value to TRUE, click on the constant with the **Operating** tool. Generally each row of a spreadsheet file contains a data array; thus in this case you want the data transposed because you want data arrays in each column.

After finishing up the wiring on the block diagram, return to the front panel and run the VI. After the data arrays have been generated, a file dialog box prompts you for the file name of the new file you are creating. Type in a file name and click on **OK**.

Do not attempt to write data in VI libraries. Doing so may result in overwriting your library and losing your previous work.

Save the VI in Users Stuff and name it Spread-sheet.vi. You now can use a spreadsheet application or a text editor to open and view the file you just created.

In this example, the data was not converted or written to file until the entire data arrays had been collected. If you are acquiring large buffers of data or would like to write the data values to disk as they are being generated, then you must use a different File I/O VI.

A working version of Spreadsheet.vi *can be found in the* Chapter 9 *folder in the* Learning *directory.* ◆

9.2.4   File I/O Express VIs

The two Express VIs that can be utilized for file I/O are the Read LabVIEW Measurement File Express VI and the Write LabVIEW Measurement File Express VI. These two Express VIs can be found on the **All Functions≫File I/O** palette as illustrated in Figure 9.16(a). They can also be readily found on the Express palettes. The Read LabVIEW Measurement File Express VI is located on the **Input** palette and the Write LabVIEW Measurement File Express VI is on the **Output** palette, as shown in Figure 9.16(b). The LabVIEW measurement Express VIs read and write .lvm files—LabVIEW measurement data files. The measurement data file (.lvm) is a tab-delimited text file that can be opened with a spreadsheet application (such as Microsoft Excel) or with a text-editing application (such as Notepad). In addition to the data an Express VI generates, the .lvm file includes information about the data, such as the date and time the data was generated.

You may often find it necessary to permanently store measurement data acquired from your DAQ device. When planning to store data to a file, remember that not all data logging applications use LabVIEW to process and analyze the stored data. Consider which applications might need to access the data, and

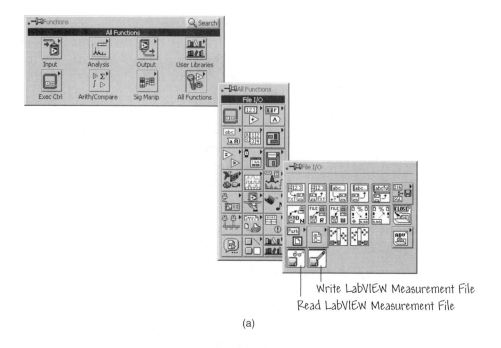

(a)

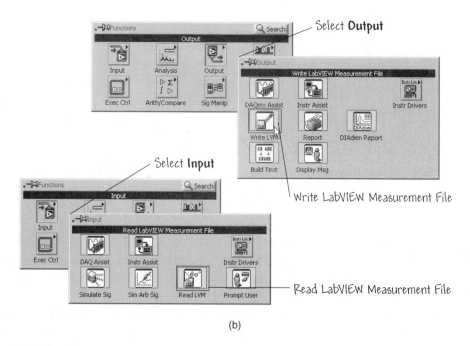

(b)

FIGURE 9.16
Locating the LabVIEW data measurement Express VIs. (a) The **File I/O** palette.
(b) The Express palettes.

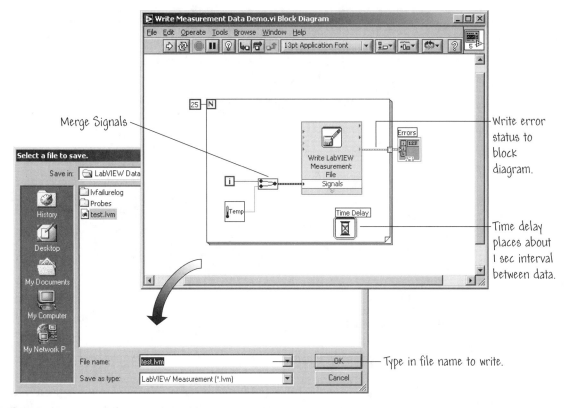

FIGURE 9.17
Writing measurement data to a file specified by the user.

write the data in an appropriate format—the data storage format defines which applications can read the file. Since LabVIEW contains standard file operation functions that exist in other languages, you have complete control over the data logging process. The LabVIEW measurement file, or .lvm file, is an ASCII text file that can is easy to create in LabVIEW and easy to read in other applications.

Practice with Writing & Reading Measurement Data Files

In this example, we practice writing and reading LabVIEW measurement files. To begin, open the Write Measurement Data Demo.vi found in Chapter 9 of the Learning directory. The block diagram is shown in Figure 9.17. Central to the process of writing measurement data files is the Write LabVIEW Measurement File Express VI, which includes the open, write, close, and error handling functions. It also handles formatting the string with either a tab or comma delimiter.

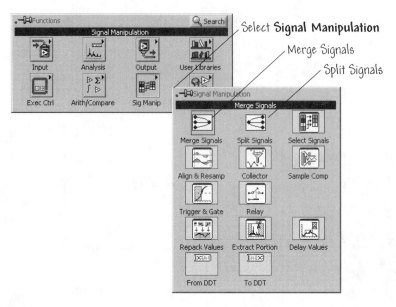

FIGURE 9.18
The Merge Signals and Split Signals functions.

The Merge Signals function found in the Write Measurement Data Demo.vi combines the iteration count and the random number into the dynamic data type for use by the Express VI (see Chapter 2 for a review of dynamic data type). The Merge Signals function is located on the **Functions≫Signal Manipulation** palette, as shown in Figure 9.18.

When the Write LabVIEW Measurement File Express VI is first placed on the block diagram, a dialog box opens automatically to configure the VI. The dialog box is shown in Figure 9.19. The configuration as shown requires the user to choose the output filename. When the VI is run, a dialog box will open, as illustrated in Figure 9.17, for specifying the output file name. This process will lead to the creation of a .lvm file that can be later opened in a spreadsheet or text editor application.

You can double-click on the Write LabVIEW Measurement File Express VI in the Write Measurement Data Demo.vi to access the dialog box.

Now we can use the Read LabVIEW Measurement File Express VI as a central component of a VI that reads the .lvm file from the Write Measurement Data Demo.vi and outputs the results to a graph. Open the Read Measurement

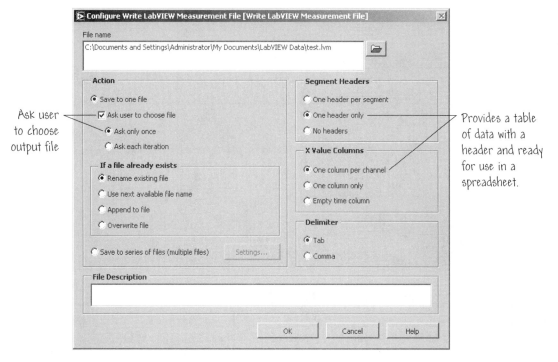

Ask user
to choose
output file

Provides a table
of data with a
header and ready
for use in a
spreadsheet.

FIGURE 9.19
The Write LabVIEW Measurement File Express VI dialog box.

Data Demo.vi found in Chapter 9 of the Learning directory. The block diagram
is shown in Figure 9.20.

Double-click on the Read LabVIEW Measurement File Express VI to access the **Configure Read LabVIEW Measurement File** dialog box, shown in
Figure 9.21. Notice that the following options have been set:

- In the **Action** section, a checkmark is placed in the **Ask user to choose file**
  checkbox.

- Set the **Segment Size** to **Retrieve segments of original size** so that all the
  data stored in the file is retrieved.

- Set **Time Stamps** to **Relative to start of measurement**. Because the dynamic
  data type stores information about the signal timing, this setting aligns the
  data with the time of the measurement.

- In the **Generic Text File** section, remove the checkmark from the **Read
  generic text files** checkbox because the data is stored in a LabVIEW measurement file.

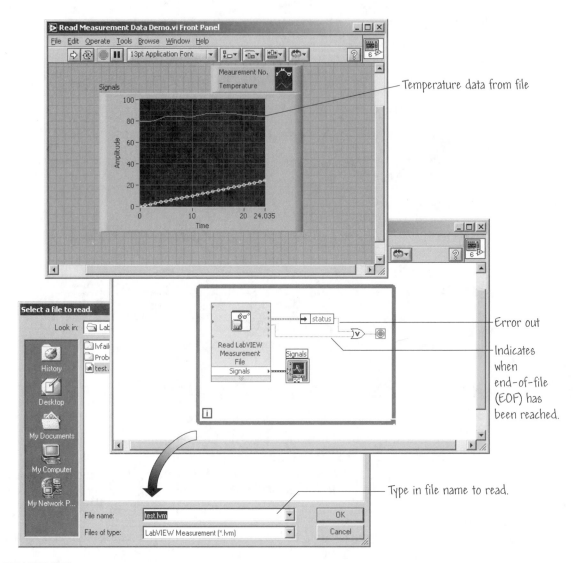

FIGURE 9.20
Reading measurement data from a file specified by the user.

Click the **OK** button to close the dialog box when you are finished. Display the front panel and run the VI. In the filename prompt that appears, select the test.lvm file that you created with the Write LabVIEW Measurement File Express VI. The temperature data that was stored in the LabVIEW measurement file appears in the waveform chart. ◆

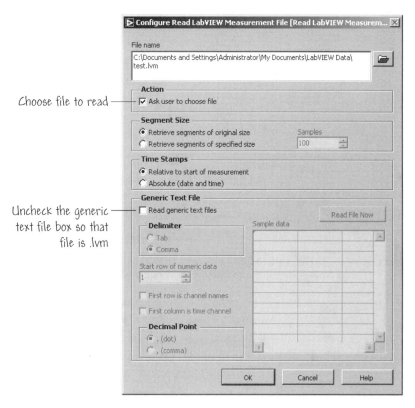

Choose file to read ——

Uncheck the generic —
text file box so that
file is .lvm

FIGURE 9.21
The Read LabVIEW Measurement File Express VI dialog box.

9.3   BUILDING BLOCKS: MEASURING VOLUME

In this exercise you will create a VI that writes data to a file. Open the Volume Chart.vi and modify the block diagram using Figure 9.22 as a guide. The Volume Chart.vi should have been saved in Users Stuff folder in the Learning directory as part of the Chapter 7 "Building Blocks" exercise. If you did not complete the exercise in Chapter 7 you can build the VI from scratch, but refer to the "Building Blocks" section in that chapter for a view of the front panel.

The block diagram uses the Write LabVIEW Measurement File Express VI. By default, the Express VI saves the data to test.lvm. You can double-click on the Express VI on the block diagram to modify the output data file name.

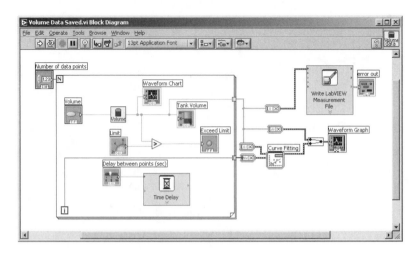

FIGURE 9.22
The Volume Data Saved.vi block diagram.

When the VI is complete, click on **Run** and watch the VI execute. When the VI is finished executing, you can view the data file in (Windows) Notepad or (Macintosh) SimpleText. Close the VI when you are done experimenting and save it as Volume Data Saved.vi in the Users Stuff folder.

9.4 RELAXED READING: USING LABVIEW IN DRIVE-BY-WIRE TESTING

The traditional mechanical design of brakes in an automobile can lead to injury to the feet and lower leg of the driver in the event of a crash. Students from Luleå University of Technology and Stanford University were challenged to work together with the Volvo Car Corporation to develop new concepts for safer brake and accelerator pedals. The project was dubbed "virtual pedals."

The challenge was to develop a control system that incorporated continuous changes of key parameters using LabVIEW Real-Time in a PXI chassis with embedded controllers. This was to be implemented as part of a drive-by-wire system in a production car. The goal was to show significant performance improvement of the braking system during test driving in a real-time environment without delay or malfunction.

Since the newly designed brake system might replace the ordinary brake and accelerator system in a working car, it was necessary to incorporate redundancy into the design. The VOLVO S80 already has an electronically controlled accelerator system. Two separate signal circuits manage its redundancy: one analog signal and a digital PWM signal. The engine PU system compares these signals and if they differ by more than 2%, the control system shuts down the servo unit in the throttle valve system.

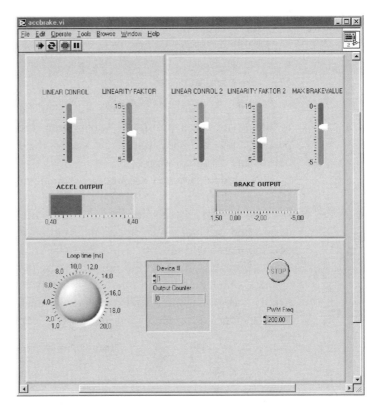

FIGURE 9.23
Student designed VI for testing an advanced drive-by-wire braking.

The student solution was to replace the brake pedal system with a DC engine controlled linear motion unit. A servo amplifier was employed to manage the high current in the DC. The hardware selected was the 1000B PXI chassis with a DC power supply, and the I/O unit selected was the NI PXI-6052E (333 kS/s) analog multifunction data acquisition card. The heart of the system was the NI PXI-8176 RT embedded controller with an Intel Pentium III processor. The controller communicates across a standard Ethernet with a Windows laptop running the LabVIEW Real-Time module.

The robust PXI chassis and the rest of the apparatus fit easily into the trunk of the car. For easy access to the driving parameters, a laptop provided access to the control panel of the system in the driving environment. The front panel of the system is shown in Figure 9.23.

LabVIEW Real-Time provided key advantages:

- It runs on its own real-time operating system, providing extremely good stability and reliability.

- The speed is outstanding because of its own operating system.

- You can modify and adjust control parameters during test driving and evaluation.

- You can easily add a CAN-bus system for communication with the vehicle onboard computer.

Another benefit of the flexible LabVIEW-based system was that it was easy to implement and test different subsystems, such as hill holding, which makes parking and starting in hills much easier for the driver without difficult and time-consuming programming.

The students created a system to evaluate and test a modern production car braking system without loss of performance and reliability. The complex interfaces developed by the students would be difficult and very cost-demanding to design and produce utilizing typical microprocessing units and their software development environments. The successful student project provides the universities a great tool for use in future projects with the automotive industry. Due to the ease of adaptability and the user-friendly interface, incoming students can readily access the system and conduct their own experiments.

For more information about this project, contact:

Holger Andersson
Korngatan 3
943 31 Öjebyn
Sweden
ho.andersson@telia.com

9.5   SUMMARY

File input and output operations store and retrieve information from a disk file. LabVIEW supplies you with simple functions that take care of almost all aspects of file I/O. You can store or retrieve data from files in ASCII byte stream, binary byte stream, or datalog files. This allows you to interact with word processing programs, spreadsheet programs, or with other VIs in LabVIEW.

KEY TERMS

ASCII: American Standard Code for Information Interchange.

Binary byte stream files: Files that store data as a sequence of bytes.

Datalog files: Files that store data as a sequence of records of a single, arbitrary data type that you specify when you create the file.

EOF: End-of-file. Character offset of the end of a file relative to the beginning of the file (that is, the EOF is the size of the file).

Hex: Hexadecimal. A base-16 number system.

Nondisplayable characters: ASCII characters that cannot be displayed, such as null space, backspace, and tab.

Not-a-path: A predefined value for the path control that means that a path is invalid.

Not-a-refnum: A predefined value that means the refnum associated with an open file is invalid.

Refnum: An identifier that G associates with a file when you open it. You use the file refnum to indicate that you want a function or VI to perform an operation on the open file.

String: A sequence of displayable or nondisplayable ASCII characters.

Text (ASCII) byte stream files: Files that store data as a sequence of ASCII characters.

EXERCISES

E9.1   Open the Learning\Chapter 8\AI Waveform.vi. Select Write to Spreadsheet File.vi from the **Functions≫File I/O** palette. Wire the array of information from AI Acquire Waveform.vi to the 1D data input of Write to Spreadsheet File.vi. Run the VI. Enter the file name you want the waveform data to be saved to. Open this file in (Windows) Notepad or (Macintosh) SimpleText to view the data. Save this VI as AI Waveform&SpreadFile.vi in Learning\Users Stuff.

E9.2   Open the Learning\Chapter 8\AI Waveform.vi. Select Write to SGL File.vi from Functions≫File I/O≫Binary File VIs. Wire the array of information from AI Acquire Waveform.vi to the 1D array input of Write to SGL File.vi. Run the VI. Enter the file name you want the waveform data to be saved to. Save this VI as AI Waveform&BinaryFile.vi in Learning\Users Stuff directory. Select Help≫Examples. . .≫Fundamentals. Click File I/O≫Binary File Examples≫Read Binary File links. Run this VI to read the file you just create with AI Waveform&Binary File.vi.

E9.3   Create a new VI that acquires analog data and writes the data to a LabVIEW Measurement File. Use the DAQ Assistant Express VI to configure the analog input operation and then use the Generate Code≫Example feature to automatically create the code for this acquisition. Take 50 measurements and write the measurements to a LabVIEW Measurement File. Refer back to Chapter 8 for more information on using the DAQ Assistant Express VI.

E9.4 Create a VI that reads the LabVIEW Measurement File created in E9.3 and displays that information on the front panel.

E9.5 Modify the VI you created in E9.4 so that data is saved to a new file every time the VI runs. Also, make it so that LabVIEW appends the next sequential number to the filename.

PROBLEMS

P9.1 Complete the crossword puzzle.

Across

4. Files that store data as a sequence of records of a single, arbitrary data type that you specify when you create the file.
6. A predefined value for the path control that means that a path is invalid.
7. Files that store data as a sequence of ASCII characters.
9. A base-16 number system.
10. An identifier that G associates with a file when you open it.

Down

1. Files that store data as a sequence of bytes.
2. American Standard Code for Information Interchange.
3. Character offset of the end of a file relative to the beginning of the file.
5. A predefined value that means the refnum associated with an open file is invalid.
8. A sequence of displayable or nondisplayable ASCII characters.

P9.2 Open the **Basic Spectral Measurements**.vi shipping example (located in the **examples\express** folder). Using the Configure Report Express VI, modify this VI to create a report that displays the graphs of both signals and includes the report title, author, company name, and comments about the report. Give the user the option to enter additional comments. Save the report as an HTML file. Run the VI and view the HTML document that you created in a Web browser or word processing application.

P9.3 Create a VI that concatenates a message string, numeric, and unit string using the Build Text Express VI and writes the data to file. Create the input variables "Number" and "Units" and configure them in the Build Text Express VI. Wire a string control to the Beginning Text input of the Build Text Express VI. Write the resulting concatenated string to a datalog file. Then create a VI that reads the datalog file you just created and displays that information on the front panel. To find out more about datalog files, refer to the online help and examples.

CHAPTER 10

Instrument Control

This chapter introduces the concept of communicating with and controlling external instruments. In this chapter we focus on the GPIB (General Purpose Interface Bus) and RS-232 (a serial interface bus standard). A brief introduction to the main components of an instrument control system are described. The application of the **Measurement & Automation Explorer (MAX)** in detecting instruments and installing instrument drivers is discussed. The notion of an instrument driver is an important topic woven throughout the chapter. The Instrument I/O Assistant is introduced.

GOALS

1. Learn about GPIB and serial instrument control.

2. Understand how to interact with your instruments using the MAX.

3. Gain some experience with instrument drivers (for the HP34401A multimeter).

4. Gain experience with the Instrument I/O Assistant.

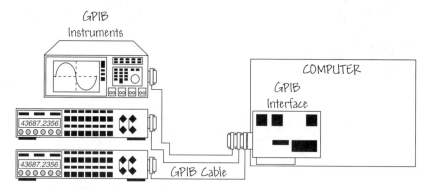

FIGURE 10.1
A typical GPIB system.

10.1   COMPONENTS OF AN INSTRUMENT CONTROL SYSTEM

LabVIEW communicates with and controls external instruments (such as oscilloscopes and digital multimeters) using GPIB (General Purpose Interface Bus), RS-232 (a serial interface bus standard), VXI (VME eXtensions for Instrumentation), and other hardware standards. In this chapter we focus on the GPIB and serial communication.

10.1.1   What Is GPIB?

Hewlett Packard developed the General Purpose Interface Bus (or **GPIB**) standard in the late 1960s to interconnect and control its line of programmable instruments. The interface bus was originally called HP-IB. In this context, a **bus** is the means by which computers and instruments transfer data. National Instruments made GPIB available to users of non-Hewlett-Packard equipment.

At the time it was developed, GPIB provided a much-needed specification and protocol to govern this communication. Figure 10.1 shows a typical GPIB system. While the GPIB is one way to bring data into a computer, it is fundamentally different from data acquisition with boards that plug into the computer. Using a special protocol, GPIB brings data that has been acquired by another computer or instrument into the computer using a "handshake," while data acquisition involves connecting a signal directly to a DAQ board in the computer.

The original purpose of GPIB was to provide computer control of test and measurement instruments. The GPIB was soon applied to intercomputer communication and control of scanners, film recorders, and other peripherals because of its 1 Mbyte/sec maximum data transfer rates. The Institute of Electrical and Electronic Engineers (**IEEE**) standardized the GPIB in 1975, and it became accepted as IEEE Standard 488-1975. The standard has since evolved into ANSI/IEEE Standard 488.2-1987. The GPIB functions for LabVIEW follow the IEEE 488.2 specification. The terms GPIB, HP-IB, and IEEE 488 are synonymous.

10.1.2   GPIB Messages

The GPIB carries two types of messages:

- **Device-dependent messages** contain device-specific information such as programming instructions, measurement results, machine status, and data files. These are often called **data messages**.

- **Interface messages** manage the bus itself, and perform such tasks as initializing the bus, addressing and unaddressing devices, and setting device modes for remote or local programming. These are often called **command messages**.

Physically, the GPIB is a digital, 24-conductor, parallel bus. It comprises 16 signal lines and 8 ground-return lines. The GPIB connector is depicted in Figure 10.2. The 16 signal lines are divided into three groups:

- **Eight data lines**: The eight data lines (denoted DIO1 through DIO8) carry both data and command messages. GPIB uses an eight-bit parallel, asynchronous data transfer scheme where whole bytes are sequentially handshaked across the bus at a speed determined by the slowest participant in the transfer. Because the GPIB sends data in bytes (1 byte = 8 bits), the messages transferred are frequently encoded as ASCII character strings. All commands and most data use the 7-bit ASCII or International Standards Organization (ISO) code set, in which case the eighth bit, DIO8, is unused or is used for parity.

- **Three handshake lines**: The three handshake lines asynchronously control the transfer of message bytes among devices. These lines guarantee that message bytes on the data lines are sent and received without transmission error.

- **Five interface management lines**: The five interface management lines manage the flow of information across the interface from device to computer.

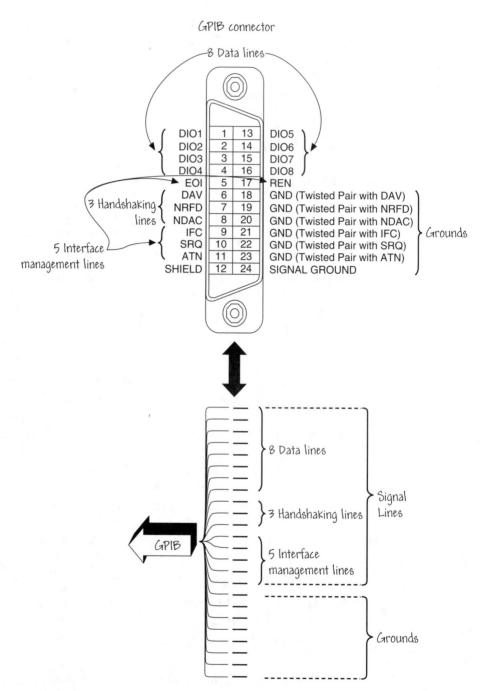

FIGURE 10.2
The GPIB connector.

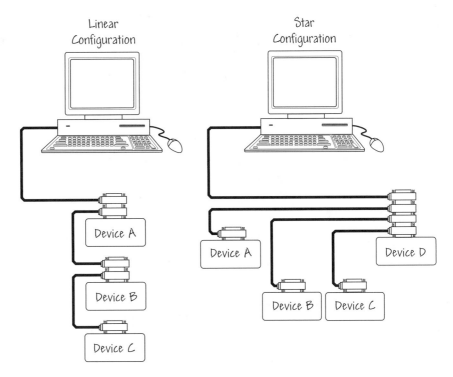

FIGURE 10.3
Typical GPIB configurations.

*The next generation in high-speed GPIB has been proposed by National Instruments, and is known as **HS488**. Speed increases up to 8 Mbyte/sec have been achieved by removing the propagation delays associated with the three handshake lines.*

10.1.3   GPIB Devices and Configurations

You can have many GPIB devices (that is, many instruments and computers) connected to the same GPIB. Typical GPIB linear and star configurations are illustrated in Figure 10.3. You can even have more than one GPIB board in your computer. A typical multiboard GPIB configuration is illustrated in Figure 10.4. GPIB devices are grouped in three categories:

- **Talkers**: send data messages to one or more listeners.

- **Listeners**: receive data messages from the talker.

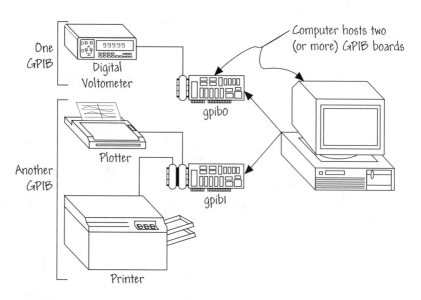

FIGURE 10.4
A typical multiboard GPIB configuration.

- **Controllers**: manage the flow of information on the GPIB by sending commands to all devices.

GPIB devices can fall into multiple categories. For example, a digital voltmeter can be both a talker and a listener.

The GPIB has one controller (usually a computer) that controls the bus. The role of the GPIB controller is similar to the role of a central processing unit in a computer. A good analogy is the switching center of a city telephone system. The communications network (the GPIB) is managed by the switching center (the controller). When a party (a GPIB device) wants to make a call (that is, to send a data message), the switching center (the controller) connects the caller (the talker) to the receiver (the listener). To transfer instrument commands and data on the bus, the controller addresses one talker and one or more listeners. The controller must address the talker and a listener before the talker can send its message to the listener. The data strings are then handshaked across the bus from the talker to the listener(s). After the talker transmits the message, the controller may unaddress both devices. LabVIEW provides VIs that automatically handle these GPIB functions.

Some bus configurations do not require a controller. For example, one device may always be a talker (called a talk-only device) and there may be one or more listen-only devices. A controller is necessary when you must change the active or addressed talker or listener. A computer usually handles the controller function. With the GPIB board and its software, the personal computer plays all three roles: Controller, Talker, and Listener.

There can be multiple controllers on the GPIB but only one controller at a time is active. The active controller is called the **controller-in-charge** (CIC). Active control can be passed from the current CIC to an idle controller. Only the system controller (usually the GPIB board) can make itself the CIC.

With GPIB, the physical distance separating the GPIB devices matters. To achieve high-rate data transfers, we must accept a number of physical restrictions. Typical numbers are as follows:

- A maximum separation of 4 meters between any two devices and an average separation of 2 meters over the entire bus. For high-speed applications, you should have at least one device per meter of cable.

- A maximum total cable length of 20 meters. A maximum of 15 meters is desirable for high-speed applications.

- A maximum of 15 devices connected to each bus, with at least two-thirds of the devices powered on. For high-speed applications, all devices should be powered on.

Bus extenders and expanders can be used to increase the maximum length of the bus and the number of devices that can be connected to the bus. You can also communicate with GPIB instruments through a TCP/IP network. For more information, refer to the National Instruments website.

10.1.4   Serial Port Communication

Serial communication is another popular means of transmitting data between a computer and another computer, or between a computer and a peripheral device. However, unlike GPIB, a serial port can communicate with only one device, which can be limiting for some applications. Serial port communication is also very slow.

Most computers and many instruments have built-in serial port(s). Since serial communication uses the built-in serial port in your computer, you can send and receive data without buying any special hardware. Serial communication uses a transmitter to send data—one bit at a time—over a single communication line to a receiver. This method works well when sending or receiving data over long distances and when data transfer rates are low. It is slower and less reliable

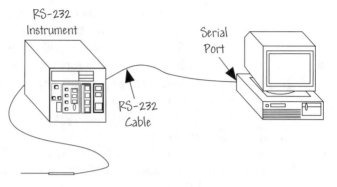

FIGURE 10.5
A typical serial communication system.

than GPIB, but you do not need a special board in your computer, and your instrument does not need to conform to the GPIB standard. Figure 10.5 shows a typical serial communication system.

There are different serial port communication standards. Developed by the Electronic Industries Association (EIA) to specify the serial interface between equipment known as Data Terminal Equipment (DTE), such as modems and plotters, and Data Communications Equipment (DCE), such as computers and terminals, the RS-232 standard includes signal and voltage characteristics, connector characteristics, individual signal functions, and recipes for terminal-to-modem connections. The most common revision to this standard is the RS-232C used in connections between computers, printers and modems.

The RS-232 serial port connectors come in two varieties: the 25-pin connector and the 9-pin connector. Both connectors are depicted in Figure 10.6. The 9-pin connector has two data lines (denoted TxD and RxD in the figure) and five handshake lines (denoted RTS, CTS, DSR, DCD, and DTR in the figure). This compact connector is occasionally found on smaller RS-232 laboratory equipment and has enough pins for the "core" set used for most RS-232 interfaces. The 25-pin connector is the "standard" RS-232 connector with enough pins to cover all the signals specified in the RS-232 standard. Only the "core" set of pins is labeled in Figure 10.6.

Serial communication requires that you specify four parameters: the baud rate of the transmission, the number of data bits encoding a character, the sense of the optional parity bit, and the number of stop bits. Each transmitted character

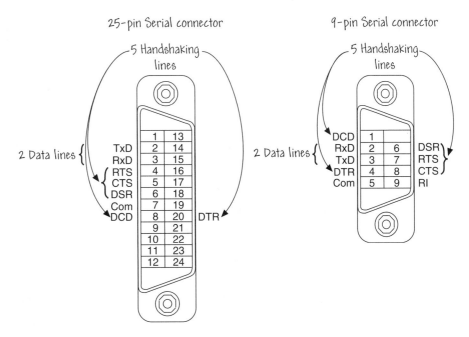

FIGURE 10.6
The RS-232 serial port connectors.

is packaged in a character frame that consists of a single start bit followed by the data bits, the optional parity bit, and the stop bit or bits. The baud rate is a measure of how fast data is moving between instruments that use serial communication.

10.2   DETECTING AND CONFIGURING INSTRUMENTS

With the **Measurement & Automation Explorer (MAX)** you can automatically detect connected instruments, install the required instrument drivers, and manage the instrument drivers already installed. The software architecture for GPIB instrument control using LabVIEW is similar to the architecture for DAQ (see Chapter 8.1). Figure 10.7 shows the software architecture on the Windows platforms. Instrument drivers are LabVIEW applications written to control a specific instrument. More on instrument drivers in the next section. You can use the instrument drivers or parts of the instrument drivers to develop your own application quickly. You can also set naming aliases for your instruments for easier instrument access. If you are using a Macintosh, the NI-488.2 Configuration Utility (available at Start≫Settings≫Control Panel configures the parameters for the

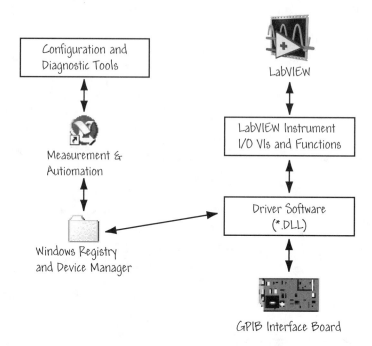

FIGURE 10.7
Software architecture for Windows platforms.

GPIB devices installed in you Macintosh computer. The Macintosh OS automatically recognizes GPIB devices. You can view or modify the default configuration settings using the NI-488.2 configuration utility.

**Practice
with the
Instrument
Wizard**

In this example, you will use the MAX to configure and test the GPIB interface. MAX interacts with the various diagnostic and configuration tools installed with the driver and also with the Windows Registry and Device Manager (see Figure 10.7). The driver-level software is in the form of a dynamically linked library (DLL) and contains all the functions that directly communicate with the GPIB board. The LabVIEW Instrument I/O VIs and functions directly call the driver software.

You configure the objects listed in the MAX by right-clicking on the item and making a selection from the shortcut menu. Figure 10.8 shows the GPIB interface board in the MAX utility and the results of pressing the **Scan for Instruments** button at the top of the window.

Select this option to enable the search for instruments.

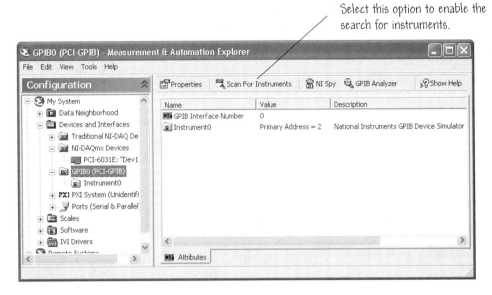

FIGURE 10.8
Using MAX to search for instruments.

The configuration utilities and hierarchy described in this example are specific to Windows platforms. If you are using a Macintosh or other operating system, refer to the manuals that came with your GPIB interface board for the appropriate information for configuring and testing that board. ◆

As discussed in Chapter 8, the MAX is a configuration utility for your software and hardware. MAX executes system diagnostics, adds new channels, interfaces, and virtual channels, and views devices and instruments connected to your system. Open MAX by double-clicking on its icon on the desktop or by selecting **Measurement & Automation Explorer** from the **Tools** menu.

The four possible selections in MAX are:

- **Data Neighborhood**—Use this selection to create virtual channels, aliases, and tags to your channels or measurements configured in Devices and Interfaces as you did in the DAQ chapter.

- **Devices and Interfaces**—Use this selection to configure resources and other physical properties of your devices and interfaces. Using this selection, you can view attributes of one or more multiple devices, such as serial numbers.

- **Scales**—Use this selection to set up simple operations to perform on your data, such as scaling.

- **Software**—Use this selection to determine which drivers and application software are installed and their version numbers.

If you do not have an instrument attached to your computer, the MAX will obviously not find any instruments. In this example, the attached instrument is a National Instruments GPIB Device Simulator, denoted by **Instrument0** in MAX as seen in Figure 10.8.

Click **Communicate with Instrument** to communicate with the instrument, as shown in Figure 10.16. By default, the string *IDN? appears in the command to send to this GPIB device. The string *IDN? is an IEEE 488.2 standard identification request to the instrument. Click **Query** to send the string. In this example the string is sent to the NI GPIB Device Simulator. The instrument replies with the string "National Instruments GPIB Device Simulator Rev A.2." The next time you launch the MAX, it will automatically contain your stored instrument configuration.

You can also examine the properties of your GPIB board. On the MAX screen, highlight the GPIB hardware and then choose **Properties**, as illustrated in Figure 10.17. A window will open which provides all the information about your GPIB board.

10.3   USING THE INSTRUMENT I/O ASSISTANT

The Instrument I/O Assistant is an Express VI that can be employed to establish communication with message-based instruments. You can communicate with an instrument that uses a serial, Ethernet, or GPIB interface. The Instrument I/O Assistant organizes instrument communication into ordered steps using a dialog window associated with the I/O Assistant Express VI. Like the DAQ Assistant (see Chapter 8), the Instrument I/O Assistant is readily configured to provide a communication link between the instrument and the computer. The Instrument I/O Assistant can be used whenever an instrument driver is not available.

To launch the Instrument I/O Assistant, place the Instrument I/O Assistant Express VI on the block diagram, as illustrated in Figure 10.9. The Instrument I/O Assistant is located on the **Functions≫Input** and **Functions≫All Functions≫Instrument I/O** palettes. Once the Instrument I/O Assistant Express VI is placed on the block diagram, the Instrument I/O Assistant configuration dialog box will automatically appear. The dialog box is shown in Figure 10.10.

If the Instrument I/O Assistant configuration dialog box does not appear, double-click the Instrument I/O Assistant icon.

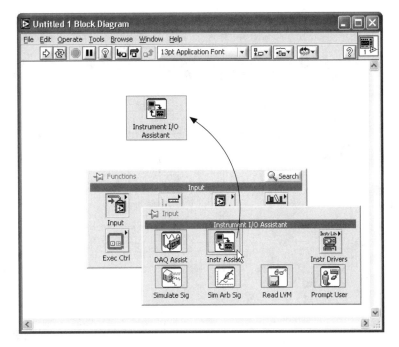

FIGURE 10.9
Placing the Instrument I/O Assistant on the Block Diagram.

The desired communication steps in the instrument communication process are specified in the Instrument I/O Assistant configuration dialog box. Once the configuration dialog box appears, the general procedure for configuring the Instrument I/O Assistant Express VI is as follows:

1. **Select an instrument.**
   Instruments that have been configured in MAX appear in the **Select an instrument** pull-down menu. In Figure 10.10, we see the National Instruments GPIB Device Simulator that was configured in MAX.

2. **Choose a Code generation type.**
   VISA code generation allows for more flexibility and modularity than GPIB code generation.

3. **Specify the communication steps.**
   Select from the following communication steps using the **Add Step** button:

   (a) **Query and Parse**—Sends a query to the instrument and parses the returned string.

   (b) **Write**—Sends a command to the instrument.

   (c) **Read and Parse**—Reads and parses data from the instrument.

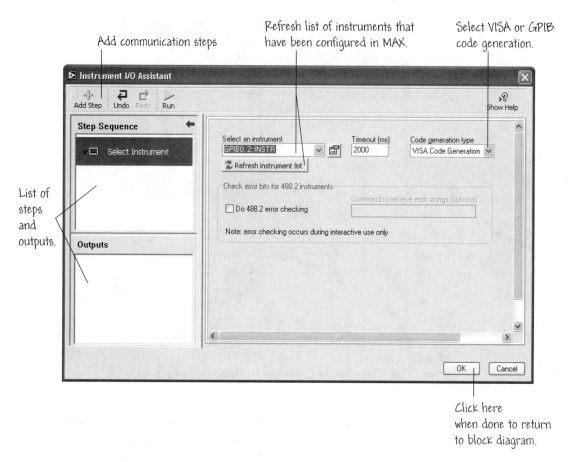

Add communication steps

Refresh list of instruments that have been configured in MAX.

Select VISA or GPIB code generation.

List of steps and outputs.

Click here when done to return to block diagram.

FIGURE 10.10
The Instrument I/O Assistant Dialog Box.

To use the Instrument I/O Assistant, you place the communication steps into a sequence. As you add steps to the sequence, they appear in the **Step Sequence** window (see Figure 10.10).

4. **Test the communication sequence.**
   After adding the desired number of steps, click the **Run** button to test the sequence of communication that you have configured for the Express VI.

5. **Return to the block diagram and complete the VI.**
   Click the **OK** button to exit the **Instrument I/O Assistant** configuration dialog box. LabVIEW adds input and output terminals to the Instrument I/O Assistant Express VI on the block diagram that correspond to the data you will receive from the instrument.

Practice with the Instrument I/O Assistant

In this example, we will construct a VI that uses the Instrument I/O Assistant to communicate with a National Instruments GPIB Device Simulator. This device uses a GPIB interface.

To begin the VI development, open a blank VI and place the Instrument I/O Assistant Express VI (located on the **Functions≫Input** palette) on the block diagram. The Instrument I/O Assistant dialog box will appear to allow you to configure the Instrument I/O Assistance Express VI.

As previously discussed, the first step is to select an instrument. In the dialog box, make sure that the National Instruments GPIB Device Simulator is selected (denoted by GPIB::2::INSTR in Figure 10.10). You can click on the **Refresh instrument list** button to make sure all the available instruments are listed.

The second step is to choose a code generation type. Since the VISA code generation allows for more flexibility and modularity than GPIB code generation, we will choose the **VISA Code Generation** from the **Code generation type** pull-down menu.

We will add two communication steps using the **Query and Parse** sequence. To accomplish this, click the **Add Step** button, and in the dialog box that appears, select **Query and Parse** to write and read from the NI GPIB Device Simulator, as illustrated in Figure 10.11. The Step Sequence window will update automatically to show the Query and Parse step. To test the communication link, type *IDN? as the command, select \n as the **Termination character**, and click the **Run this step** button. If no error warning appears in the lower half of the dialog box, this step has successfully completed. To parse the data received, click the **Auto parse** button. Notice that Token now appears in the **Outputs** pane on the left side of the dialog box. This value represents the string returned from the identification query—the name of the instrument.

Now, repeat the previous configuration to include a second communication step, as illustrated in Figure 10.12. Click the **Add Step** button and select **Query and Parse**. Enter the command MEAS:DC? and click the **Run this step** button to test the communication. As before, to parse the data received, click the **Auto parse** button. The data returned is a random numeric value, which in this case is 2.4323 as shown in Figure 10.12. This is a simulated voltage reading. The name Token2 now appears in the **Outputs** pane on the left side of the dialog box.

Generally we should rename the outputs to more accurately represent the variables. Right-click on the output **Token** and select **Rename** to change the name of the output, as illustrated in Figure 10.13. Rename **Token** to **IO String** and rename **Token2** to **Voltage**.

Click the **OK** button to exit the Instrument I/O Assistant and return to the block diagram. The configuration of the Instrument I/O Assistant Express VI is complete.

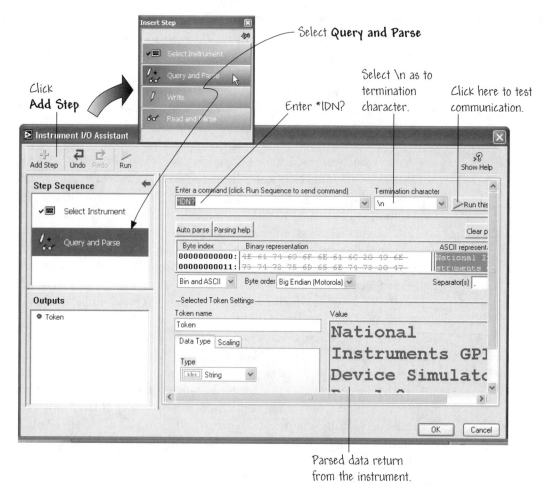

FIGURE 10.11
Configuring the Instrument I/O Assistant Express VI.

To view the code generated by the Instrument I/O Assistant, right-click the Instrument I/O Assistant icon and select **Open Front Panel** from the shortcut menu. When asked if you want to convert to a subVI select **Convert**. This converts the Express VI to a subVI. Switch to the block diagram to see the code generated. Figure 10.14 shows the code generated by the Instrument I/O Assistant after completing the above steps.

Once an Express VI has been converted to a subVI, it cannot be converted back.

Add second communication step by entering MEAS:DC.

Select **Run this Step** to test the communication link.

Add step

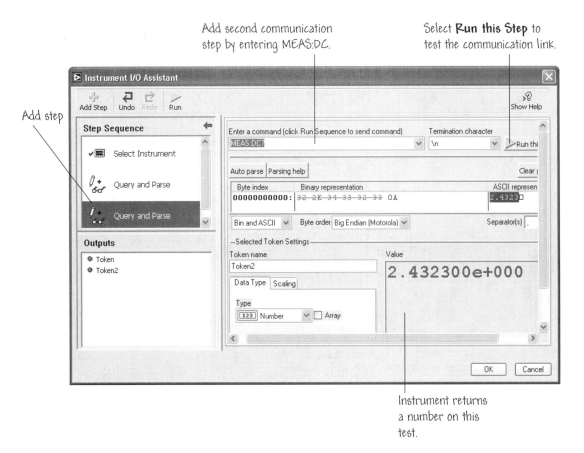

Instrument returns a number on this test.

FIGURE 10.12
Adding a second communication step.

The final step in the development of the VI is to wire appropriate controls and indicators on the block diagram. In our case, we need to wire an indicator to the **ID String** and another to the **Voltage** output. Right-click the ID String output and select Create≫Indicator from the shortcut menu. Similarly, right-click the **Voltage** output and select Create≫Indicator from the shortcut menu. The appropriate indicators will automatically wire to the outputs of the Instrument I/O Assistant Express VI, as illustrated in Figure 10.15. Wire the **Error Out** output to the Simple Error Handler VI.

Now we finally test the VI. Display the front panel and select **Run**. You will most likely need to resize the string indicator to accommodate the instrument identification. As seen in Figure 10.15, the NI GPIB Device Simulator returns the identification **National Instruments GPIB Device Simulator Rev A.2**, and on this test it returned a voltage reading of **0.95**.

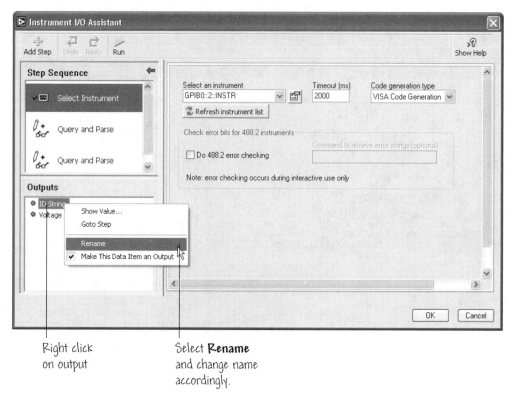

FIGURE 10.13
Renaming the outputs.

Save the VI as IO Assistant Demo.vi in the Users Stuff folder in the Learning directory. ◆

10.4 INSTRUMENT DRIVERS

An **instrument driver** is a piece of software (that is, a VI) that controls a particular instrument. Instrument drivers eliminate the need to learn the complex, low-level programming commands for each instrument. LabVIEW is ideally suited for creating instrument drivers since the VI front panel can simulate the operation of an instrument's front panel. The block diagram sends the necessary commands to the instrument to perform the various operations specified on the front panel. When using an instrument driver, you do not need to remember the commands necessary to control the instrument—this is specified via input on the front panel.

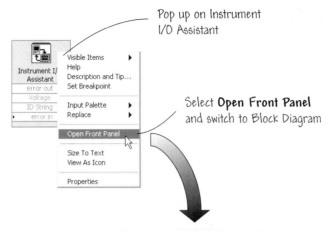

Pop up on Instrument
I/O Assistant

Select **Open Front Panel**
and switch to Block Diagram

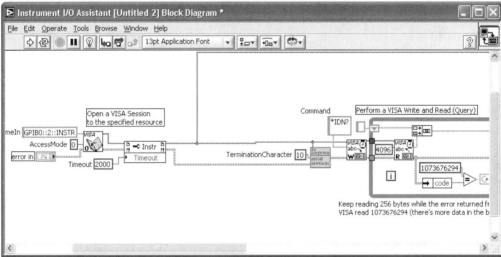

FIGURE 10.14
Viewing the code generated by the Instrument I/O Assistant.

LabVIEW provides many VIs that can be used in the development of an instrument driver for your hardware. These VIs can be grouped into the following categories:

- Standard VISA I/O functions
- Traditional GPIB functions and added capability via the GPIB 488.2
- Serial port communication functions

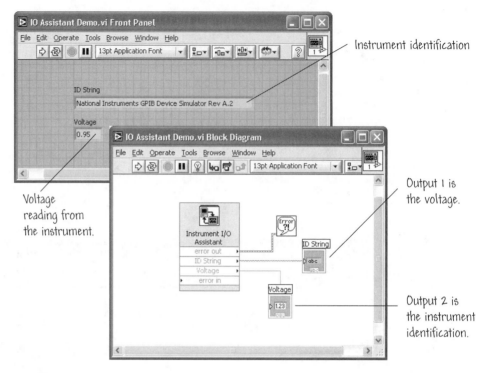

FIGURE 10.15
Completing the instrument VI.

VISA stands for Virtual Instrument Software Architecture. In essence, VISA is a VI library for controlling GPIB, serial, or VXI instruments and making the appropriate calls depending on the type of instrument. VISA by itself does not provide instrumentation programming capability—it is a high-level application programming interface (API) that calls lower-level code to control the hardware. Each VISA instrument driver VI corresponds to a programmatic operation, such as configuring, reading from, writing to, and triggering an instrument. The GPIB and serial port functions provide similar capabilities.

Two questions come to mind:

1.   When should you attempt to develop your own instrument driver from "scratch" using the VISA, GPIB, or serial port functions?

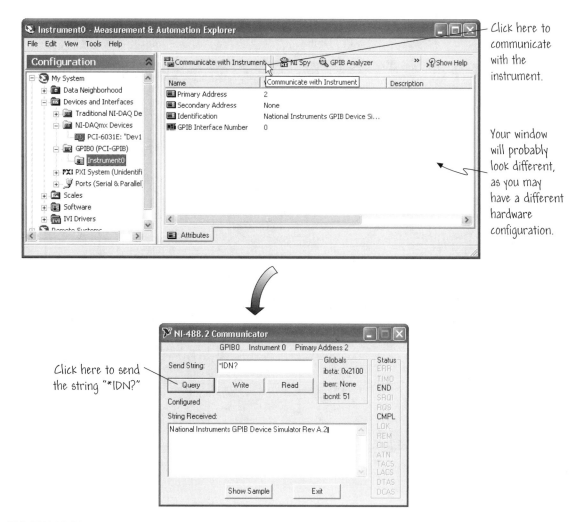

FIGURE 10.16
A National Instruments GPIB device simulator is detected.

And if you must develop your own instrument driver,

2. When should you use the VISA functions, and when should you use the GPIB functions?

The answer to the first question is simple—students new to LabVIEW should not attempt to develop their own instrument drivers! They should use the instrument drivers developed by National Instruments, rather than attempt to develop drivers from "scratch." Instrument drivers can be downloaded from the National

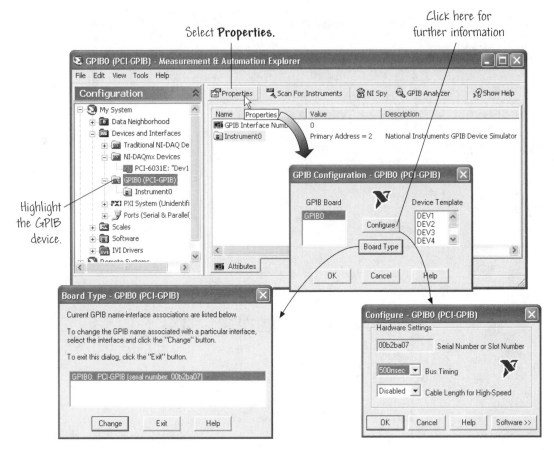

FIGURE 10.17
Examining the properties of the GPIB devices.

Instruments website using the Instrument Driver Network. To access the driver network, connect to

http://www.ni.com/idnet

directly or use Tools≫Instrumentation≫Instrument Driver Network... to automatically connect to the network as illustrated in Figure 10.18. If an Instrument Driver is not available, use the Instrument I/O Assistant.

The LabVIEW instrument driver library contains instrument drivers for a variety of programmable instruments that use the GPIB, serial, or VXI interfaces. You can use an instrument driver from the library to control your instrument, or you can customize the instrument drivers, since they are distributed with their

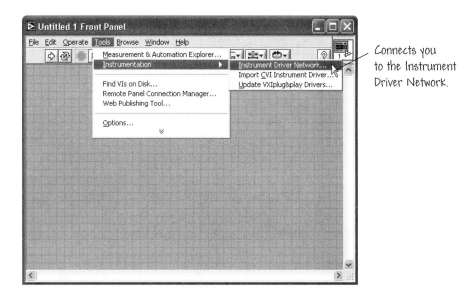

FIGURE 10.18
Link automatically to the National Instruments Instrument Driver Network.

block diagram source code. Once you have properly installed the appropriate instrument drivers using the MAX, you can access the instrument drivers on the **Functions** palette, as shown in Figure 10.19.

   If you have a Web browser installed, you can link automatically to the National Instruments Instrument Driver Network, as shown in Figure 10.18. The Instrument Driver Network provides the complete library of available instrument drivers for LabVIEW. In this Web page, you can search for your instrument among over 2200 drivers available with free source code. You then can download the instrument driver you need and install it into the instr.lib folder in the LabVIEW root directory.

If you decide to build your own instrument driver, VISA is the standard API throughout the instrumentation industry. In addition, the VISA functions can control a suite of instruments of different types, including GPIB, serial, or VXI. In other words, VISA provides interface independence. Students that need to program instruments for different interfaces only need to learn one API.

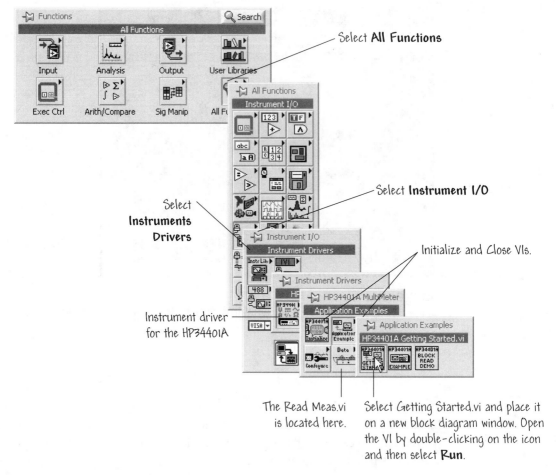

FIGURE 10.19
Locating the installed instrument drivers on the **Functions** palette.

The HP34401A Instrument Driver

In this example we will take a look at the HP34401A instrument driver. The HP 34401 A digital multimeter instrument driver is installed automatically as part of the *LabVIEW Student Edition*. Begin by first opening a new VI. Using Figure 10.19 as a guide to locating the HP34401A Getting Started VI, drop the VI on the block diagram. Then double-click on the VI, which should open up the front panel. The Getting Started VIs are used to verify communication with your instrument and test a typical programmatic instrument operation. The HP34401A Getting Started VI front panel is shown in Figure 10.20.

Address field

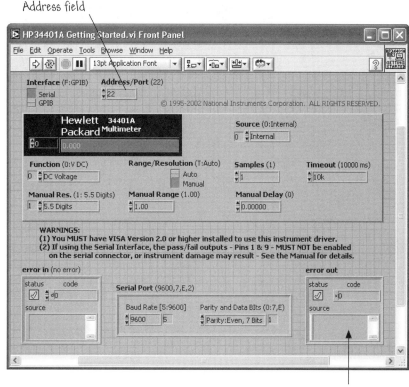

Any errors detected
will be listed here.

FIGURE 10.20
The HP34401a Getting Started VI front panel.

With the exception of the address field, the defaults for most controls on the front panel will be sufficient for your first run. You will need to set the address appropriately. If you do not know the address of your instrument, refer to the MAX for help. After running the VI, check to see that reasonable data was returned and an error was not reported in the error cluster (bottom right corner of the front panel).

If your Getting Started VI does not work, you need to check that

- NI-VISA is installed. If you did not choose this as an option during your Lab-VIEW installation, you will need to install it before rerunning your Getting Started VI.

- The instrument address is correct.

- The selected instrument driver supports the exact instrument model you are using.

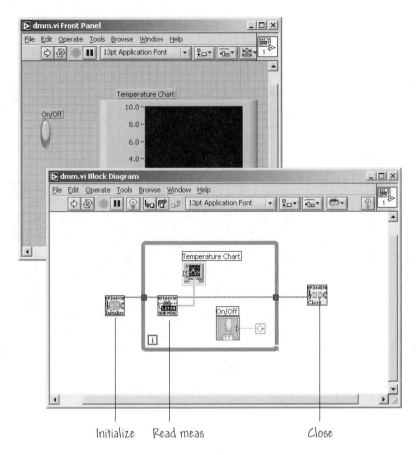

FIGURE 10.21
A simple HP34401A instrument driver.

Once you have verified basic communication with your instrument using the Getting Started VI, you may want to customize instrument control for your particular needs. If your application needs are similar to the Getting Started VI, the simplest means of creating a customized VI is to save a copy of the Getting Started VI by selecting **Save As...** from the **File** menu. You can change the default values on the front panel by selecting **Make Current Values Default** from the **Operate** menu.

The block diagram of an instrument driver generally consists of three VIs: the Initialize VI, an application function VI, and the Close VI. A simple instrument driver can be assembled from the VIs provided by LabVIEW for the HP34401A. Observing Figure 10.19, we see that the three VIs we need are HP34401A Initialize.vi, HP34401A Read Meas.vi, and HP34401A Close.vi. A straightforward instrument driver comprised of the three basic VIs is shown in Figure 10.21. Build the block diagram shown in Figure 10.21, if you have the HP34401A or do not have an instrument. If you are using another instrument driver, build a VI similar to Figure 10.21 using the Initialize, Read Meas, and Close VIs for that instrument. ◆

10.4.1   Developing Your Own Instrument Driver

Make sure to check thoroughly for existing instrument drivers for your instrument before starting to build your own from scratch. You should check first at the National Instruments website as discussed in the previous sections. If that search fails, you should then search the website of the manufacturer of your equipment—they may have developed the instrument driver already. During your website searches, be on the lookout for instrument drivers that support a similar instrument, since instruments from the same model series often have similar command sets. If you find any such drivers, download them and assess the similarity of their command sets to that of your instrument—they may work directly or with minor modifications. For instruments from the same model series, you might need to contact the manufacturer and ask for details on the differences between the command sets.

If an instrument driver for your particular instrument does not exist, you can do one of several things (in rank order):

1. Use a driver for a similar instrument. Often similar instruments from the same manufacturer have similar if not identical command sets. The degree to which an instrument driver will need to be changed will depend on how similar the instruments and their command sets are. If the command sets are very different, you may be better off starting from scratch.

2. Use the Instrument I/O Assistant.

3. Create a simple instrument driver using the guidelines in the *Developing a Simple On-line help Driver* topic found in the LabVIEW **Online Reference**.

4. Develop a complete, fully functional instrument driver. To develop a National Instruments quality driver, you can download **Application Note AN006** *Developing a LabVIEW Instrument Driver* from the National Instruments website at

http://www.ni.com/appnotes.nsf/

This application note will help you to develop a complete instrument driver. To aid in the development of your instrument driver, National Instruments has created standards for instrument driver structure, device management, instrument I/O, and error reporting. The **Application Note AN006** describes these standards, as well as the purpose of a LabVIEW instrument driver, its components, and the integration of these components. In addition, this application note suggests a process for developing useful instrument drivers.

As to the question of whether to use VISA, GPIB, or serial functions, the recommendation is to rely on VISA. There are four major reasons for using VISA over GPIB. VISA

- is the industry standard,

- provides interface independence,

- provides platform independence, and

- will be easily adaptable in future instrumentation control applications.

VISA uses the same operations to communicate with instruments regardless of the interface type. VISA is designed so that programs written using VISA function calls are easily portable from one interface type to another like GPIB, serial or VXI. To ensure platform independence, VISA strictly defines its own data types. The VISA function calls and their associated parameters are uniform across all platforms, so that software can be ported to other platforms and then recompiled. A LabVIEW program using VISA can be ported to any platform supporting LabVIEW. A final advantage of using VISA is that it is an object-oriented API that will easily adapt to new instrumentation interfaces as they are developed in the future, making application migration to the new interfaces easy.

 If you must develop your own instrument driver, you should use VISA functions rather than GPIB because of the versatility of VISA.

10.5   FUTURE OF INSTRUMENT DRIVERS AND INSTRUMENT CONTROL

Current instrument drivers use VISA, which is a common software interface for controlling GPIB, serial, or VXI instruments. At the present time and into the future, there will be one instrument driver for all oscilloscopes, for instance, no matter the manufacturer, model, or hardware interface (such as GPIB). These new instrument drivers are called Interchangeable Virtual Instruments (IVI) drivers and are supported by the IVI Foundation. The IVI Foundation is comprised of end-user test engineers and system integrators with many years of experience building

GPIB and VXI-based test systems. By defining a standard instrument driver model that enables engineers to swap instruments without requiring software changes, the IVI Foundation members believe that significant savings in time and money will result because of

- Software that does not change when instruments become obsolete.

- A single software application that can be used on a system with different instrument hardware, which maximizes existing resources.

- Portable code that can be developed in test labs and hosted on different instrument in the production environment.

The following instrument types or classes have been defined by the IVI Foundation: oscilloscope, DMM, arbitrary waveform generator, switch, and power supply. More instrument types will be defined in the future. For more information on the IVI drivers and the IVI Foundation, refer to the **Application Note AN 121** "Using IVI Drivers to Build Hardware-Independent Test Systems with LabVIEW and LabWindows/CVI" on the National Instruments website, www.ni.com/appnotes.nsf. You may also want to visit the IVI Foundation website at

http://www.ivifoundation.org/

BUILDING BLOCK

10.6   BUILDING BLOCKS: DEMO SCOPE

If you do not have instrument I/O hardware installed, you can use the Demo Scope VI as a substitute to learn about instrument control. The Demo Scope VI is the demonstration equivalent of a Getting Started VI for an actual instrument driver.

Open the Demo Scope VI instrument driver located in the library

vi.lib\tutorial.llb

Click on **Run** to begin the process of acquiring data from one or two channels on your oscilloscope, as shown in Figure 10.22. You can play around with changing the time and volts per division settings. When you are finished acquiring data, click the **Stop[F4]** button to stop the VI, as shown in Figure 10.22.

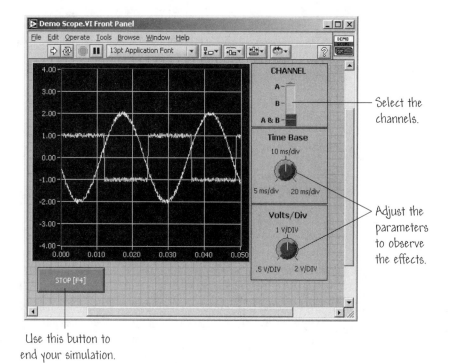

FIGURE 10.22
The Demo Scope VI front panel.

Switch to the block diagram window, shown in Figure 10.23. Notice that the first function is the initialization, followed by the commands to send to the instrument in the DemoScope_Application_Example.vi. The DemoScope_Close.vi then closes communication with the instrument. LabVIEW instrument drivers follow this model—initializing the instrument, then calling the functions to control the instrument, and finally closing the instrument communication.

10.7 RELAXED READING: AUTOMATING THE SAN FRANCISCO BAY MODEL

The San Francisco Bay Model is a physical model of the bay used to simulate tidal conditions and river flows and measure hydraulic conditions. Sensors and transducers placed around the model gather velocity and salinity readings every two seconds. The data is analyzed to determine the environmental impacts of dredging new shipping channels and constructing new levees and dams.

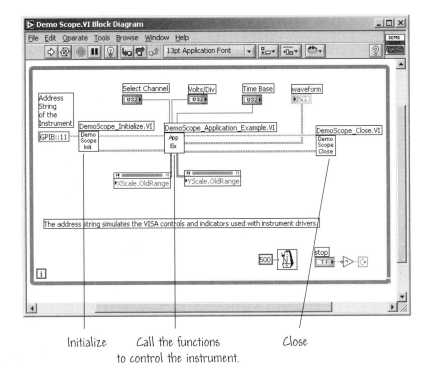

FIGURE 10.23
The Demo Scope VI block diagram.

Three areas of automation on the bay model were investigated:

1. **Tidal Control**—precise water levels at the Golden Gate Bridge must be reproduced to within 0.01 in.

2. **High Channel Sound Data Acquisition**—more than 150 sensors located around the model acquire information then made available on the Internet.

3. **Distributed Control**—flow controllers located around the model simulate river flows from the Sacramento Delta.

The tide control system is comprised of a 75 hp, three-phase pump used to drive water into the model at a constant flow rate. In the Pacific Ocean area, the model has three butterfly valves that control the flow of water out of the model and three ultrasonic level sensors to monitor the water level at three different locations.

The system monitors the water level, filters the data, and determines the proper outflow for the desired tide. Using National Instruments motion control technology, LabVIEW converts the desired outflow to motor positions on the servo-controlled butterfly valves. The PID controller is provided with a setpoint profile that represents the desired tide; LabVIEW handled the rest.

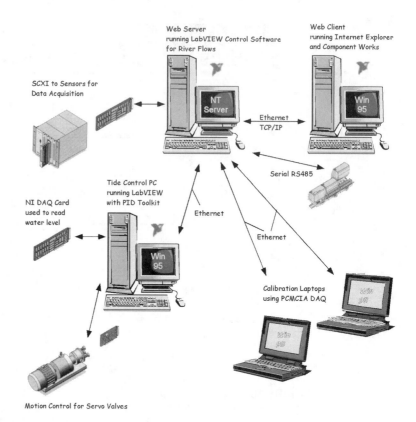

FIGURE 10.24
System configuration.

The data acquisition portion of the system is housed in a server machine containing data acquisition and signal conditioning hardware. LabVIEW scans 256 channels at a rate of 30 Hz, collects the data, scales it appropriately, and stores it in an MS Access database. An application was developed using National Instruments ComponentWorks to post the data on the Internet.

A 166 MHz Pentium laptop equipped with portable DAQ hardware was employed to perform the calibrations and send back to the server via wireless Ethernet. LabVIEW controlled the data collection and relayed the calibration information back to the data acquisition server.

Because the simulated river flows are located in distant locations around the model, they are excellent applications for National Instruments FieldPoint distributed I/O. Analog input modules monitor the flow rates and feed the data through multiple PID loops controlled with LabVIEW. FieldPoint varies the flow controllers to achieve the desired flow rates.

The biggest challenge of the entire project was the tide control system. In this large hydraulic system, controllers tend to oscillate as waves bounce off walls and reverberate throughout the system. With the analysis features of LabVIEW, integrated with the PID and motion control features, digital finite impulse response (FIR) filtering was applied to the data to eliminate undesired frequencies before feeding the data into the PID control loop. LabVIEW was invaluable in performing rapid prototyping and quick modifications of algorithms and parameters. The options were abundant and easily implemented.

For more information, contact:

Cal-Bay Systems, Inc.
3070 Kerner Blvd., Suite B
San Rafael, CA 94901
Web: www.calbay.com

10.8   SUMMARY

As we already know from the chapter on data acquisition, LabVIEW can communicate with external devices. As discussed in Chapter 8, you can use DAQ boards in conjunction with LabVIEW to read and generate analog input, analog output, and digital signals. In this chapter we learned that LabVIEW can also control external instruments (such as digital voltmeters and oscilloscopes) over the GPIB bus or through a serial port. MAX is used to detect instruments and to install instrument drivers. An instrument driver is a VI that controls a particular instrument. Students should use the instrument drivers developed by National Instruments rather than attempt to develop drivers from "scratch." The Instrument I/O Assistant provides a relatively straightforward way to build instrument communication VIs. Instrument drivers can be downloaded from the National Instruments website using the Instrument Driver Network.

KEY TERMS

Bus: The means by which computers and instruments transfer data.

Controller: A GPIB device that manages the flow of information on the GPIB by sending commands to all devices. A computer usually handles the controller function.

Controller-in-charge: The active controller in a GPIB system.

Device-dependent messages: Messages that contain device-specific information such as programming instructions, measurement results, machine status, and data files. These are often called **data messages**.

GPIB: General Purpose Interface Bus is the common name for the communications interface system defined in ANSI/IEEE Standard 488-1975 and 488.2-1987.

IEEE: Institute for Electrical and Electronic Engineers.

Instrument driver: A set of LabVIEW VIs that communicate with an instrument using standard VISA I/O functions.

Interface messages: Messages that manage the bus itself, and perform such tasks as initializing the bus, addressing and unaddressing devices, and setting device modes for remote or local programming. These are often called **command messages**.

Listener: A GPIB device that receives data messages from the talker.

MAX: A utility that provides information about connected instruments, and installs and manages instrument drivers.

Serial communication: A popular means of transmitting data between a computer and a peripheral device by sending data one bit at a time over a single communication line.

Talker: A GPIB device that sends data messages to one or more listeners.

VISA: Virtual Instrument Software Architecture. VISA is a VI library for controlling GPIB, serial, or VXI instruments.

EXERCISES

E10.1   A more sophisticated oscilloscope than the one discussed in "Building Blocks" is called Two Channel Oscilloscope.vi and is located in Examples\Apps\

demos.llb. Open and run the oscilloscope example. You can select between displaying channels A, B, or both A and B concurrently. Once the channel(s) is selected, you can set the time base, volts.div, trigger source, slope, and level. This VI is intended to demonstrate the flexibility of LabVIEW for instrument control.

E10.2 From the LabVIEW startup screen, select a **New VI**. Select Tools≫Instrumentation≫Instrument Driver Network.... Once you have connected to the website, navigate around until you find and click on **Download Driver**. Locate an instrument you are using or have used and download the driver. Run the executable to install the VIs in the proper location. Close LabVIEW, then open LabVIEW. Select **New VI** and select Functions≫Instrument Drivers and the subpalette of the driver you just downloaded. You will notice that the VI organization is very similar to that of the HP34401A instrument driver discussed in this chapter.

E10.3 Set a VISA alias for your instrument in MAX. In MAX, right-click your instrument and select **VISA Properties** from the shortcut menu to set the VISA alias for your instrument.

PROBLEMS

P10.1 Open Frequency Response.vi located in Examples\Apps\freqresp.llb. This VI simulates an application that uses GPIB instruments to perform a frequency response test. What brand of digital multimeter is being simulated in this example? Switch to the block diagram and verify that For Loops, Formula Nodes, graphs, and arrays are some of the LabVIEW objects used in the code. Run the VI in continuous run mode and observe the effects of varying the amplitude and frequency of the signal.

P10.2 Open a new VI and launch the Instrument I/O Assistant Express VI (found on the **Functions≫Input** palette). Select the instrument to communicate with. Add a **Query and Parse** step, enter the command string, and run the step. Use the Auto parse feature to parse the returned data and enter a token name for the returned data. Next, use the **Write** step to send another command and the **Read and Parse** step to read data from the instrument. Again, use the Auto parse feature and specify a token name for the data. Return to LabVIEW and display the returned data on the front panel.

P10.3 Create a VI that communicates with an instrument from LabVIEW using VISA. Use the VISA Open, VISA Write, VISA Read, and VISA Close functions. Use the VISA alias you created in Exercise 10.3 to address the instrument. Display the write string (i.e., *IDN?) and the data to be read on the front panel. (*Hint*: You need to specify the byte count on the VISA Read in order to return data.)

P10.4 Complete the crossword puzzle.

Across

1. A popular means of transmitting data between a computer and a peripheral device by sending data one bit at a time over a single communication line.
5. A GPIB device that manages the flow of information on the GPIB by sending commands to all devices.
7. A GPIB device that receives data messages from the talker.
8. A GPIB device that sends data messages to one or more listeners.
9. A set of LabVIEW VIs that communicate with an instrument using standard VISA I/O functions.
11. The common name for the communications interface system defined in ANSI/IEEE Standard 488–1975 and 488.2–1987.
12. The means by which computers and instruments transfer data.

Down

2. Messages that manage the bus itself, and perform such tasks as initializing the bus and addressing devices.
3. The active controller in a GPIB system.
4. Institute for Electrical and Electronic Engineers.
6. A VI library for controlling GPIB, serial, or VXI instruments.
10. A utility that provides information about connected instruments, and installs and manages instrument drivers.

CHAPTER 11

Analysis

LabVIEW is an excellent environment for analysis of signals and systems. The G programming language is ideally suited to developing programs for solving linear algebraic systems of equations, curve fitting, integrating ordinary differential equations, computing function zeroes, computing derivatives of functions, integrating functions, generating and analyzing signals, computing discrete Fourier transforms, and filtering signals. The Express VIs for curve fits, filters, signal generation, and spectral analysis are described. This chapter presents an overview of these topics.

GOALS

1. Introduce some of LabVIEW's capabilities for analyzing signals and systems.

2. Study some of the mathematical analysis VIs made available by LabVIEW, especially the analysis Express VIs.

11.1   LINEAR ALGEBRA

LabVIEW provides a number of important VIs dedicated to solving systems of linear algebraic equations. These types of systems arise frequently in engineering and scientific applications. An entire branch of mathematics is dedicated to the study of *linear algebra*. We will only be able to touch on a few important subjects within the linear algebra arena in the space available to us here.

11.1.1   Review of Matrices

The basic element used in the computations is the so-called *matrix*. A matrix is represented by an $m \times n$ array of numbers:

$$
\mathbf{A} =
\begin{bmatrix}
a_{0,0} & a_{0,1} & \cdots & a_{0,n-1} \\
a_{1,0} & a_{1,1} & \cdots & a_{1,n-1} \\
\cdots & \cdots & \cdots & \cdots \\
a_{m-1,0} & a_{0,1} & \cdots & a_{m-1,n-1}
\end{bmatrix},
$$

where n is the number of columns, and m is the number of rows. When $m \neq n$, the matrix is called a *rectangular* matrix; conversely when $m = n$, the matrix is called a *square* matrix. An $m \times 1$ matrix is called a *column vector*, and a $1 \times n$ matrix is called a *row vector*. Other special forms of the matrix are the *diagonal* matrix, the *zero* matrix, and the *identity* matrix. Examples of these three types of matrices are

$$
\mathbf{A} =
\begin{bmatrix}
2 & 0 & 0 \\
0 & 4 & 0 \\
0 & 0 & -6
\end{bmatrix}, \quad
\mathbf{0} =
\begin{bmatrix}
0 & 0 & 0 \\
0 & 0 & 0 \\
0 & 0 & 0
\end{bmatrix}, \quad
\mathbf{I} =
\begin{bmatrix}
1 & 0 & 0 \\
0 & 1 & 0 \\
0 & 0 & 1
\end{bmatrix},
$$

respectively. The elements of a matrix, denoted by $a_{i,j}$, can be real or complex numbers.

Addition of two matrices is performed element by element. For example,

$$
\begin{bmatrix} 2 & 3 & 1 \\ -7 & 4 & 5 \\ 2 & 0 & -6 \end{bmatrix} + \begin{bmatrix} 2 & -7 & 2 \\ 2 & 4 & 1 \\ 1 & 5 & -1 \end{bmatrix} = \begin{bmatrix} 4 & -4 & 3 \\ -5 & 8 & 6 \\ 3 & 5 & -7 \end{bmatrix}.
$$

You can easily check that $\mathbf{A} + \mathbf{B} = \mathbf{B} + \mathbf{A}$ and that $(\mathbf{A} + \mathbf{B}) + \mathbf{C} = \mathbf{A} + (\mathbf{B} + \mathbf{C})$. Therefore, matrix addition is commutative and associative.

If we multiply a matrix by a scalar, α, the result is obtained by multiplying each element of the matrix by the scalar, yielding

$$
\alpha \mathbf{A} = \begin{bmatrix} \alpha a_{0,0} & \alpha a_{0,1} & \cdots & \alpha a_{0,n-1} \\ \alpha a_{1,0} & \alpha a_{1,1} & \cdots & \alpha a_{1,n-1} \\ \cdots & \cdots & \cdots & \cdots \\ \alpha a_{m-1,0} & \alpha a_{0,1} & \cdots & \alpha a_{m-1,n-1} \end{bmatrix}.
$$

Multiplication of matrices $\mathbf{AB}$ requires the two matrices to be of compatible dimensions—the number of columns of $\mathbf{A}$ must be equal to the number of rows of $\mathbf{B}$. Thus if $\mathbf{A}$ is an $m \times n$ matrix, and $\mathbf{B}$ is an $n \times p$ matrix, the product $\mathbf{C} = \mathbf{AB}$ is an $m \times p$ matrix. The element $c_{i,j}$ of the matrix $\mathbf{C}$ is given by

$$
c_{i,j} = a_{i,1}b_{1,j} + a_{i,2}b_{2,j} + \cdots + a_{i,m}b_{m,j}.
$$

In general, matrix multiplication is not commutative; that is,

$$
\mathbf{AB} \neq \mathbf{BA}.
$$

The *transpose* of a real matrix (that is, a matrix comprised of only real numbers) is formed by interchanging the rows and columns. For example, if

$$
\mathbf{A} = \begin{bmatrix} 2 & 3 & 1 \\ -7 & 4 & 5 \end{bmatrix}, \quad \text{then} \quad \mathbf{A}^T = \begin{bmatrix} 2 & -7 \\ 3 & 4 \\ 1 & 5 \end{bmatrix},
$$

where the matrix $\mathbf{A}^T$ is the transpose of $\mathbf{A}$. A real matrix is called a *symmetric* matrix if $\mathbf{A}^T = \mathbf{A}$. If the elements of the matrix $\mathbf{C}$ are complex numbers, then we extend the notion of a transpose to *complex conjugate transpose*. This means that we transpose the matrix and then replace each element with its own complex conjugate. We denote the complex conjugate transpose as $\mathbf{C}^H$. A complex matrix is called a *Hermitian* matrix if $\mathbf{C}^H = \mathbf{C}$.

For square matrices, we can define the operations **trace**, **determinant**, and **inversion**. The trace of an $n \times n$ square matrix $\mathbf{A}$ is the sum of the diagonal elements,

$$\text{tr } \mathbf{A} = a_{1,1} + a_{2,2} + \cdots + a_{n,n}.$$

The determinant of a 2×2 matrix is given by

$$\det \mathbf{A} = \left| \begin{bmatrix} a_{1,1} & a_{1,2} \\ a_{2,1} & a_{2,2} \end{bmatrix} \right| = a_{1,1}a_{2,2} - a_{1,2}a_{2,1}.$$

If the determinant is identically equal to zero, then we say that the matrix is *singular*. In general, the determinant can be computed as a function of the *minors* and *cofactors* of the matrix. The minor of an element $a_{i,j}$ is the determinant of an $n - 1 \times n - 1$ matrix formed by removing the ith row and the jth column of the original matrix $\mathbf{A}$. For example, if

$$\det \mathbf{A} = \left| \begin{bmatrix} a_{1,1} & a_{1,2} & a_{1,3} \\ a_{2,1} & a_{2,2} & a_{2,3} \\ a_{3,1} & a_{3,2} & a_{3,3} \end{bmatrix} \right|,$$

then the minor of the $a_{2,3}$ element is

$$M_{2,3} = \left| \begin{bmatrix} a_{1,1} & a_{1,2} \\ a_{3,1} & a_{3,2} \end{bmatrix} \right| = a_{1,1}a_{3,2} - a_{1,2}a_{3,1}.$$

The cofactor of $a_{i,j}$ is defined as

$$\alpha_{i,j} = \text{cofactor } a_{i,j} = (-1)^{i+j}M_{i,j}.$$

In general, we can compute the determinant of an $n \times n$ square matrix $\mathbf{A}$ as

$$\det \mathbf{A} = \sum_{j=1}^{n} a_{i,j}\alpha_{i,j}$$

for any row i. Similarly, we can compute the determinant as

$$\det \mathbf{A} = \sum_{i=1}^{n} a_{i,j}\alpha_{i,j}$$

for any column j.

The *adjoint matrix* of an $n \times n$ square matrix $\mathbf{A}$ is formed by transposing the matrix and replacing each element $a_{i,j}$ with the cofactor $\alpha_{i,j}$. Therefore, we have

$$\text{adjoint } \mathbf{A} = \begin{bmatrix} \alpha_{1,1} & \alpha_{1,2} & \alpha_{1,3} \\ \alpha_{2,1} & \alpha_{2,2} & \alpha_{2,3} \\ \alpha_{3,1} & \alpha_{3,2} & \alpha_{3,3} \end{bmatrix}^{T} = \begin{bmatrix} \alpha_{1,1} & \alpha_{2,1} & \alpha_{3,1} \\ \alpha_{1,2} & \alpha_{2,2} & \alpha_{3,2} \\ \alpha_{1,3} & \alpha_{2,3} & \alpha_{3,3} \end{bmatrix}.$$

The matrix inverse is denoted by $\mathbf{A}^{-1}$ and can be computed as

$$\mathbf{A}^{-1} = \frac{\text{adjoint } \mathbf{A}}{\det \mathbf{A}}.$$

The matrix inverse must satisfy the relationship

$$\mathbf{A}^{-1}\mathbf{A} = \mathbf{A}\mathbf{A}^{-1} = \mathbf{I}.$$

The matrix inverse does not exist (that is, it is singular) when $\det \mathbf{A} = 0$. If the matrix $\mathbf{A}$ is singular, then there exists a nonzero vector $\mathbf{v}$ such that $\mathbf{A}\mathbf{v} = \mathbf{0}$.

11.1.2 System of Algebraic Equations

Suppose that we want to solve the following system of algebraic equations:

$$4x_1 + 6x_2 + x_3 = 4$$
$$x_1 + 2x_2 + 3x_3 = 0$$
$$5x_2 - x_3 = 1$$

The unknowns are the variables x_1, x_2, and x_3. We can identify the two column vectors $\mathbf{x}$ and $\mathbf{b}$ as

$$\mathbf{x} = \begin{bmatrix} x_1 \\ x_2 \\ x_3 \end{bmatrix} \quad \text{and} \quad \mathbf{b} = \begin{bmatrix} 4 \\ 0 \\ 1 \end{bmatrix}.$$

Then we can write the system of algebraic equations as

$$\mathbf{A}\mathbf{x} = \mathbf{b},$$

where

$$\mathbf{A} = \begin{bmatrix} 4 & 6 & 1 \\ 1 & 2 & 3 \\ 0 & 5 & -1 \end{bmatrix}.$$

Thus, we have rewritten the problem in a compact matrix notation. Can we solve for the vector $\mathbf{x}$? In this case, the matrix $\mathbf{A}$ is invertible (i.e., the inverse exists), so the solution is readily obtained as

$$\mathbf{x} = \mathbf{A}^{-1}\mathbf{b} = \begin{bmatrix} 0.9123 \\ 0.1228 \\ -0.3860 \end{bmatrix}.$$

If the matrix $\mathbf{A}$ is singular, then the number of solutions depends on the vector $\mathbf{b}$. If a solution does exist in the singular case, it is not unique!

For any given $n \times n$ square matrix $\mathbf{A}$, we would like to know if there exists a scalar λ and a corresponding vector $\mathbf{v} \neq \mathbf{0}$ such that

$$\lambda \mathbf{v} = \mathbf{A}\mathbf{v}.$$

This scalar λ is called the *eigenvalue*, and the corresponding vector $\mathbf{v}$ is called the *eigenvector*. Rearranging yields

$$\lambda \mathbf{v} - \mathbf{A}\mathbf{v} = (\lambda \mathbf{I} - \mathbf{A})\,\mathbf{v} = \mathbf{0}.$$

Therefore, a solution exists (for $\mathbf{v} \neq \mathbf{0}$) if and only if

$$\det\,(\lambda \mathbf{I} - \mathbf{A}) = 0.$$

If $\mathbf{A}$ is an $n \times n$ matrix, then $\det\,(\lambda \mathbf{I} - \mathbf{A}) = 0$ is an nth-order polynomial (known as the *characteristic equation*) whose solutions are called the *characteristic roots* or *eigenvalues*. The eigenvalues of a square matrix are not necessarily unique and may be complex numbers (even for a real-valued matrix!). Given the eigenvalues of a matrix, we can compute the trace and determinant as

$$\text{tr } \mathbf{A} = \sum_{i=1}^{n} \lambda_i$$

$$\det \mathbf{A} = \prod_{i=1}^{n} \lambda_i$$

We see that if any eigenvalue is zero, then the determinant is zero, and thus it follows that the matrix is singular.

An interesting fact regarding eigenvalues is that if a matrix is real and symmetric, then its eigenvalues will be real. If we compute the "square" of a real matrix $\mathbf{A}$ according to $\mathbf{B} = \mathbf{A}\mathbf{A}^T$, then $\mathbf{B}$ is real and symmetric. In fact, the matrix $\mathbf{B}$ is nonnegative, in the sense that, for any nonzero column vector $\mathbf{v}$ it follows that the scalar value $\mathbf{v}^T \mathbf{B}\mathbf{v} \geq 0$. Now if we compute the eigenvalues of $\mathbf{B}$, we find that they are all nonnegative and real. Taking the square root of each eigenvalue yields quantities known as *singular values*, and denoted here by β_i for $n = 1, 2, \ldots, n$. An $n \times n$ matrix has n nonnegative singular values. Two important singular values are the maximum and minimum values, β_{max} and β_{min}, respectively. Singular values are important in computational linear algebra because they tell us something about how close a matrix is to being singular. We know that if the determinant of a matrix is zero, then it is a singular matrix. But what about if you compute the determinant numerically with the computer and determine that the determinant is 10^{-10}? Is this close enough to zero to say that the matrix is singular? The answer to this is provided by the **condition number** of a matrix. It can be defined in different ways, but it is commonly defined as

$$\text{cond } \mathbf{A} = \frac{\beta_{max}}{\beta_{min}}.$$

The condition number can vary from 0 to ∞. A matrix with a condition number near 1 is closer to being nonsingular than a matrix with a very large condition number. The condition number is useful in assessing the accuracy of solutions to systems of linear algebraic equations.

As a final practical note, it is not generally a good idea to explicitly compute the matrix inverse when solving systems of linear algebraic equations, because inaccuracies are associated with the numerical computations—especially when the condition number is high. The preferred solution technique involves using *matrix decompositions*. Popular techniques include

- Singular Value Decomposition (SVD)
- Cholesky decomposition (or QR)

The idea is to decompose the matrix into component matrices that have "nice" numerical properties.

For example, suppose we decompose the matrix

$$\mathbf{A} = \mathbf{QR},$$

where $\mathbf{Q}$ is an *orthogonal* matrix (that is, $\mathbf{Q}^T \mathbf{Q} = \mathbf{I}$), and $\mathbf{R}$ is upper triangular (all the elements below the diagonal are zero). Then,

$$\mathbf{Ax} = \mathbf{QRx} = \mathbf{b}.$$

Multiplying both sides by $\mathbf{Q}^T$ yields

$$\mathbf{Rx} = \mathbf{Q}^T \mathbf{b}.$$

Then you use the fact that $\mathbf{R}$ is upper triangular to *back-substitute* and solve for $\mathbf{x}$ without ever computing a matrix inverse.

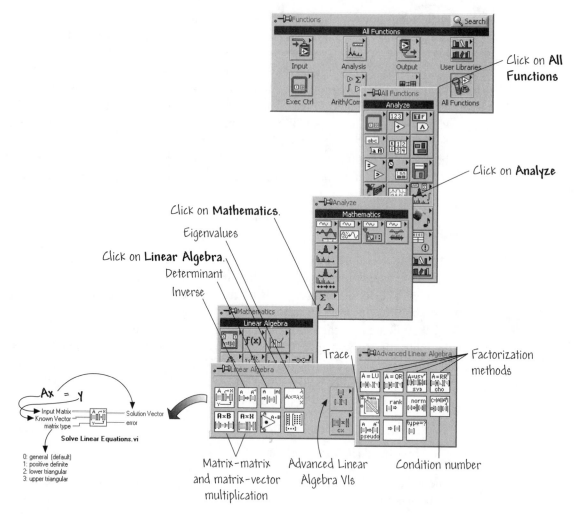

FIGURE 11.1
The linear algebra VIs.

11.1.3 Linear System VIs

The *Student Edition of LabVIEW* comes with a complete set of VIs that can be used to perform all the matrix computations discussed previously (and much more!). You can find the VIs on the palette shown in Figure 11.1.

The Linear Algebra Calculator.vi is a useful VI that can be used to perform a variety of linear algebra operations on a matrix **A** or to solve systems of linear algebraic equations. This VI can take data as input on the front panel or read the data from a file (that is, from a spreadsheet file). The next example gives you the opportunity to experiment with linear algebra computations.

Linear Algebra Calculator

Open Linear Algebra Calculator.vi located in Examples\Analysis\ linaxmpl.llb. The front panel and block diagram are shown in Figure 11.2.

By default, the VI is set up to solve the system of linear algebraic equations

$$\mathbf{Ax} = \mathbf{b},$$

where

$$\mathbf{A} = \begin{bmatrix} 4 & 2 & -1 \\ 1 & 4 & 1 \\ 0.1 & 1 & 2 \end{bmatrix}, \quad \text{and} \quad \mathbf{b} = \begin{bmatrix} 2 \\ 12 \\ 10 \end{bmatrix}.$$

Choose **Solve Linear Equations** in the lower left side of the front panel and run the VI. Verify that you obtain the solution

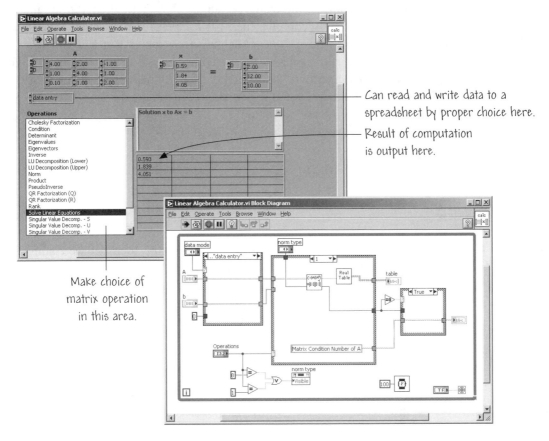

FIGURE 11.2
The linear algebra calculator.

$$\mathbf{x} = \begin{bmatrix} 0.59 \\ 1.84 \\ 4.05 \end{bmatrix}.$$

Use the calculator to compute the determinant of the matrix $\mathbf{A}$. You should get the result det $\mathbf{A} = 23.6$. Compute the condition number, the inverse, and the trace of the matrix $\mathbf{A}$ (condition number $= 4.418$ and trace $= 10$). The VI runs until the stop button is pressed.

Modify the input matrix to be

$$\mathbf{A} = \begin{bmatrix} 4 & 2 & -1 \\ 1 & 4 & 1 \\ 2 & 8 & 2 \end{bmatrix}.$$

Compute the condition number. You should find that the condition number is very, very large! Now compute the matrix inverse (by selecting **Inverse** from the menu). What happens? An error message appears that lets you know that the system of equations cannot be solved because the input matrix is singular. Vary the (3,3) element and observe the effect on the condition number. When the condition number reduces to less than 10, compute the matrix inverse again. Does it work in this case? It should!

Set the (3,3) term of the matrix to 2.01:

$$\mathbf{A} = \begin{bmatrix} 4 & 2 & -1 \\ 1 & 4 & 1 \\ 2 & 8 & 2.01 \end{bmatrix}.$$

Verify that the determinant is 0.14. Now compute the condition number—it should be quite high (above 2,500). What does this result lead you to conclude about the advisability of solving the system of linear algebraic equations by inverting the $\mathbf{A}$ matrix? Basically, when the condition number is high, it is not advisable to solve the system of equations by matrix inversion.   ◆

11.2 CURVE FITTING

Curve fitting is a common technique used in science, engineering, business, medicine, and other fields in the analysis of data. The technique involves extracting a set of curve parameters (or coefficients) from the data set to obtain a functional description of the data set. Using curve fitting, digital data can be represented by a continuous model. For example, you may want to fit the data with a straight line model. The curve-fitting procedure would provide values for the linear curve fit in terms of slope and axis offset.

11.2.1   Curve Fits Based on Least Squares Methods

The main algorithm used in the curve-fitting process is known as the least squares method. Define the error as

$$e(\mathbf{a}) = [f(x, \mathbf{a}) - y(x)]^2,$$

where $e(\mathbf{a})$ is a measure of the difference between the actual data and the curve fit, $y(x)$ is the observed data set, $f(x, \mathbf{a})$ is the functional description of the data set (this is the curve-fitting function), and $\mathbf{a}$ is the set of curve coefficients that best describes the curve. For example, let $\mathbf{a} = (a_0, a_1)$. Then the functional description of a line is

$$f(x, \mathbf{a}) = a_0 + a_1 x.$$

The least squares algorithm finds $\mathbf{a}$ by solving the Jacobian system

$$\frac{\partial e(\mathbf{a})}{\partial \mathbf{a}} = 0.$$

The curve-fitting VIs solve the Jacobian system automatically and return the set of coefficients that best describes the input data set. The automatic nature of this process provides the opportunity to concentrate on the results of the curve fitting rather than dealing with mechanics of obtaining the curve-fit parameters.

When we curve-fit data, we generally have available two input sequences, Y and X. The sequence X is usually the independent variable (e.g., time) and the sequence Y is the actual data. A point in the data set is represented by (x_i, y_i), where x_i is the ith element of the sequence X, and y_i is the ith element of the sequence Y. Since we are dealing with samples at discrete points, the VIs calculate the mean-square error (MSE), which is a relative measure of the residuals between the expected curve values and the actual observed values, using the formula

$$\text{MSE} = \frac{1}{n} \sum_{i=0}^{n-1} (f_i - y_i)^2,$$

where f_i is the sequence of fitted values, y_i is the sequence of observed values, and n is the number of input data points.

LabVIEW offers a number of curve-fitting algorithms, including:

■   Linear Fit:

$$y_i = a_0 + a_1 x_i$$

- Exponential Fit:

$$y_i = a_0 e^{a_1 x_i}$$

- General Polynomial Fit:

$$y_i = a_0 + a_1 x_i + a_2 x_2 + \cdots$$

- General Linear Fit:

$$y_i = a_0 + a_1 f_1(x_i) + a_2 f_2(x_i) + \cdots$$

where y_i is a linear combination of the parameters $a_0, a_1, \ldots$. This type of curve fit provides user-selectable algorithms (including SVD, Householder, Givens, LU, and Cholesky) to help achieve the desired precision and accuracy.

- Nonlinear Levenberg-Marquardt Fit:

$$y_i = f(x_i, a_0, a_1, a_2, \ldots)$$

where $a_0, a_1, a_2, \ldots$ are the parameters. This method does not require y to have a linear relationship with $a_0, a_1, a_2, \ldots$. Although it can be used for linear curve fitting, the Levenberg-Marquardt algorithm is generally used for nonlinear curve fits.

The curve-fitting VIs can be found on the **Curve Fit** palette, as illustrated in Figure 11.3.

Practicing with Curve Fitting

Open Regressions Demo.vi located in Examples\Analysis\regressn.llb. The front panel is shown in Figure 11.4. The VI generates noisy data samples that are approximately linear, exponential, or polynomial and then uses the appropriate curve-fitting VIs to determine the best parameters to fit the given data. You can control the noise level with the knob on the front panel. You can also select an algorithm and the number of samples to fit. For the polynomial curve fits, the order of the polynomial can be varied via front panel input.

Select *Linear* in the **Algorithm Selector** control and set the **Noise Level** to around 0.05. Run the VI and make a note of the computed error displayed in the **mse** indicator. Increase the noise level to 0.1 and again make a note of the computed error. Continue this process for noise levels of 0.15, 0.2, and 0.25. Did you detect any trends? You should have seen the MSE increase as the noise level increased.

Select Polynomial in the **Algorithm Selector** control and set the **Noise Level** to around 0.05. Set the **Order** control to 2. Run the VI and make a note of the computed error. Follow the same procedure as above. Did you obtain the same

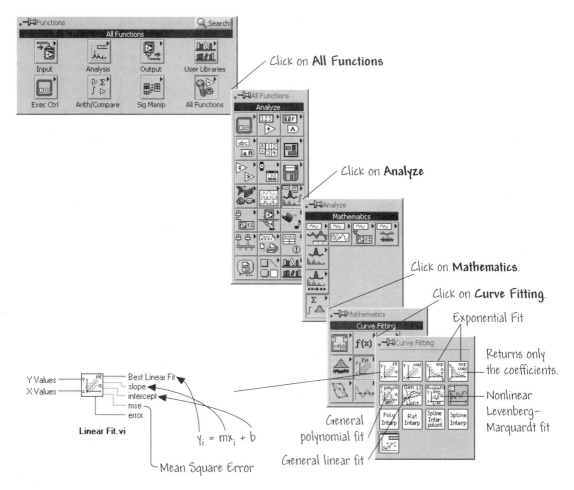

FIGURE 11.3
The curve-fitting VIs.

trends as the noise increased? Reduce the noise level to 0.15. This time run a computer experiment with the noise level fixed but increase the polynomial order from 2 to 6. Did you detect any trends as the polynomial order increased? In this case, as you increase the polynomial order, the computed error fluctuates, but remains basically on the same order of magnitude.

With the **Algorithm Selector** control set to Polynomial, run the VI with the polynomial order set to 0. Then change the polynomial order to 1 and run the VI. With polynomial order equal to 0 and 1, the fitted curve is a horizontal line and a straight line with a (generally) nonzero slope, respectively. Experiment with the Regressions Demo VI and see if you can discover new trends. Consider comparing the linear fit with the exponential fit.   ◆

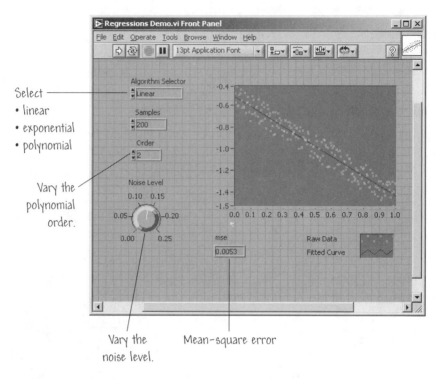

Select —
• linear
• exponential
• polynomial

Vary the polynomial order.

Vary the noise level.   Mean-square error

FIGURE 11.4
A demo to investigate curve fitting with linear, exponential, and polynomial fits.

In Examples\Analysis\regressn.llb you will find other VIs that can be used to experiment with curve fits. The general linear fit and nonlinear Levenberg-Marquardt fit demonstration VIs are called General LS Fit Example.vi and Nonlinear Lev-Mar Exponential Fit.vi, respectively.

11.2.2 Fitting a Curve to Data with Normal Distributions

Real-world data is very often normally (or Gaussian) distributed. The mathematical description of a normal distribution is

$$f(x) = \frac{1}{\sigma\sqrt{2\pi}} \exp\left[-\frac{1}{2}\left(\frac{x-m}{\sigma}\right)^2\right],$$

where m is the mean and σ is the standard deviation. Figure 11.5 shows the normal distribution with $m = 0$ and $\sigma = 0.5$, 1.0, and 2.0. As seen in the figure, the normal distribution is bell-shaped and symmetric about the mean, m. The area under the bell shaped curve is unity! The two parameters that completely describe the normally distributed data are the mean, m, and the standard deviation, σ. The peak of the bell-shaped curve occurs at m. The smaller the value of σ, the higher the peak at the mean and the narrower the curve.

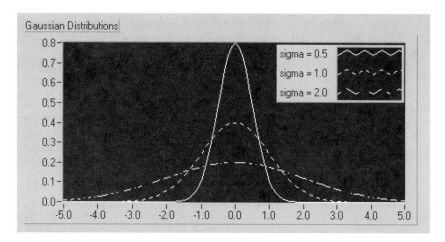

FIGURE 11.5
The normal distribution shape for three values of σ: 0.5, 1.0, and 2.0.

This normal distribution is illustrated in the Figure 11.6. The standard deviation is an important parameter that defines the "bounds" within which a certain percentage of the data values are expected to occur. For example:

- About two-thirds of the values will lie between $m - \sigma$ and $m + \sigma$.

- About 95% of the values will lie between $m - 2\sigma$ and $m + 2\sigma$.

- About 99% of the values will lie between $m - 3\sigma$ and $m + 3\sigma$.

Therefore, an interpretation of these values is that the probability that a normally distributed random value lies outside $\pm 2\sigma$ is approximately 0.05 (or 5%).

Normal Distributions

In the folder **Chapter 11** in the **Learning** directory you will find a VI called **Normal (Gaussian) Fit.vi**. This VI generates a random data set and then plots the distribution. The front panel is shown in Figure 11.7. If you want to experiment with normally distributed data, open and run the VI. You can vary the number of samples and the standard deviation (that is, the σ). If you set $\sigma = 0.5$, 1.0, and 2.0, you will be able to duplicate the graphs shown in Figure 11.5.

Run the VI and vary the number of samples from 10 to 10,000. You should notice that as the number of samples increases, the shape of the data distribution becomes more and more bell-shaped. Does this exercise relate in any way to your experience with grade distributions in class? Would you expect to have a bell-shaped grade distribution in a class of 5 students? 100 students? 500 students? (Answers: no, maybe, yes)

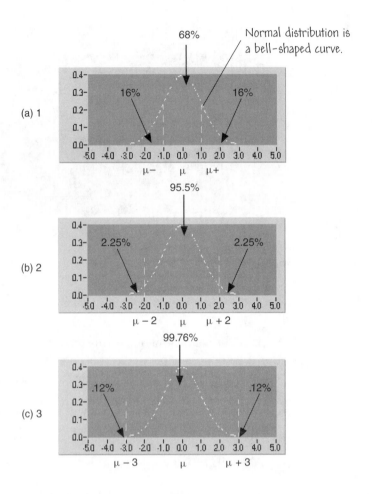

FIGURE 11.6
The normal distribution for 1σ, 2σ, and 3σ.

Using **Normal (Gaussian) Fit.vi** as a starting point, construct a VI to curve-fit the normal distribution and to compute the mean and sigma from the curve-fit parameters. ◆

11.2.3 The Curve Fitting Express VI

LabVIEW 7.0 provides a significant number of Express VIs designed for analysis purposes. The **Signal Analysis** Express palette is shown in Figure 11.8. We will discuss only a subset of the Express VIs: Curve Fitting, Spectral, Filter, and Simulate Signals. In this section, we present the Curve Fitting Express VI.

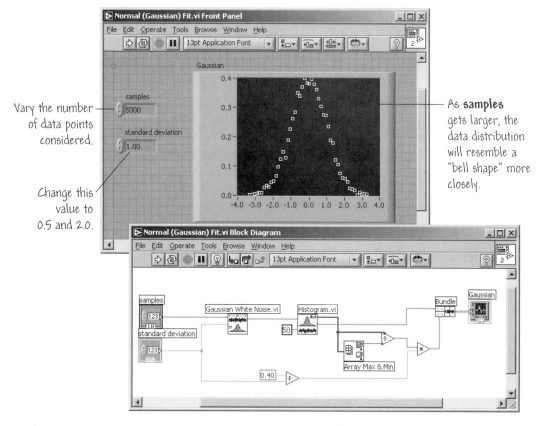

FIGURE 11.7
Normally distributed data.

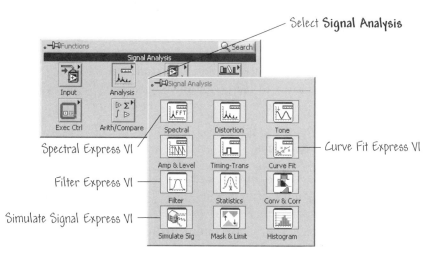

FIGURE 11.8
The Signal Analysis Express VIs.

To use the Curve Fitting VI, you first place it on the block diagram. As illustrated in Figure 11.9, once the Curve Fitting Express VI is placed on the block diagram, a dialog box automatically appears to configure the VI. The curve fitting options include a linear fit, a quadratic fit, a spline fit, a polynomial fit (of order specified by the user) and a general least squares fit. In Figure 11.9, a 5th-order polynomial fit is selected. Once the Curve Fitting VI is configured, click **OK** to return to the block diagram.

On the block diagram, the Express VI will expand to display the inputs and outputs. The key input is the signal, and the key output is the best fit signal. You can also access the curve fit residuals and mean squared error. These are useful in quantifying the accuracy of the curve fit itself. The Curve Fitting VI is now ready for inclusion in a VI for signal analysis.

Practice with the Curve Fitting Express VI

Consider the VI shown in Figure 11.10. Notice on the block diagram that two Express VIs—the Curve Fitting Express VI and the Simulate Signal Express VI—are utilized. This VI was developed to assist in investigating the impact of selecting various curve fitting strategies to fit a noisy sine wave signal.

The Curve Fitting Express Demo.vi is located in Chapter 11 of the Learning directory.

There are four elements on the block diagram:

- The Curve Fitting Express VI—this can be configured by double-clicking on the icon to access the configuration dialog box.

- The Simulate Signal Express VI—this VI will be described in Section 11.7 in more detail. It has been configured here to provide a sine wave signal at a frequency of 10.1 Hz, and normally distributed white noise is added to the signal.

- The Merge Signals function—this function was described in Section 9.2.5. Its purpose is to merge the raw signal and the best fit signal (after curve fitting) into a format acceptable to the waveform graph.

- The waveform graph—this is used to graphically display the smoothed signal and the raw signal.

Run the Curve Fitting Express Demo.vi and observe the results on the waveform graph. Recall that the Curve Fitting Express VI was configured to use a 5th-order polynomial fit.

Double-click the Curve Fitting Express VI to access the configuration dialog box. Reduce the order of the polynomial fit to 2nd-order. You should find that the curve fit is now much poorer than before with the 5th-order polynomial. Try the linear fit and see what happens.

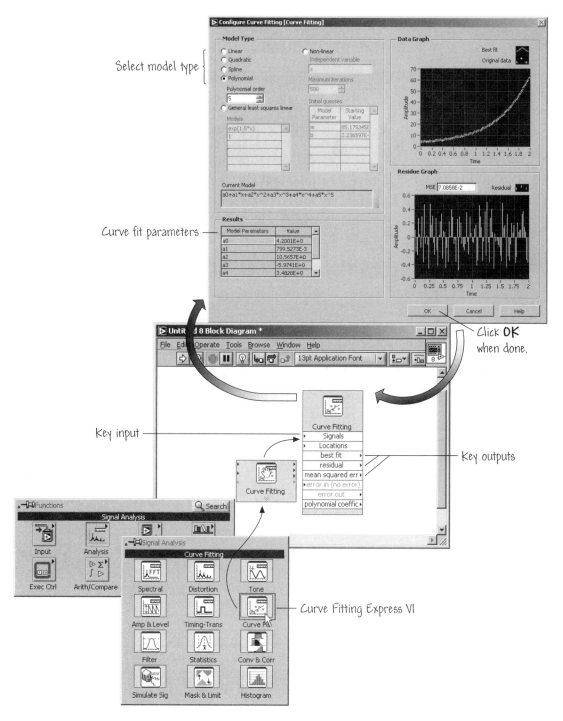

FIGURE 11.9
Configuring the Curve Fitting Express VI.

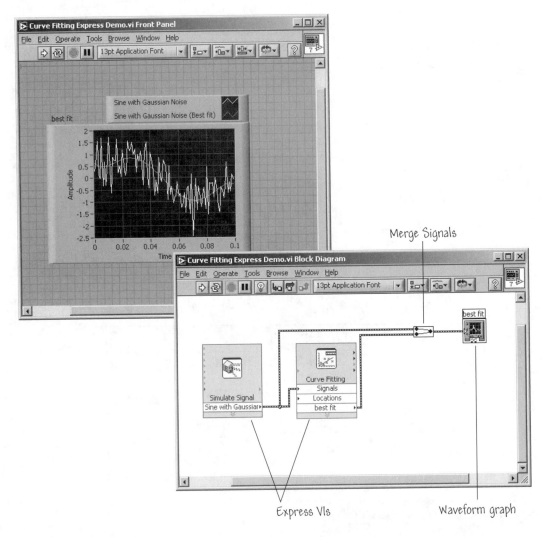

FIGURE 11.10
The Curve Fitting Express Demo.vi for investigating noisy sine wave signals.

11.3   DISPLAYING FORMULAS ON THE FRONT PANEL

In Chapter 5 we discussed Formula Nodes. Recall that a Formula Node is a resizable box placed on the block diagram (similar to the Sequence Structure, Case Structure, For Loop, and While Loop) containing one or more formula statements. The formula statements use a syntax similar to most text-based programming languages for arithmetic expressions. In this section we will learn about placing formulas on the front panel (rather than on the block diagram, as with Formula Nodes) and then gaining access to a family of analysis VIs for optimization, integration, graphing, and many more. The formulas retain the familiar text-based programming language syntax.

There are a few differences between using parser VIs to place formulas on the front panel and Formula Nodes. The Formula Nodes have access to the binary functions—max, min, mod, rem—and the parser VIs do not have this access. However, the parser VIs can access more complex functions—Gamma, spike, Legendre elliptic integral 1st kind, square, and more—and have access to most of the same standard functions, such as $\sin x$, e^x, and $\tanh x$, as do the Formula Nodes. The Formula Nodes have access to a variety of logical, conditional, inequality, and equality operators, such as $<, >, ==, \geq, \leq$, while the parser VIs do not. The parser VIs and the Formula Nodes play different roles in LabVIEW. We can learn about the parser VIs by working an example.

Practice with Parser VIs

Consider the VI shown in Figure 11.11, which shows how you can enter a formula on the front panel, evaluate it, and graph the results. Open a new VI and construct a VI using Figure 11.11 as a model. The parser VI at the center of this VI is the Eval Single-Variable Array.vi. It calculates an array of function values at given points in a given interval; that is,

$$y_i = f(x_i),$$

where $i = 1, \ldots, n$. Examining the block diagram in Figure 11.11, you can see that the For Loop is used to generate the x_i values, and $n = 1,000$. The points range from 0 to 9.99 in increments of 0.01.

After building the VI, return to the front panel and type in the expression $\sin(x)$ in the formula control and run the VI. The VI waveform will show the usual sine wave curve. The function has only one variable, which we called x—we could just as easily have input $\sin(z)$ or $\sin(y)$.

Change the formula to $3 * x^2 + x * \log(x)$ and run the VI. Try another expression, such as $\text{step}(y) + \cos(y) * \sin(y)$. Then type in another expression of your choice in the formula control and run the VI. When you are finished, save the VI in User's Stuff and call it EvalVal.vi.

A working version of EvalVal.vi can be found in the **Chapter 11** *folder in the* **Learning** *directory.* ◆

The Substitute Variables VI is another important VI in the parser library. This VI is used to substitute formulas for parameters in other formulas, resulting in more complex functions. This provides a great degree of control over the formulas themselves. Suppose that we define a "generic" formula as

$$\sin(A) + e^B.$$

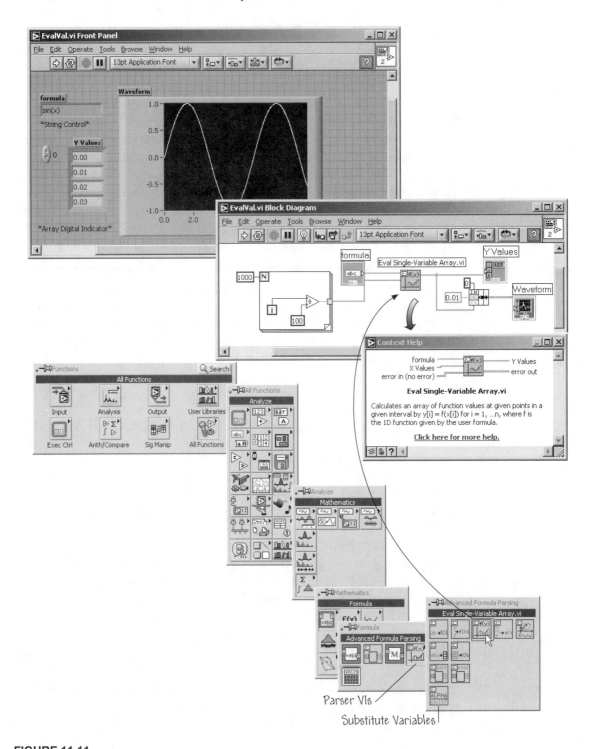

FIGURE 11.11
A VI that allows you to enter a formula on the front panel.

We can then "substitute" for A and B to construct a more sophisticated function. We might, for example, use the Substitute Variables VI to obtain

$$\sin(\ln(x)) + e^{\cos x}.$$

The substitutions in this case are $A \rightarrow \ln(x)$ and $B \rightarrow \cos(x)$.

Substitute Variables

Consider the VI shown in Figure 11.12. This VI is similar to the VI that you developed in the previous example called EvalVal.vi, but includes the Substitute Variables VI. This allows you to enter a formula on the front panel, evaluate it, and graph the results. You can substitute formulas for the function parameters in the original formulas. Open a new VI and construct a VI using Figure 11.12 as a model. You might also consider opening EvalVal.vi and using it as a starting point for this new VI.

The parser VI at the center of this VI is Substitute Variables.vi; see Figure 11.11 to find it on the **Parser VIs** palette . This VI substitutes a formula string by given rules, as illustrated in Figure 11.12. The input **original formula** to the Substitute Variables.vi is the formula that you input on the front panel control. The input **Substitution Rules** specifies the substitutions to be made for

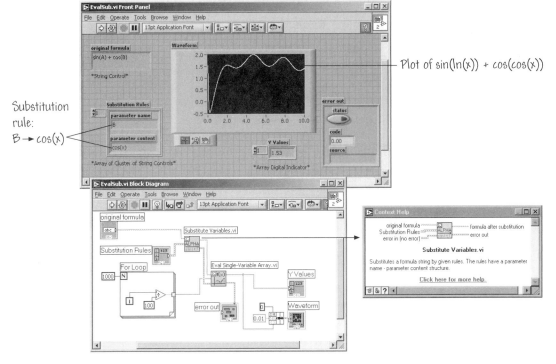

FIGURE 11.12
A VI that allows you to substitute formulas.

the parameters in the original formula. Each element in the **Substitution Rules** array specifies a parameter and its associated substitution rule. The output **formula after substitution** is the resulting formula after the parameter substitutions specified in **Substitution Rules** have been made.

Include a For Loop to generate the x_i values, where $i = 1, \ldots, 1,000$. The points at which you evaluate the function should range from 0 to 9.99 in increments of 0.01. After building the VI, return to the front panel and type in the equation $\sin(A) + \cos(B)$ in **original formula**. Make the substitutions $A \rightarrow \ln(x)$ and $B \rightarrow \cos(x)$:

- In element 0 of the **Substitution Rules** control, type in
  - **parameter name**: A
  - **parameter content**: $\ln(x)$
- In element 1 of the **Substitution Rules** control, type in
  - **parameter name**: B
  - **parameter content**: $\cos(x)$

Run the VI and observe the waveform on the graph display.

- In element 0 of the **Substitution Rules** control, type in
  - **parameter name**: A
  - **parameter content**: step $(x - 1)$
- In element 1 of the **Substitution Rules** control, type in
  - **parameter name**: B
  - **parameter content**: square (x)

Run the VI and observe the waveform on the graph display. Try a few substitutions of your choosing.

When you are finished, save the VI in User's Stuff and call it EvalSub.vi.

 A working version of EvalSub.vi can be found in the Chapter 11 *folder in the* Learning *directory.* ◆

11.4 DIFFERENTIAL EQUATIONS

In LabVIEW, you can solve linear and nonlinear ordinary differential equations (ODEs) using one of seven VIs for solving first- and higher-order differential equations. Figure 11.13 shows the palette with the available ODE VIs.

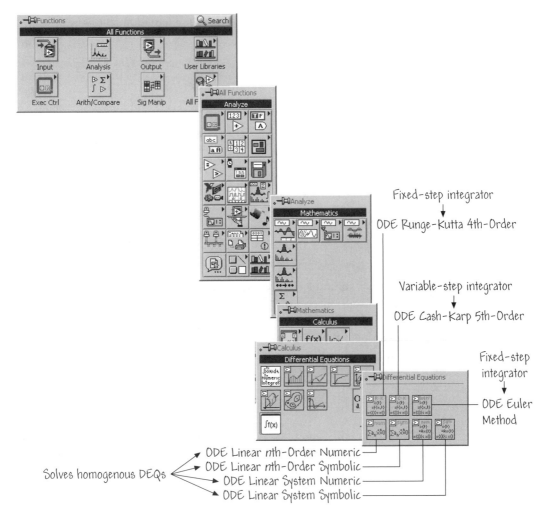

FIGURE 11.13
Solving differential equations using LabVIEW VIs.

The order of a differential equation is the order of the highest derivative in the differential equation.

Suppose that we have a set of first-order ordinary differential equations:

$$\dot{\mathbf{x}}(t) = \mathbf{f}(\mathbf{x}(t), \mathbf{u}(t)),$$

where $\mathbf{x}$ is the vector $(x_1 x_2 \cdots x_n)^T$, sometimes known as the *state vector*, and $\mathbf{u}$ is the vector $(u_1 u_2 \cdots u_m)^T$ of inputs to the system. When $\mathbf{u} = 0$, the differential equation is termed a *homogeneous* differential equation, and when

$$\mathbf{f}(\mathbf{x}(t), \mathbf{u}(t)) = \mathbf{A}\mathbf{x}(t) + \mathbf{B}\mathbf{u}(t),$$

the system is a system of *linear* ordinary differential equations.

An example of a *nonhomogeneous* system is

$$\dot{x}_1(t) = x_1^2(t) + \sin x_2(t) + u_1(t)$$
$$\dot{x}_2(t) = x_1(t) + x_2^3(t) + 3u_2(t),$$

where we write

$$\mathbf{x} = \begin{pmatrix} x_1 \\ x_2 \end{pmatrix} \quad \mathbf{u} = \begin{pmatrix} u_1 \\ u_2 \end{pmatrix} \quad \mathbf{f}(\mathbf{x}(t), \mathbf{u}(t)) = \begin{pmatrix} x_1^2(t) + \sin x_2(t) + u_1(t) \\ x_1(t) + x_2^3(t) + 3u_2(t) \end{pmatrix}$$

To compute a solution we need to specify the initial conditions. For a system with n first-order differential equations, we need to specify n initial conditions, $x_1(0), x_2(0), \ldots, x_n(0)$.

We can also represent physical systems by higher-order differential equations. An example of a second-order mass-spring-damper system is

$$m\frac{d^2 y(t)}{dt} + b\frac{dy(t)}{dt} + ky(t) = g(t),$$

where the system's parameters are m = mass, b = damping coefficient, k = spring constant, and the input function is $g(t)$. When $g(t) = 0$, the system is homogeneous. To compute the solution, we need two initial conditions: $y(0)$ and $\dot{y}(0)$.

You can describe an nth-order differential equation equivalently by n first-order differential equations. Consider the second-order DEQ presented above and define

$$x_1 = y \quad \text{and} \quad x_2 = \dot{y} \quad \text{and} \quad u = g(t).$$

Taking time-derivatives of x_1 and x_2 yields

$$\dot{x}_1 = \dot{y} = x_2$$
$$\dot{x}_2 = \ddot{y} = -\frac{b}{m}\dot{y} - \frac{k}{m}y + \frac{1}{m}g(t) = -\frac{b}{m}x_2 - \frac{k}{m}x_1 + \frac{1}{m}u(t),$$

or,

$$\dot{\mathbf{x}}(t) = \mathbf{A}\mathbf{x}(t) + \mathbf{B}u(t) = \begin{bmatrix} 0 & 1 \\ -\frac{k}{m} & -\frac{b}{m} \end{bmatrix} \mathbf{x}(t) + \begin{bmatrix} 0 \\ \frac{1}{m} \end{bmatrix} u(t)$$

LabVIEW has five VIs to solve sets of first-order differential equations, and two VIs to solve nth-order differential equations. Each VI uses a different numerical method for solving the differential equations. Each method has an associated *step size* that defines the time intervals between solution points. To solve DEQs of the form

$$\dot{\mathbf{x}}(t) = \mathbf{f}(\mathbf{x}(t), \mathbf{u}(t)),$$

use one of the following VIs:

- ODE Cash-Karp 5th-Order: Variable-step integrator that adjusts the step size internally to reduce numerical errors.

- ODE Euler Method: A very simple fixed-step integrator—that is, it executes fast, but the integration error associated with the Euler method is usually unacceptable in situations where solution precision is important.

- ODE Runge-Kutta 4th-Order: Fixed-step integrator that provides much more precise solutions than the Euler method.

To solve DEQs of the form

$$\dot{\mathbf{x}}(t) = \mathbf{A}\mathbf{x}(t)$$

where the coefficients of the matrix $\mathbf{A}$ are constant, use one of the following VIs:

- ODE Linear System Numeric: Generates a numerical solution of a homogeneous linear system of differential equations.

- ODE Linear System Symbolic: Generates a symbolic solution of a homogeneous linear system of differential equations.

By a symbolic solution, we mean that the output of the VI is actually a formula rather than a array of numbers. The solution is presented as a formula displayed on the VI front panel.

For solving homogeneous, higher-order differential equations of the form

$$a_0 \frac{d^n y}{dt^n} + a_1 \frac{d^{n-1} y}{dt^{n-1}} + \cdots + a_{n-1} y = 0,$$

LabVIEW has two VIs:

- ODE Linear nth-Order Numeric: Generates a numeric solution of a linear system of nth-order differential equations.

- ODE Linear nth-Order Symbolic: Generates a symbolic solution of a linear system of nth-order differential equations.

Which VI should you use? A few general guidelines follow:

- Use the ODE Euler Method VI sparingly and only for very simple ODEs.

- For most situations, choose the ODE Runge-Kutta 4th-Order or the ODE Cash-Karp 5th-Order VI.

  ▫ If you need the solution at equal intervals, choose the ODE Runge-Kutta 4th-Order VI.

  ▫ If you are interested in a global solution and fast computation, choose the ODE Cash-Karp 5th-Order VI.

The Pendulum

The objective of this exercise is to build a VI that solves a second-order differential equation that models the motion of a pendulum. You will see how incorporation of the Substitute Variables VI allows you to vary the model parameters. The pendulum model is given by

$$\frac{d^2\theta}{dt^2} + \frac{c}{ml}\frac{d\theta}{dt} + \frac{g}{l}\sin\theta = 0,$$

where m is the mass of the pendulum, l is the length of the rod, $g = 9.8$ is the acceleration due to gravity, and θ is the angle between the rod and a vertical line passing through the point where the rod is fixed (that is, the equilibrium position).

The pendulum model is a homogeneous, nonlinear equation (notice the $\sin\theta$ term!)—you have three choices of VIs:

- ODE Cash-Karp 5th-Order

- ODE Euler Method

- ODE Runge-Kutta 4th-Order

The first step is to formulate the pendulum model as two first-order differential equations. All three integration VIs listed above are for first-order systems. This is achieved by making the following substitution:

$$x_1 = \theta \quad \text{and} \quad x_2 = \dot{\theta}.$$

Go ahead and convert the second-order DEQ to two first-order DEQs. You should end with the resulting system:

$$\dot{x}_1 = x_2$$
$$\dot{x}_2 = -\frac{c}{ml}x_2 - \frac{g}{l}\sin x_1$$

Construct a VI to simulate the motion of the pendulum. Use the front panel of the VI shown in Figure 11.14 as a guide.

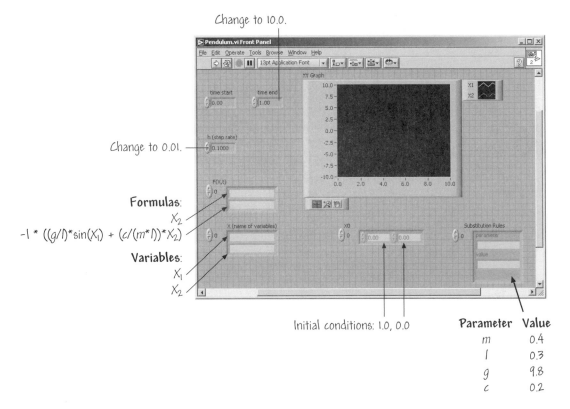

FIGURE 11.14
Simulating the motion of a pendulum—front panel.

A block diagram is shown in Figure 11.15 that can be used as a guide in the VI development. To begin the development, use the ODE Euler Method VI. The inputs that you will need follow:

- **X**: An array of strings listing the dependent variables (x_1 and x_2).
- **time start**: The point in time at which to start the calculations ($t_0 = 0$).
- **time end**: The point in time at which to end the calculations ($t_f = 10$).
- **h**: The time increment at which to perform the calculations ($h = 0.01$).
- **X0**: The initial conditions ($x_1(0) = 1$, and $x_2(0) = 0$).

The pendulum differential equations are typed in the **F(x,t)** control. The VI uses the Substitute Variables VI as a vehicle for varying the pendulum parameters. This allows you to enter the pendulum model on the front panel and then to substitute numerical values for m, l, g, and c.

When the VI is ready to accept inputs, enter the right side of the pendulum model in the **F(X,t)** control. You will enter x_2 and $-(g/l)\sin x_1 - (c/ml)x_2$. For the substitution rules use

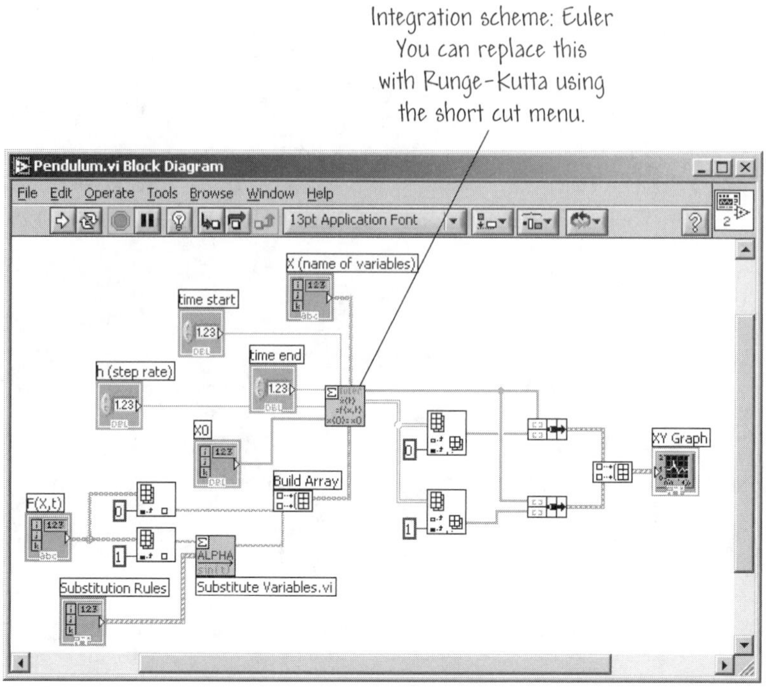

Integration scheme: Euler
You can replace this
with Runge-Kutta using
the short cut menu.

FIGURE 11.15
Simulating the motion of a pendulum—block diagram.

- In element 0 of the **Substitution Rules** control, type in
  - **parameter name**: m
  - **parameter content**: 0.4
- In element 1 of the **Substitution Rules** control, type in
  - **parameter name**: l
  - **parameter content**: 0.3
- In element 2 of the **Substitution Rules** control, type in
  - **parameter name**: g
  - **parameter content**: 9.8
- In element 3 of the **Substitution Rules** control, type in
  - **parameter name**: c
  - **parameter content**: 0.2

Run the VI and observe the waveform on the graph display. You should see a nice, stable response damping out around 6 seconds.

On the front panel, change the **h (step rate)** from 0.01 to 0.1. Run the VI. Did you detect a problem? The system response is no longer stable! Nothing has changed with the physical model, so this must be due to the Euler integration scheme. Now, switch to the block diagram and pop up on the ODE Euler Method VI and **Replace** it with the ODE Runge-Kutta 4th-Order VI. Both VIs have the same inputs and outputs. Run the VI. What happens? The expected smooth, stable response is obtained. This demonstrates the benefit of the Runge-Kutta method over the Euler method—you can run with larger time steps and obtain more accurate solutions.

Back on the front panel, switch the time step back to **h (step rate)** $= 0.01$. Run the VI and verify that the results remain essentially the same. Investigate the effect that varying the pendulum mass has on the response.

When you are finished, save the VI in User's Stuff and call it Pendulum.vi.

A working version of Pendulum.vi can be found in the Chapter 11 folder in the Learning directory. If you run this VI, make sure to verify that all the input parameters are correct! ◆

11.5   FINDING ZEROES OF FUNCTIONS

LabVIEW provides six VIs that can be used to compute zeroes of functions. The VIs can be used to determine the zeroes of general functions of the following form:

$$f(x, y) = 0$$
$$g(x, y) = 0$$

For example, you could use the VIs to find the zeroes of $\sin(x) + \cos(x)$ in the range $-10 \le x \le 10$.

Very often in the course of studying mathematics, engineering, business, and science it is necessary to compute the *zeroes* of a polynomial. A nth-order polynomial has the form

$$f(x) = x^n + a_{n-1}x^{n-1} + a_{n-2}x^{n-2} + \cdots + a_1x + a_0 = 0.$$

The zeroes of the polynomial are the values of x such that $f(x) = 0$. The zeroes are also known as the *roots* of the polynomial. For example, we discussed in previous sections that the characteristic equation associated with a $n \times n$ matrix **A** is an nth-order polynomial, and the zeroes of the characteristic equation are the eigenvalues of the matrix. Eigenvalues can be real and imaginary. But in this discussion, when we talk about zeroes of a function, we mean the real roots.

LabVIEW VIs are shown in Figure 11.16. There are seven VIs:

- Find All Zeroes of f(x): This VI determines all the zeroes of a 1D function in a specified interval.

- Newton-Raphson Zero Finder: Uses derivatives to assist in determining a zero of a 1D function in a specified interval.

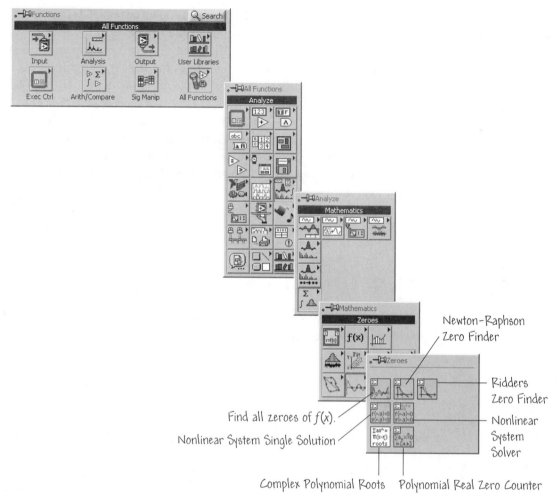

FIGURE 11.16
VIs for finding zeroes of functions.

- Nonlinear System Single Solution: Computes the zeroes of a nonlinear function, where an approximation is provided as input.

- Nonlinear System Solver: Computes the zeroes of a set of nonlinear functions.

- Polynomial Real Zero Counter: Determines the number of real zeroes of a polynomial in an interval without actually computing the zeroes.

- Ridders Zero Finder: This VI computes a zero of a function in a given interval, but the function must be continuous and when evaluated at the edges of the interval, the function must have different signs.

- Complex Polynomial Roots: Finds the complex roots of a complex polynomial.

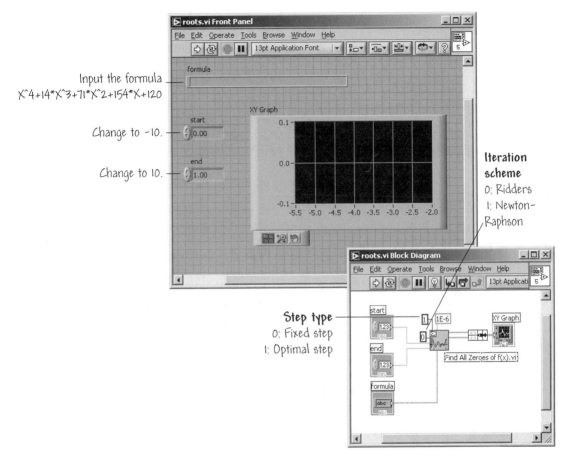

FIGURE 11.17
Computing the zeroes of a polynomial.

Finding Zeroes of a Polynomial Function

The objective of this exercise is to build a VI to compute the zeroes of a polynomial function in a given interval. Construct a VI using the front panel and block diagram shown in Figure 11.17 as a guide.

Use the VI **Find All Zeroes of f(x).vi** to compute the zeroes. This VI is based on a numerical scheme that iterates on the interval to converge on the zeroes. Since it is an iterative scheme, it utilizes a search algorithm. You have two choices for the search—Ridders method and Newton-Raphson. As the search algorithm iterates, it can use either uniformly spaced function values or an optimal step size. These two inputs are wired as constants on the block diagram shown in Figure 11.17—you can change that to have them input from the front panel. You should experiment with the different search and step size possibilities to see which you prefer. The inputs to the VI are:

- **formula**: Type in the formula.

- **start** and **end**: The endpoints of the search interval.

When your VI is ready to accept inputs, enter the following formula:

$$x^4 + 14x^3 + 71x^2 + 154x + 120.$$

Set the endpoints to -10 and 10. Run the VI. Where are the zeroes? You should determine them to be $x = -5, -4, -3$, and -2. When you are finished, save the VI in User's Stuff and call it roots.vi.

A working version of roots.vi can be found in the Chapter 11 *folder in the* Learning *directory.*   ◆

11.6   INTEGRATION AND DIFFERENTIATION

LabVIEW VIs for integration and differentiation are shown in Figure 11.18. The figure shows that there are many other VIs available for working with 1D functions—unfortunately we can not cover all the analysis capabilities of LabVIEW in this book.

A straightforward implementation of the integration and differentiation VIs is shown in Figure 11.19. You can find the VI shown in Figure 11.19 in the Chapter 11 folder within the Learning directory—it is called Derivative and Integration.vi. Open the VI and enter the formula $\sin(x)$ and run the VI. You will find that three plots appear on the graph, including the function, the derivative of the function at a number of points defined by the **number of points** input, and a plot of the integral of the function. Change the value of **end** from the default value of 1 to 6.28 and run the VI again. This time you should see one complete cycle of the sine wave, and the plot of the derivative of the sine function should appear as a cosine function. If you run the VI in **Run Continuously** mode, you can vary the parameter **number of points** and watch the sine function become smoother as the number increases and, conversely, become less smooth as the number of points decreases.

11.7   SIGNAL GENERATION

You can use LabVIEW to generate signals for testing and other purposes when real-world signals are not available. You may also want to generate signals rather than rely on signals acquired from the real world whenever you need to accurately control the signal characteristics (such as magnitude, frequency and phase of periodic signals, and so on). In this section we discuss some of the possibilities for generating signals using VIs. The discussion begins by considering the notion of normalized frequency.

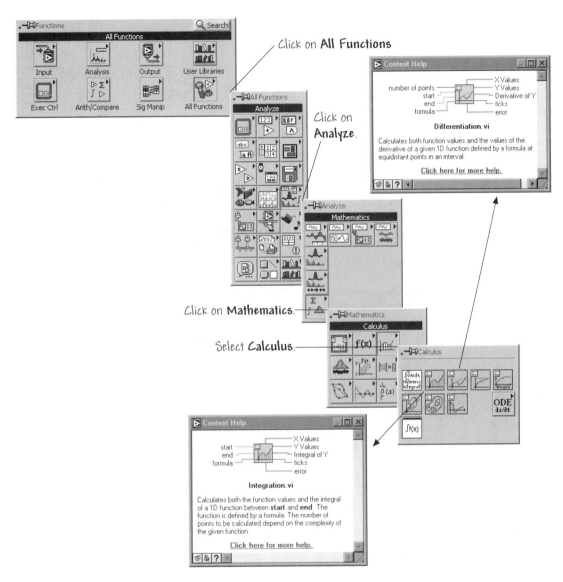

FIGURE 11.18
VIs for integration and differentiation.

11.7.1 Normalized Frequency

In the digital signal world (and with many Signal Generation VIs) we often use the so-called **digital frequency** or **normalized frequency** (in units of cycles/sample) computed as

$$f = \text{digital frequency} = \frac{\text{analog frequency}}{\text{sampling frequency}}.$$

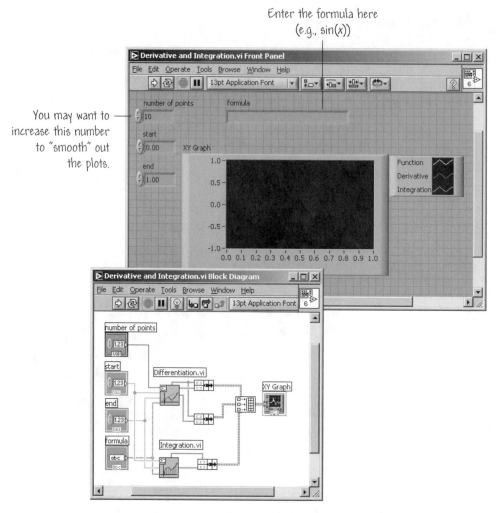

FIGURE 11.19
A VI for integration and differentiation of a 1D function.

The analog frequency is generally measured in units of Hz (or cycles per second) and the sampling frequency in units of samples per second. The normalized frequency is assumed to range from 0.0 to 1.0 corresponding to a frequency range of 0 to the sampling frequency, denoted by f_s. The normalized frequency wraps around 1.0, so that a normalized frequency of 1.2 is equivalent to 0.2. As an example, a signal sampled at the **Nyquist frequency** (that is, at $f_s/2$) is sampled twice per cycle (that is, two samples/cycle). This corresponds to a normalized frequency of $1/2$ cycles/sample = 0.5 cycles/sample. Therefore, we see that the reciprocal of the normalized frequency yields the number of times that the signal is sampled in one cycle (more on sampling in Sec. 11.8.4).

The following VIs utilize frequencies given in normalized units:

1. Sine Wave
2. Square Wave
3. Sawtooth Wave
4. Triangle Wave
5. Arbitrary Wave
6. Chirp Pattern

When using these VIs, you will need to convert the frequency units given in the problem to the normalized frequency units of cycles/sample. The VI depicted in Figure 11.20 illustrates how to generate two cycles of a sine wave and then convert cycles to cycles/sample.

In the example shown in Figure 11.20, the number of cycles (2) is divided by the number of samples (50), resulting in a normalized frequency of $f = 2/50$ cycles/sample. This implies that it takes 50 samples to generate two cycles of the sine wave. What if the problem specifies the frequency in units of Hz

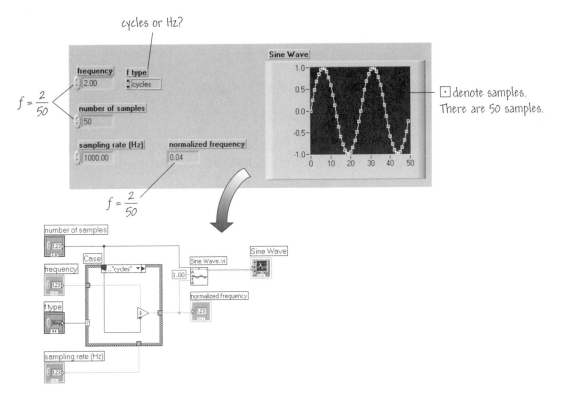

FIGURE 11.20
Generating two cycles of a sine wave and converting cycles to cycles/sample.

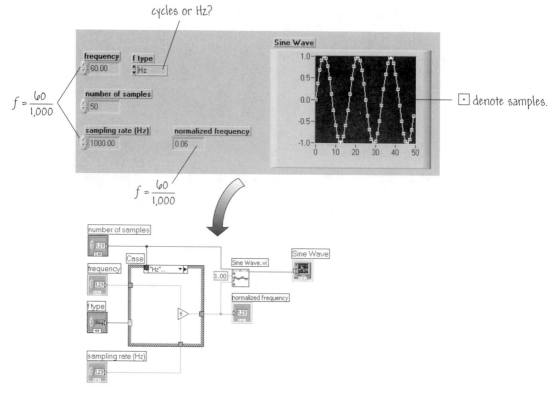

FIGURE 11.21
Generating a 60 Hz sine wave and computing normalized frequency.

(cycles/sec)? In this case, if you divide the frequency in Hz (cycles/sec) by the sampling rate given in Hz (samples/sec), you obtain units of cycles/sample:

$$\frac{\text{cycles/sec}}{\text{samples/sec}} = \frac{\text{cycles}}{\text{sample}}.$$

The illustration in Figure 11.21 shows a VI used to generate a 60 Hz sine signal and to compute the normalized frequency when the input is in Hz. The normalized frequency is found by dividing the frequency of 60 Hz by the sampling rate of 1,000 Hz to get the normalized frequency of $f = 0.06$ cycles/sample:

$$f = \frac{60}{1,000} = 0.06 \, \frac{\text{cycles}}{\text{sample}}.$$

Therefore, we see that it takes almost 17 samples to generate one cycle of the sine wave. The number 17 comes from computing the reciprocal of $f = 0.06$.

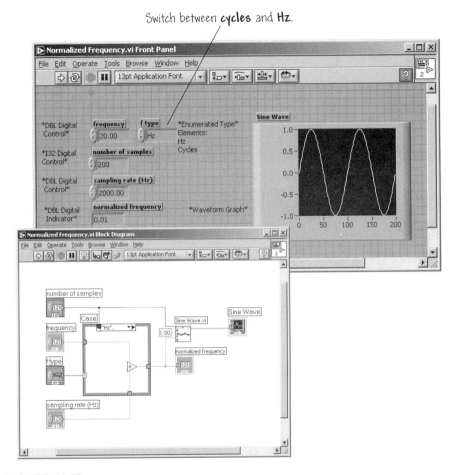

FIGURE 11.22
A VI to compute the normalized frequency.

Normalized Frequencies

Open the VI called Normalized Frequency.vi located in the Chapter 11 folder in the Learning directory. The front panel and block diagram are shown in Figure 11.22. You can use this VI to experiment with calculating the normalized frequency when the input is in cycles and in Hz. Make sure that the **f type** is selected as Hz and run the VI. You should find that the normalized frequency is $f = 0.01$ with the default VI input values.

Manually compute the normalized frequency for **f type** = Hz, **frequency** = 10, and **sampling rate (Hz)** = 1,000. Modify the VI input parameters accordingly and verify that you obtain the same answer as by hand. The answer should be $f = 0.01$, computed as

$$f = \frac{10}{1,000} = 0.01 \; \frac{\text{cycles}}{\text{sample}}.$$

You can also work in cycles (rather than Hz) by selecting **f type** to be cycles. Then, in this situation, the normalized frequency is computed as a ratio of **frequency** to **number of samples**.   ◆

11.7.2   Wave and Pattern VIs

The basic difference in the operation of the Wave or Pattern VIs is whether or not the particular VI keeps track of the phase of its own generated signal internally. The Wave VIs keep track of phase internally; the Pattern VIs do not. You can distinguish between the two types of VIs by recognizing that the VI names contain either the word *wave* or *pattern*, as illustrated in Figure 11.23.

The Wave VIs operate with normalized frequencies in units of cycles/sample. The only Pattern VI that uses normalized units is the Chirp Pattern VI.

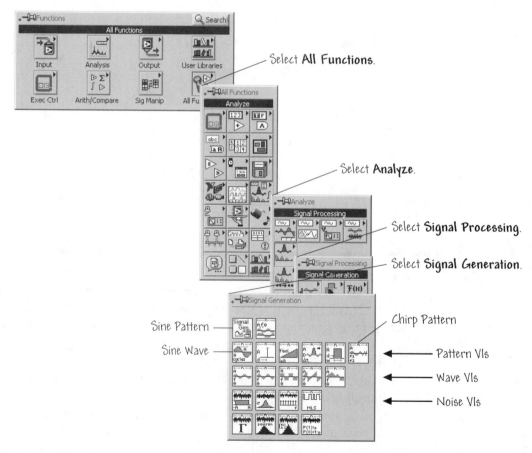

FIGURE 11.23
Signal generation VIs.

Since the Wave VIs can keep track of the phase internally, they allow the user to control the value of the initial phase. The **phase in** control specifies the initial phase (in degrees) of the first sample of the generated waveform and the **phase out** indicator specifies the phase of the next sample of the generated waveform. In addition, a **reset phase** control dictates whether or not the phase of the first sample generated when the wave VI is called is the phase specified at the **phase in** control, or whether it is the phase available at the **phase out** control when the VI last executed. A TRUE value of reset phase sets the initial phase to **phase in**—a FALSE value sets it to the value of **phase out** when the VI last executed.

Practice with Signal Generation

In this exercise you will construct a VI that uses the two types of signal generation VIs. The front panel and block diagram for the VI is shown in Figure 11.24. The two main functions used on the block diagram are the Sine Pattern.vi and Sine Wave.vi. These functions are located on the **Signal Generation** palette, as shown in Figure 11.23. The VI also calculates and displays the normalized frequency.

Construct the VI using Figure 11.24 as a guide and run the VI when it is ready. Vary the VI inputs, paying particular attention to the effect that changing the control **phase in (degrees)** has on the Sine Pattern waveform graph. You should notice that the waveform begins to shift left and right as you vary the phase up and down. When you set the **reset phase** to TRUE (ON), the initial phase is reset to the value specified by **phase in** each time the VI is called in the loop; otherwise, the initial phase is set to the previous phase output. While the VI is running, set the **reset phase** to the ON position. The sine wave should be rendered stationary, whereas with the reset button set to the OFF position, the sine wave varies with the varying phase.

Stop the execution of the VI using the **Stop** button located at the bottom left of the VI front panel in Figure 11.24. When you are finished experimenting with the VI, save it as Signal Generation.vi in the Users Stuff folder in the Learning directory.

You can find a working version of Signal Generation.vi *in the* Chapter 11 *folder in the* Learning *directory.* ◆

11.7.3   The Simulate Signal Express VI

The Simulate Signal Express VI is located on the **Signal Analysis** Express palette shown in Figure 11.8. This VI simulates sine waves, square waves, triangular waves, sawtooth waves, and noise signals. In Section 11.2.3, we used the Signal Analysis Express VI to simulate a noisy sine wave.

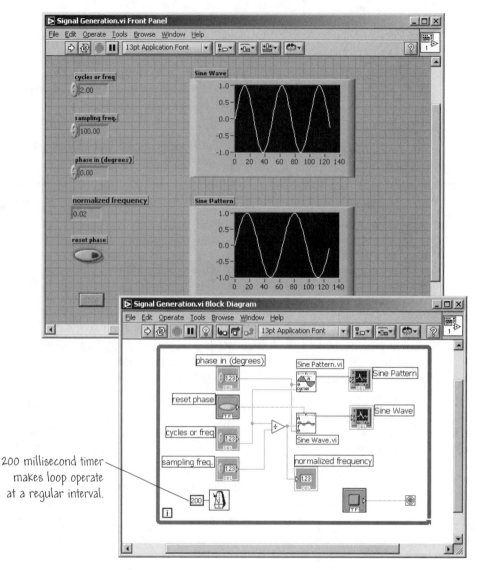

FIGURE 11.24
Using the Sine Wave and Sine Pattern VIs to construct a signal generation VI.

As with all Express VIs, to use the Simulate Signal Express VI, you first place it on the block diagram. As illustrated in Figure 11.25, once the Simulate Signal Express VI is placed on the block diagram, a dialog box automatically appears to configure the VI. In Figure 11.25, a sawtooth wave is selected. The **frequency** is set to 10.1 Hz, although in the forthcoming example we will wire a control to the corresponding frequency input to programmatically vary the frequency from

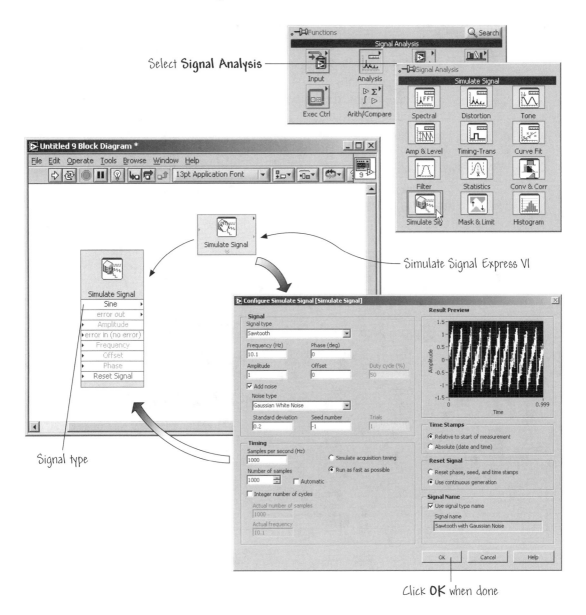

FIGURE 11.25
Configuring the Simulate Signal Express VI.

the front panel. We do the same with the **amplitude**, set to 1 in Figure 11.25,
but ultimately wired to a control to the corresponding input to programmatically
vary the amplitude. We do the same as well for the **Offset** variable. Once the
Simulate Signal Express VI is configured as desired, click **OK** to return to the
block diagram.

Consider the VI shown in Figure 11.26. The Simulate Signal Express Demo.vi was developed to assist in investigating various signals generated by the Simulate Signal Express VI. Notice that the Simulate Signal Express VI is the central element on the block diagram in Figure 11.26. The Express VI is configured to generate a sawtooth signal. The remaining elements on the block diagram are placed there to permit easy access to key parameters of the sawtooth signal: frequency, amplitude, and offset.

The Simulate Signal Express Demo.vi is located in Chapter 11 of the Learning directory.

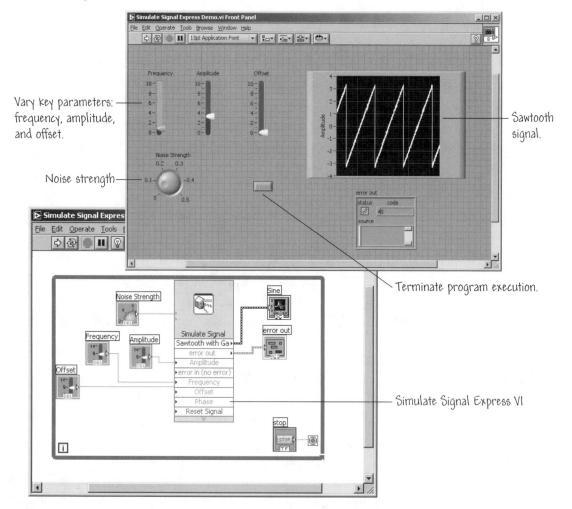

FIGURE 11.26
The Simulate Signal Express Demo.vi for investigating signal generation.

Run the Simulate Signal Express Demo.vi and observe the results on the waveform chart. Vary the frequency, amplitude, and offset and examine the resulting impact on the sawtooth signal.

Double-click the Simulate Signal Express VI to access the configuration dialog box. Select a different signal, such as the DC signal and repeat the above investigation. What happens? You should find that the Broken Run indicator appears and there are broken wires on the block diagram. Remember that when you change the signal type, the inputs also change. In the case of a DC signal, it makes no sense to input a frequency; hence, the wire to that input generates an error. ◆

11.8 SIGNAL PROCESSING

In this section we discuss three main topics: the Fourier transform (including the discrete Fourier transform and the fast Fourier transform), smoothing windows, and a brief overview of filtering.

11.8.1 The Fourier Transform

In Chapter 8 we covered the subject of data acquisition, where we discussed the fact that the samples of a measured signal obtained from the DAQ system are a time-domain representation of the signal, giving the amplitudes of the sampled signal at the sampling times. A significant amount of information is coded into the time-domain representation of a signal—maximum amplitude, maximum overshoot, time to settle to steady-state, and so on. The signal contains other useful information that becomes evident when the signal is transformed into the frequency domain. In other words, you may want to know the frequency content of a signal rather than the amplitudes of the individual samples. The representation of a signal in terms of its individual frequency components is known as the **frequency-domain representation** of the signal.

A common practical algorithm for transforming sampled signals from the time domain into the frequency domain is known as the **discrete Fourier transform**, or DFT. The relationship between the samples of a signal in the time domain and their representation in the frequency-domain is established by the DFT. This process is illustrated in Figure 11.27.

If you apply the DFT to a time-domain signal represented by N samples of the signal, you will obtain a frequency-domain representation of the signal of length N. We denote the individual components of the DFT by $X(i)$. If the signal is sampled at the rate f_s Hz, and if you collect N samples, then you can compute the frequency resolution as $\Delta f = f_s/N$. This implies that the ith sample of the DFT occurs at a frequency of $i \Delta f$ Hz. We let the pth element $X(p)$ correspond to the Nyquist frequency. Regardless of whether the input signal is real or complex,

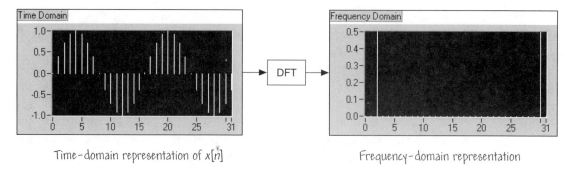

Time-domain representation of $x[n]$ Frequency-domain representation

FIGURE 11.27
The DFT establishes the relationship between the samples of a signal in the time-domain and their representation in the frequency-domain.

the frequency-domain representation is always complex and contains two pieces of information—the amplitude and the phase.

For real-valued time-domain signals (denoted here by $x(i)$), the DFT is symmetric about the index $N/2$ with the following properties:

$$|X(i)| = |X(N-i)| \quad \text{and} \quad \text{phase}(X(i)) = -\text{phase}(X(N-i)).$$

The magnitude of $X(i)$ is **even symmetric**, that is, symmetric about the vertical axis. The phase of $X(i)$ is **odd symmetric**, that is, symmetric about the origin. This symmetry is illustrated in Figure 11.28. Since there is repetition of information contained in the N samples of the DFT (due to the symmetry properties), only half of the samples of the DFT need to be computed, since the other half can be obtained from symmetry.

Figure 11.29a depicts a *two-sided transform* for a complex sequence with $N = 8$ and $p = N/2 = 4$. Since $N/2$ is an integer, the DFT contains the Nyquist frequency. When N is odd, $N/2$ is not an integer, and thus there is no component at the Nyquist frequency. Figure 11.29b depicts a two-sided transform when $N = 7$.

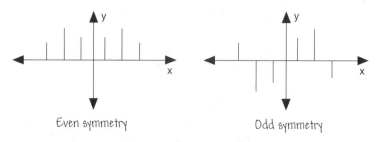

Even symmetry   Odd symmetry

FIGURE 11.28
Even and odd symmetric signals.

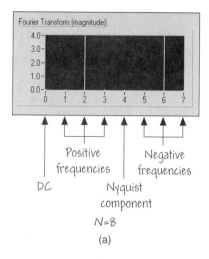 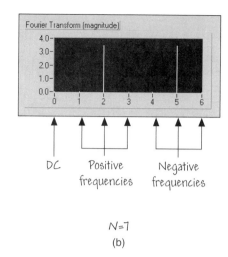

N=8

(a)

N=7

(b)

FIGURE 11.29
A two-sided transform representation of a complex sequence.

The computationally intensive process of computing the DFT of a signal with N samples requires approximately N^2 complex operations. However, when N is a power of 2, that is, when

$$N = 2^m \quad \text{for} \quad m = 1 \text{ or } 2 \text{ or } 3 \text{ or } \cdots,$$

you can implement the so-called **fast Fourier transforms** (FFTs), which require only approximately $N\log_2(N)$ operations. In other words, the FFT is an efficient algorithm for calculating the DFT when the number of samples (N) is a power of 2.

Practice with FFT

In this example you will practice with FFTs by opening an existing VI and experimenting with the input parameters. Locate and open FFT_2sided.vi. You will find this VI in the Chapter 11 folder in the Learning directory. The front panel is shown in Figure 11.30.

The VI demonstrates how to use the FFT VI to analyze a sine wave of user-specified frequency. The VI block diagram, shown in Figure 11.31, contains three main VIs:

- Real FFT.vi: This VI computes the fast Fourier transform (FFT) or the discrete Fourier transform (DFT) of the input sequence. Real FFT.vi will execute FFT routines if the size of the input sequence is a power of 2. If the size of the input sequence is not a power of 2, then an efficient DFT routine is called.

- Sine Wave.vi: This VI generates an array containing a sine wave. The VI is located in Analyze≫Signal Processing≫Signal Generation subpalette of the **Functions** palette.

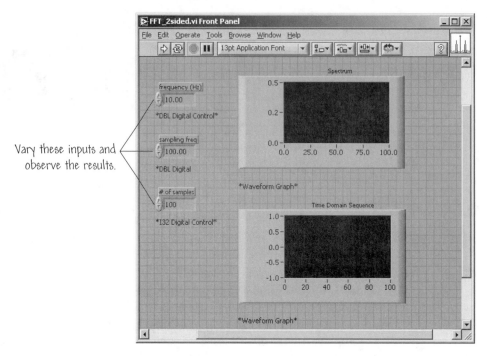

Vary these inputs and observe the results.

FIGURE 11.30
The front panel of FFT_2sided.vi.

- Complex To Polar.vi: Separates a complex number into its polar components represented by magnitude and phase. The input can be a scalar number, a cluster of numbers, an array of numbers, or an array of clusters. In the case of Figure 11.31, the input is an array of numbers.

Run the VI and experiment with the input parameters. Run several numerical experiments using the **Run Continuously** mode. What happens to the spectrum when you vary the signal frequency? For example, set the input signal frequency to 10 Hz, the sampling frequency to 100 Hz, and the number of samples to 100. In this case, $\Delta f = 1$ Hz. The spectrum should have two corresponding peaks. Check this using the VI.

For another experiment, set the signal frequency to 50 Hz, the sampling frequency to 100 Hz, and the number of samples to 100, and run the VI. With the VI running in **Run Continuously** mode, set the sampling frequency to 101 and observe the effects on the time-domain sequence waveform and the corresponding spectrum. Now, slowly increase the sampling frequency and see what happens! When you are finished, close the VI and do not save any changes.   ◆

11.8.2   Smoothing Windows

When using discrete Fourier transform methods to analyze a signal in the frequency domain, it is assumed that the available data of the time-domain signal

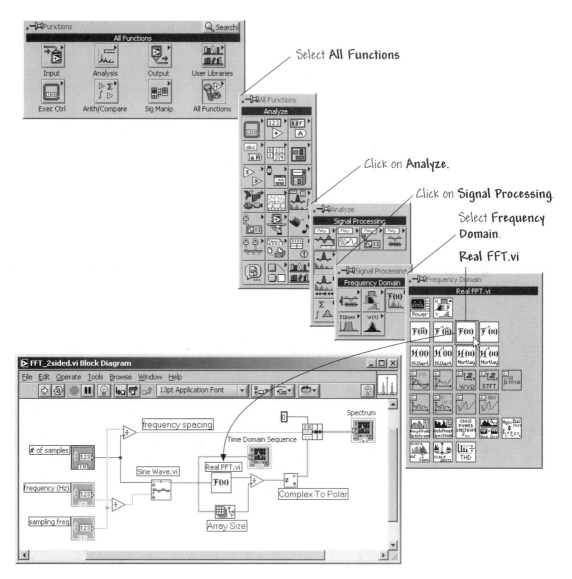

FIGURE 11.31
The block diagram of FFT_2sided.vi.

represents at least a single period of a periodically repeating waveform. Unfortunately, in most realistic situations, the number of samples of a given time-domain signal available for DFT analysis is limited and this can sometimes lead to a phenomenon known as **spectral leakage**. To see this, consider a periodic waveform created from one period of a sampled waveform, as illustrated in Figure 11.32.

The first period shown in Figure 11.32 is the sampled portion of the waveform. The sampled waveform is then repeated to produce the periodic waveform. Sampling a noninteger number of cycles of the waveform results in discontinuities

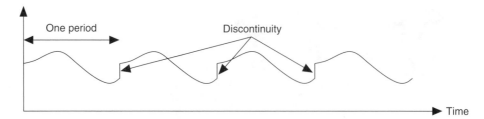

FIGURE 11.32
A periodic waveform created from one period of a sampled waveform.

between successive periods! These discontinuities induced by the process of creating a periodic waveform lead to very high frequencies (higher than the Nyquist frequency) in the spectrum of the signal—frequencies that were not present in the original signal. Therefore, the spectrum obtained with the DFT will not be the true spectrum of the original signal. In the frequency domain, it will appear as if the energy at one frequency has "leaked out" into all the other frequencies, leading to what is known as spectral leakage.

Figure 11.33 shows a sine wave and its corresponding Fourier transform. The sampled time-domain waveform is shown in Graph 1 in the upper left corner. In

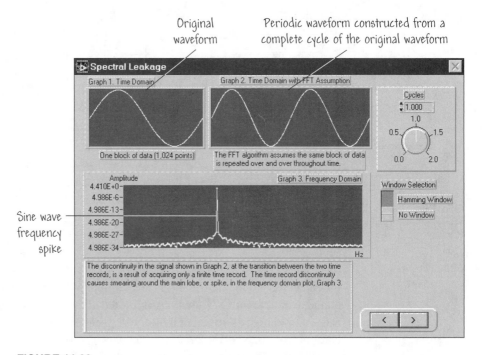

FIGURE 11.33
One complete period of a sine wave is repeated to obtain a periodic signal with no discontinuities. The corresponding Fourier transform shows no leakage.

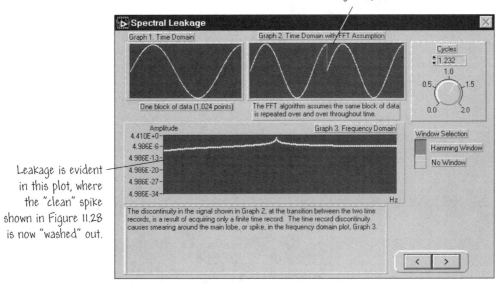

FIGURE 11.34
A portion of a sine wave period is repeated to obtain a periodic signal with discontinuities. The corresponding Fourier transform shows leakage.

this case, the sampled waveform is an integer number of cycles of the original sine wave. The sampled waveform can be repeated in time, and a periodic version of the original waveform thereby constructed. The constructed periodic version of the waveform is depicted in Graph 2 in the upper middle section of Figure 11.33. The constructed periodic waveform does not have any discontinuities because the sampled waveform is an integer number of cycles of the original waveform. The corresponding spectral representation of the periodic waveform is shown in Graph 3. Because the time record in Graph 2 is periodic, contains no discontinuities, and is an accurate representation of the true waveform, the computed spectrum is correct.

In Figure 11.34 a spectral representation of another periodic waveform is shown. However, in this case a noninteger number of cycles of the original waveform is used to construct the periodic waveform, resulting in the discontinuities in the waveform shown in Graph 2. The corresponding spectrum is shown in Graph 3. The energy is now "spread" over a wide range of frequencies—compare this result to Graph 3 in Figure 11.33. The smearing of the energy is called spectral leakage, as mentioned earlier.

Leakage results from using only a finite time sample of the input signal. One (unpractical) solution to the leakage problem is to obtain an infinite time record, from $-\infty$ to $+\infty$, yielding an ideal FFT solution. In practice, however,

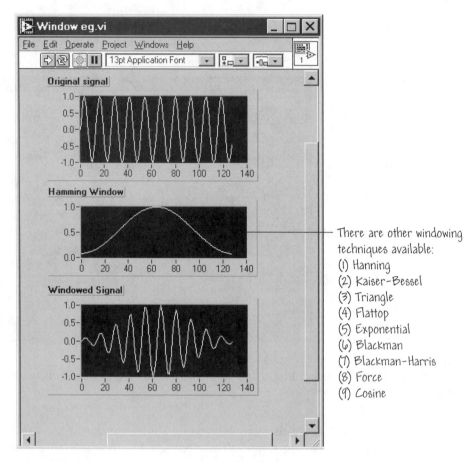

FIGURE 11.35
A sinusoidal signal windowed using a Hamming window.

we are limited to working with a finite time record. A practical approach to the problem of spectral leakage is the so-called **windowing** technique. Since the amount of spectral leakage depends on the amplitude of the discontinuity, the larger the discontinuity, the more the leakage. Windowing reduces the amplitude of the discontinuities at the boundaries of each period by multiplying the sampled original waveform by a finite length window whose amplitude varies smoothly and gradually towards zero at the edges.

One such windowing technique uses the *Hamming window*, as illustrated in Figure 11.35. The sinusoidal waveform of the windowed signal gradually tapers to zero at the ends—see the bottom graph in Figure 11.35. When computing the discrete Fourier transform on data of finite length, you can use the windowing technique applied to the sampled waveform to minimize the discontinuities of the constructed periodic waveform. This approach will minimize the spectral leakage.

The Hamming Window

Open a new front panel and place four waveform graphs and one digital control, as shown in Figure 11.36.

Construct a block diagram using Figure 11.37 as a guide. In the block diagram, we use three main VIs:

- Hamming Window.vi: This VI applies the Hamming window to the input sequence. It is located in Functions≫Analyze≫Signal Processing≫Windows, as shown in Figure 11.37. If we denote the input sequence as $\mathbf{X}$ (with n elements) and the output sequence of the Hamming window as $\mathbf{Y}$, then

$$\mathbf{Y}(i) = \mathbf{X}(i)\,[0.54 - 0.46\cos\omega]$$

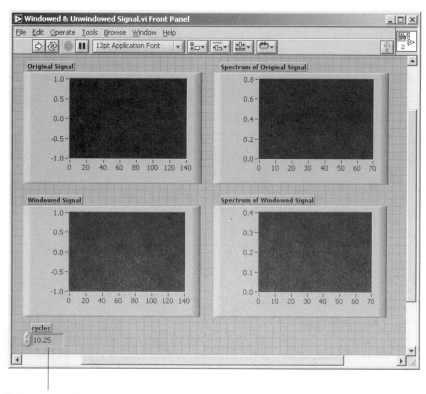

Set value equal to noninteger value to examine the positive effects of the Hamming window.

FIGURE 11.36
The front panel for a VI to investigate the use of windows.

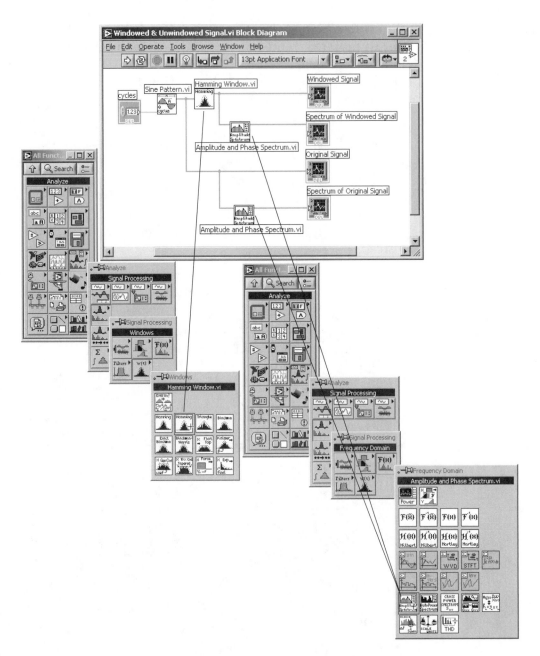

FIGURE 11.37
The block diagram for a VI to investigate the effect of windowing.

where

$$\omega = \frac{2\pi i}{n}.$$

- The **Amplitude and Phase Spectrum.vi** computes the amplitude spectrum of the windowed and nonwindowed input waveforms. You can find this particular VI in the palette Functions≫Analyze≫Signal Processing≫ Frequency Domain, as shown in Figure 11.37.

- The **Sine Pattern.vi** generates a sine wave with the number of cycles specified in the control labeled **cycles**. It is located in Functions≫Analyze≫ Signal Processing≫Signal Generation.

As interesting numerical experiment, make the following two runs:

- Set the control **cycles** to 10. Since this is an integer number, when you repeat the waveform to construct a periodic waveform, you will not have any discontinuities. You should observe that the spectrum of the windowed and the nonwindowed waveforms are both centered at 10, and that the spectrum of the original signal displays no spectral leakage.

- Set the control **cycles** to 10.25. Since this is not an integer, you should observe that the spectrum of the windowed and the nonwindowed waveforms are different. The nonwindowed spectrum should show distinct signs of spectral leakage due to the discontinuities when constructing the periodic waveform. The windowed waveform, while not a perfect spike centered at 10, displays significantly less leakage.

Save your VI as **Windowed & Unwindowed Signal.vi** in the **Users Stuff** folder in the **Learning** directory.

You can find a working version of **Windowed & Unwindowed Signal.vi** *in* **Chapter 11** *of the* **Learning** *directory.* ◆

11.8.3   The Spectral Measurements Express VI

The **Signal Analysis** Express palette shown in Figure 11.8 contains the Spectral Measurements Express VI. This Express VI performs spectral measurements, such as spectral power density, on signals.

To use the Spectral Measurements VI, you first place it on the block diagram. As illustrated in Figure 11.38, once the Spectral Measurements Express VI is placed on the block diagram, a dialog box automatically appears to configure the VI. The spectral measurements options include choosing the measurement type, such as magnitude (peak or RMS), power spectrum, and power spectral density, and then selecting the windowing from among Hanning, Hamming, Blackman-Harris, and Low Sidelobes. In Figure 11.38, the magnitude RMS is selected as the

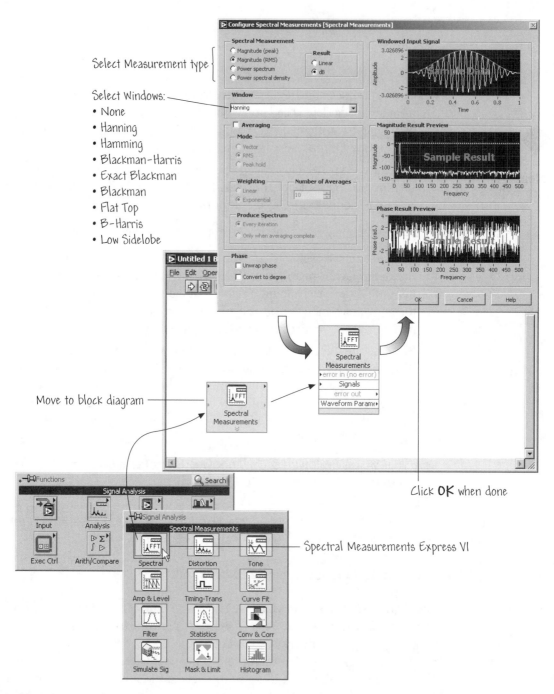

Select Measurement type

Select Windows:
- None
- Hanning
- Hamming
- Blackman-Harris
- Exact Blackman
- Blackman
- Flat Top
- B-Harris
- Low Sidelobe

Move to block diagram

Click OK when done

Spectral Measurements Express VI

FIGURE 11.38
Configuring the Spectral Measurements Express VI.

measurement type and the Hamming window is selected for windowing. Once the Spectral Measurements VI is configured, click **OK** to return to the block diagram.

On the block diagram, you can expand the Express VI to display the inputs and outputs. The key input is the signal, and the key output is the spectral measurement, such as the power spectrum. The Spectral Measurements VI is now ready for inclusion in a VI for signal analysis.

**Practice
with the
Spectral
Measurements
Express VI**

Consider the VI shown in Figure 11.39. Notice on the block diagram that two Express VIs, the Spectral Measurements Express VI and the Simulate Signal Express VI, are utilized. This VI was developed to assist in investigating the power spectral density of a sine wave signal.

The Spectral Express Demo.vi is located in Chapter 11 *of the* Learning *directory.*

There are four elements on the block diagram:

- The Spectral Measurements Express VI—this can be configured by double-clicking on the icon to access the configuration dialog box.

- The Simulate Signal Express VI—this VI was described in Section 11.7. It has been configured here to provide a sine wave signal at a frequency of 10.25 Hz.

- The waveform graph—this is used to graphically display the power spectral density of the sine wave signal, both the windowed and unwindowed versions.

- The Error Status Indicators—contains the error status of the Express VIs.

Run the Spectral Express Demo.vi and observe the results on the waveform graphs. Double-click the Simulate Signal Express VI to access the configuration dialog box. Change the frequency of sine wave signal to 50 Hz, run the Spectral Express Demo.vi and observe the effects on the power spectral density. You should observe the peak of the power spectral density shift to 50 Hz. Test the VI using different windowing methods.   ◆

11.8.4   Filtering

There are two main types of filters: analog and digital. In this section we will consider digital filters only. Why digital filters? Because digital filters

- are software programmable,

- are stable and predictable,

- do not drift with changes in external environmental conditions, and

- generally have superior performance-to-cost ratios compared to their analog counterparts.

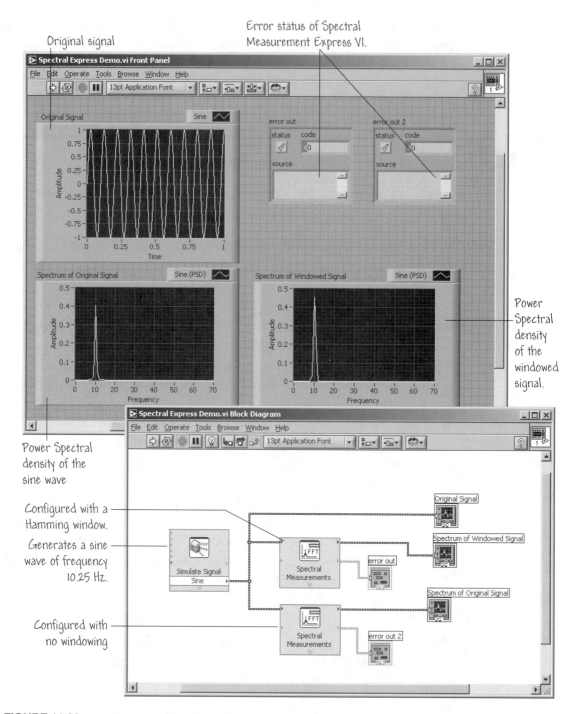

FIGURE 11.39
The Spectral Express Demo.vi for investigating the power spectral density of sine wave signals.

LabVIEW can be used to control digital filter parameters (such as filter order, cutoff frequency, stopband and passbands, amount of ripple, and stopband attenuation). You can envision a LabVIEW-based DAQ system wherein data is acquired from external sources (such as an accelerometer sensor) and filtered in the VI software. The results can be easily analyzed and studied using the graphics provided by the G programming language and written to a spreadsheet (as we will discuss in the next chapter). The key is that by using LabVIEW you can utilize digital filters, allowing the VIs to handle the design issues, computations, memory management, and the actual data filtering.

The theory of filters is a rich and interesting subject, and one that cannot be dealt with here in any depth. Please refer to other reference materials for in-depth coverage of filtering theory.[1] But a brief discussion of terms is needed to give you a better understanding of the filter parameters and how they relate to the VI inputs.

The **sampling theorem** states that a continuous-time signal can be reconstructed from discrete, equally spaced samples if the sampling frequency is at least twice that of the highest frequency in the time signal. The sampling interval is often denoted by δt. The **sampling frequency** is computed as the inverse of the sampling interval:

$$f_s = \frac{1}{\delta t}.$$

Thus, according to the sampling theorem, the highest frequency that the filter can process—the **Nyquist frequency**—is

$$f_{nyq} = \frac{f_s}{2}.$$

As an example, suppose that you have a system with sampling interval $\delta t = 0.01$ second. Then the sampling frequency is $f_s = \frac{1}{0.01} = 100$ Hz. From the sampling theorem we find that the highest frequency that the system can process is $f_{nyq} = \frac{f_s}{2} = 50$ Hz. If we expect that the signals that we need to process have components at frequencies higher than 50 Hz, then we must upgrade the system to allow for shorter sampling intervals, say for example, $\delta t = 0.001$ second. What is f_{nyq} in this case?

One main use of filters is to remove unwanted noise from a signal—usually if the noise is at high frequencies. Depending on the frequency range of operation, filters either pass or attenuate input signal components. Filters can be classified into the following types:

---

1. A good source of information for LabVIEW users is *LabVIEW Signal Processing* by Mahesh L. Chugani, Abhay R. Samant, and Michael Cerna, Prentice Hall, Upper Saddle River, New Jersey, 1998.

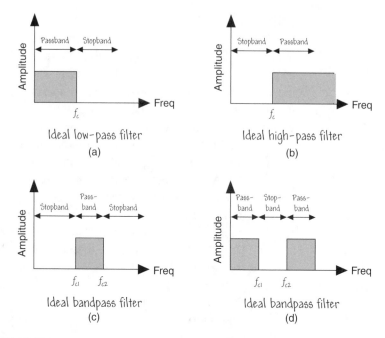

FIGURE 11.40
The ideal frequency response of common ideal filters.

1. A **low-pass filter** passes low frequencies and attenuates high frequencies. The ideal low-pass filter passes all frequencies below the *cutoff frequency* f_c.

2. A **high-pass filter** passes high frequencies and attenuates low frequencies. The ideal high-pass filter passes all frequencies above f_c.

3. A **bandpass filter** passes a specified band of frequencies. The ideal bandpass filter only passes all frequencies between f_{c1} and f_{c2}.

4. A **bandstop filter** attenuates a specified band of frequencies. The ideal band-stop filter attenuates all frequencies between f_{c1} and f_{c2}.

The ideal frequency response of these filters is illustrated in Figure 11.40.

The frequency points f_c, f_{c1} and f_{c2} are known as the cutoff frequencies and can be viewed as filter design parameters. The range of frequencies that is passed through the filter is known as the **passband** of the filter. An ideal filter has a gain of one (0 dB) in the passband—that is, the amplitude of the output signal is the same as the amplitude of the input signal. Similarly, the ideal filter completely attenuates the signals in the stopband—that is, the stopband attenuation is $-\infty$ dB. The low-pass and high-pass filters have one passband and one stopband. The range of frequencies that do not pass through the filter is known as the **stopband** of the filter. The stopband frequencies are rejected or attenuated by the filter. The passband and the stopband for the different types of filters are

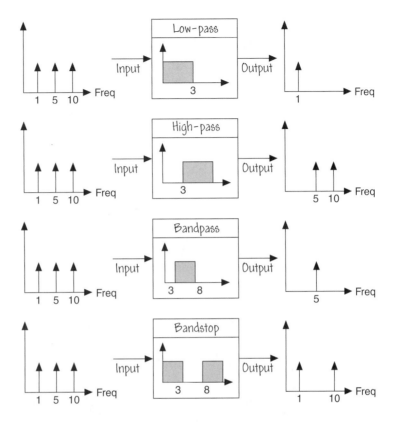

FIGURE 11.41
The output of the common filter in the case where the input signal contains component frequencies of 1 Hz, 5 Hz, and 10 Hz.

shown in Figure 11.40. The bandpass filter has one passband and two stopbands. Conversely, the bandstop filter has two passbands and one stopband.

Suppose you have a signal containing component frequencies of 1 Hz, 5 Hz, and 10 Hz. This input signal is passed separately through low-pass, high-pass, bandpass, and bandstop filters. The low-pass and high-pass filters have a cutoff frequency of 3 Hz, and the bandpass and bandstop filters have cutoff frequencies of 3 Hz and 8 Hz. What frequency content of the signal can be expected? The output of the filter in each case is shown in Figure 11.41. The low-pass filter passes only the signal at 1 Hz because this is the only component of the input signal lower than the 3 Hz cutoff. Conversely, the high-pass filter attenuates the 1 Hz component and passes a signal with components at 5 and 10 Hz. The bandpass filter passes only the signal component at 5 Hz, and the bandstop filter filters out the signal component at 5 Hz and passes a signal with frequency content at 1 and 10 Hz.

Ideal filters are not achievable in practice. It is not possible to have a unit gain (0 dB) in the passband and a gain of zero ($-\infty$ dB) in the stopband— there is always a *transition region* between the passband and stopband. A more

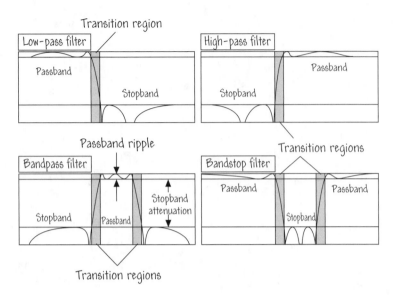

FIGURE 11.42
A realistic frequency response of a filter.

realistic filter will have the passband, stopband, and transition bands as depicted in Figure 11.42. The variation in the passband is called the *passband ripple*; see the bandpass filter in Figure 11.42. The *stopband attenuation*, also depicted in Figure 11.42, cannot be infinite (as it would be for an ideal filter).

If we view the filter as a linear system, then we can consider the response of the system (that is, the filter) to various types of inputs.[2] One interesting input is the *impulse*. If the input to the digital filter is the sequence $x(0), x(1), x(2), \ldots$, then the impulse is given by $x(0) = 1, x(1) = x(2) = \cdots = 0$. The **impulse response** of a filter (that is, the output of the filter when the input is an impulse) provides another classification system for filters. The Fourier transform of the impulse response is known as the **frequency response**. The frequency response of a system provides a wealth of information about the system, including how it will respond to periodic inputs at different frequencies. Therefore, the frequency response will tell us about the filter characteristics: How does the filter respond in the passbands and stopbands? How accurately does the filter cut off and attenuate high-frequency components?

We can classify a filter based on its impulse response as either a **finite impulse response** (FIR) filter or an **infinite impulse response** (IIR) filter. For an IIR filter, the impulse response continues indefinitely (in theory), and the output depends on current and past values of the input signal and on past values of the output.

---

2. A good source of information on systems and system response to various inputs is *Modern Control Systems*, by Richard C. Dorf and Robert H. Bishop, Prentice Hall, Upper Saddle River, NJ, 2000.

In practical applications, the impulse response for stable IIR filters decays to near zero in a finite time. For an FIR filter the impulse response decays to zero in a finite time, and the output depends only on current and past values of the input signal. Take the example of processing noisy range measurements (that is, distance measurements) to a fixed target. Suppose that you want to determine the distance to the fixed object and you have available a ranging device (e.g., a laser ranging device) that is corrupted by random noise (as is the case for most realistic sensors!). One way to estimate the range is to take a series of range measurements $x(0), x(1), x(2), \ldots, x(k)$ and filter them by computing a running average:

$$x_{\text{ave}}(k) = \frac{1}{k} \sum_{i=1}^{k} x(i).$$

The output of the filter is $x_{\text{ave}}(k)$. This is an FIR filter because the output depends only on previous values of the input $(x(0), x(1), x(2), \ldots, x(k-1)$ and on the current value of the input $(x(k))$. Now, we can rewrite the filter as

$$x_{\text{ave}}(k) = \frac{k-1}{k} x_{\text{ave}}(k-1) + \frac{1}{k} x(k).$$

This is an IIR filter because the output depends on current and previous values of the input as well as on previous values of the output (that is, the $x_{\text{ave}}(k-1)$ term). Mathematically, the two filters provide the same output, but they are implemented differently. The FIR filter is sometimes referred to as a *nonrecursive* filter; the IIR filter is known as a *recursive* filter.

One disadvantage of IIR filters is that the phase response is nonlinear. You should use FIR filters for situations where a linear phase response is needed. A strong advantage of IIR filters are that they are recursive, thus reducing the memory storage requirements. The well-known Kalman filter, which can be viewed as an IIR filter (it is actually a bit more complicated to implement than an IIR filter), was used successfully to filter navigation data acquired by the Apollo spacecraft to find its way to the moon and to rendezvous around the moon for the long journey home.[3] LabVIEW provides many different types of filters, as shown in Figure 11.43.

The IIR filter VIs available in LabVIEW include the following:

- Butterworth: Provides a smooth response at all frequencies and a monotonic decrease from the specified cutoff frequencies. Butterworth filters are maximally flat—the ideal response of unity in the passband and zero in the stopband—but do not always provide a good approximation of the ideal filter response because of the slow rolloff between the passband and the stopband.

---

3. A well-told story on the use of Kalman filters during Apollo can be found in *An Introduction to the Mathematics and Methods of Astrodynamics*, by R. H. Battin, AIAA Education Series, 1987.

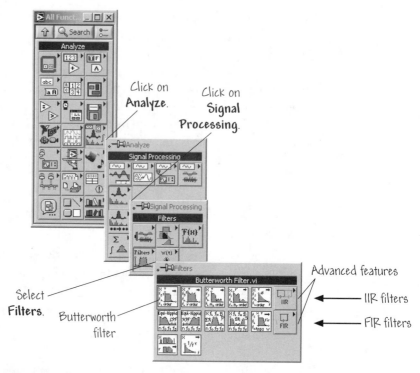

FIGURE 11.43
IIR and FIR filter choices are located in the **Functions** palette.

- Chebyshev: Minimizes the peak error in the passband by accounting for the maximum absolute value of the difference between the ideal filter and the filter response you want (the maximum tolerable error in the passband). The frequency response characteristics of Chebyshev filters have an equiripple magnitude response in the passband, monotonically decreasing magnitude response in the stopband, and a sharper rolloff than for Butterworth filters.

- Inverse Chebyshev: Also known as Chebyshev II filters. They are similar to Chebyshev filters, except that inverse Chebyshev filters distribute the error over the stopband (as opposed to the passband), and are maximally flat in the passband (as opposed to the stopband). Inverse Chebyshev filters minimize peak error in the stopband by accounting for the maximum absolute value of the difference between the ideal filter and the filter response you want. The frequency response characteristics are equiripple magnitude response in the stopband, monotonically decreasing magnitude response in the passband, and a rolloff sharper than for Butterworth filters.

- Elliptic: Minimize the peak error by distributing it over the passband and the stopband. Equiripples in the passband and the stopband characterize the magnitude response of elliptic filters. Compared with the same-order Butterworth or Chebyshev filters, the elliptic design provides the sharpest transition between the passband and the stopband. For this reason, elliptic filters are widely used.

- Bessel: Can be used to reduce nonlinear phase distortion inherent in all IIR filters. Bessel filters have maximally flat response in both magnitude and phase. Furthermore, the phase response in the passband of Bessel filters, which is the region of interest, is nearly linear. Like Butterworth filters, Bessel filters require high-order filters to minimize the error.

The FIR filter VIs available in LabVIEW include the following:

- Windowed: The simplest method for designing linear-phase FIR filters is the window design method. You select the type of windowed FIR filter you want—low-pass, high-pass, bandpass, or bandstop—via input to the FIR Windowed Filter.vi.

- Optimum filters based on the Parks-McClellan algorithm: Offers an optimum FIR filter design technique that attempts to design the best filter possible for a given filter complexity. Such a design reduces the adverse effects at the cutoff frequencies. It also offers more control over the approximation errors in different frequency bands—control that is not possible with the window method. The VIs available include

  - Equiripple Low-pass

  - Equiripple High-pass

  - Equiripple Bandpass

  - Equiripple Bandstop

Which filter is best suited for your application? Obviously, the choice of filter depends on the problem at hand. Figure 11.44 shows a flowchart that can serve as a guide for selecting the best filter for your needs. Keep in mind that you will probably use the flowchart to determine several candidate filters, and you will have to experiment to make the final choice.

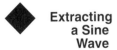

Extracting a Sine Wave

Open a new front panel and construct a front panel similar to the one shown in Figure 11.45. You will need to place one digital control, two vertical slides, and two waveform graphs. Label them according to the scheme shown in the figure.

Construct a block diagram similar to the one shown in Figure 11.45. In the block diagram, we use three main VIs:

- Butterworth.vi: This VI is used to filter the noise. It is located in Functions ≫Analyze≫Signal Processing≫Filters, as shown in Figure 11.43.

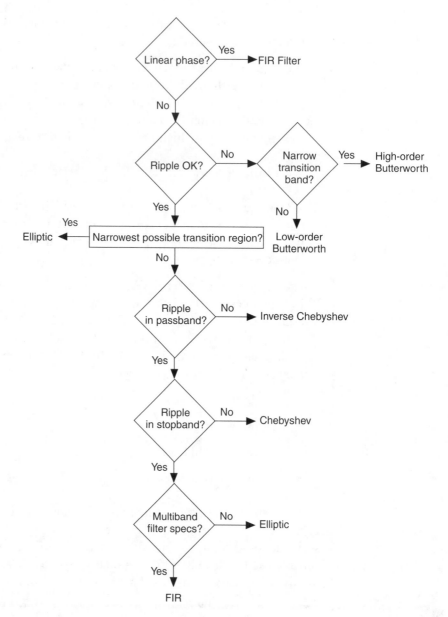

FIGURE 11.44
Flowchart that can serve as a guide for selecting the best filter.

- The Uniform White Noise.vi generates a white noise that is added to the sinusoidal signal. You can find this particular VI in the palette Functions≫ Analyze≫Signal Processing≫Signal Generation.

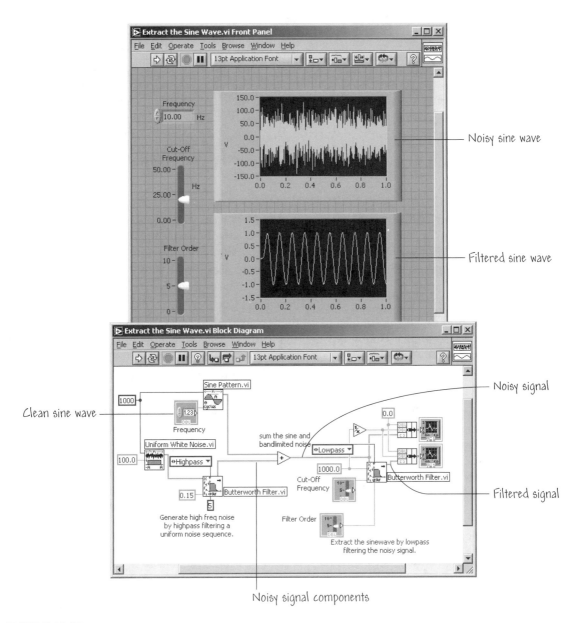

FIGURE 11.45
The front panel and block diagram for a VI to filter a noisy sine wave.

- The Sine Pattern.vi generates a sine wave of the desired frequency. It is located in Functions≫Analyze≫Signal Processing≫Signal Generation.

With this VI you are generating 10 cycles of a sine wave (this value can be varied on the front panel), and there are 1,000 samples. Select a cutoff frequency of 25 Hz and a filter order of 5. Note that we did not previously discuss filter order—it is a measure of filter complexity and is related to the number of terms retained in the filter. Run the VI. Vary the cutoff frequency and observe the effects. What happens when the cutoff frequency is set to 50? Does the filtered signal contain noise components? When you are finished exploring, save your VI as **Extract the Sine Wave.vi** in the **Users Stuff** folder.

You can find a working version of the VI above in folder **Chapter 11** *in the* **Learning** *directory. It is called* **Extract the Sine Wave.vi**. ◆

11.8.5 The Filter Express VI

The Filter Express VI is located on the **Signal Analysis** Express palette, shown in Figure 11.8. The Filter Express VI processes signals through filters and windows. The filter options include lowpass, highpass, bandpass, bandstop, and smoothing. To use the Filter Express VI, you first place it on the block diagram. As illustrated in Figure 11.46, once the Filter Express VI is placed on the block diagram, the **Configure Dialog** box automatically appears to configure the VI. Figure 11.46 show a lowpass filter selected with a cutoff frequency of 100 Hz. Once the Filter Express VI is configured, click **OK** to return to the block diagram. The Filter Express VI is now ready for inclusion in a VI for signal analysis. On the block diagram, you can expand the Filter Express VI to display the inputs and outputs. The key input is the signal and the key output is the filtered signal.

Practice with the Filter Express VI

Consider the VI shown in Figure 11.47. Notice on the block diagram that three Express VIs are used: the Filter Express VI, the Simulate Signal Express VI and the Spectral Measurements Express VI. The **Filter Express Demo.vi** was developed to assist in investigating the lowpass filtering of a sinusoidal signal.

The **Filter Express Demo.vi** *is located in Chapter 11 of the* **Learning** *directory.*

The four Simulate Signal Express VIs have been configured to provide sine wave signals at frequencies of 1, 10, and 100 Hz and a DC signal of magnitude 10. The sinusoidal signal that is to be filtered is a sum of the four individual signals from the Simulate Signal Express VIs. The Filter Express VI has been configured to provide a lowpass filter with default cutoff frequency of 25 Hz. A numeric control is employed to vary the cutoff frequency directly from the front panel. The Spectral Measurements Express VIs appearing in Figure 11.47 have been configured to provide power spectral densities of the incoming signals. The waveform graphs display the power spectral density of the filtered and unfiltered sinusoidal signals.

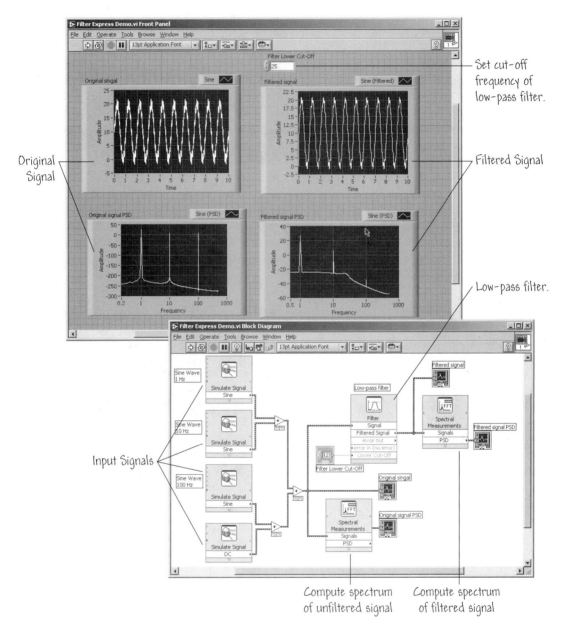

FIGURE 11.46
Configuring the Filter Express VI.

Run the Filter Express Demo.vi and observe the results on the waveform graphs. When the cutoff frequency is set to the default value of 25 Hz, you should observe that the filtered signal has a significantly reduced frequency component at 100 Hz. This is because the lowpass filer has essentially removed the 100 Hz component of the sinusoidal signal.

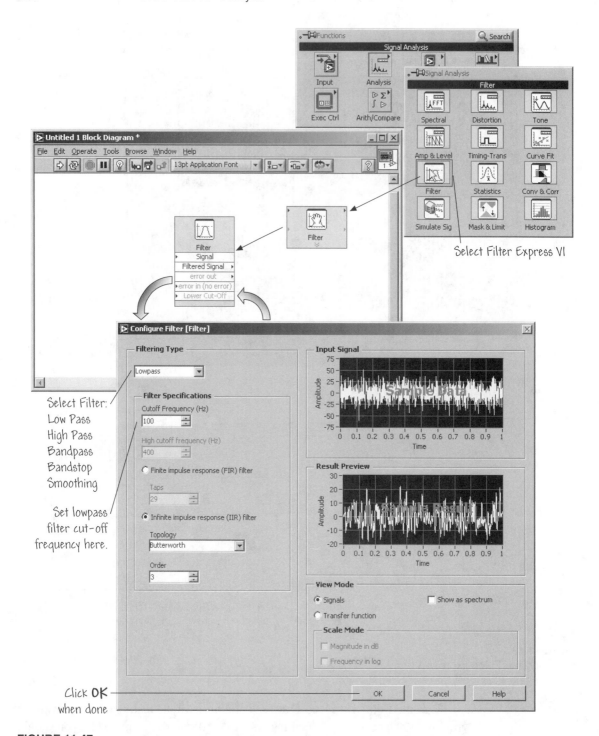

Select Filter Express VI

Select Filter:
Low Pass
High Pass
Bandpass
Bandstop
Smoothing

Set lowpass
filter cut-off
frequency here.

Click OK
when done

FIGURE 11.47
The Filter Express Demo.vi for investigating the filtering of a sinusoidal signal containing three main frequencies at 1, 10, and 100 Hz.

Let's investigate the effect of varying the lowpass filter cutoff frequency on the filtering process. The numeric control located on the front panel will be employed to vary the lowpass filter cutoff frequency. Start the VI by selecting the **Run Continuously** button. Slowly reduce the cutoff frequency and observe the power spectral density graph of the filtered signal (lower right hand-side of the block diagram). Notice that as the cutoff frequency is reduced to a value under 10 Hz, the component of the filtered sinusoidal signal at 10 Hz is correspondingly reduced.

Experiment with increasing the lowpass filter cutoff frequency above 100 Hz. Can you find a value of the cutoff frequency for which the filtered signal power spectral density shows the 100 Hz component?   ◆

BUILDING BLOCK

11.9   BUILDING BLOCKS: ANALYZING VOLUME

In the Chapter 9 "Building Block" exercise, you constructed a VI named Volume Data Saved.vi. In this exercise you will modify Volume Data Saved.vi by adding several analysis VIs. The VI that you developed in Chapter 9 should be have been saved in the Users Stuff folder.

The target front panel and block diagram are shown in Figure 11.48. The two additions are the Spectral Measurements Express VI and the Statistics Express VI, which compute the mean and power spectral density of the measured volume data, respectively. The mean value is displayed in a numeric indicator and the power spectral density is shown on a waveform graph.

Enter a value for the **Number of data points** on the front panel. A value of 100 is a reasonable number. Run the VI and observe the results—the curve fit of the measured volume is superimposed on the raw data. Notice that the curve fit is much smoother than the raw volume data. Since the measured data is noisy, each time you run the VI, you will get different results. When you are finished with the VI, save it in the Users Stuff folder and name it Volume Analysis.vi.

*The Statistics Express VI is located in the **Functions≫Signal Analysis** palette. It provides many statistical quantities of the incoming signal, such as mean, median, RMS, and standard deviation.*

11.10   RELAXED READING: MOTORCYCLE BRAKE TESTING

A tier-one supplier to large motorbike manufacturers in India wanted an on-road data acquisition system to test motorbike front disk brakes. The brakes were to be tested by measuring stopping distance during various road trials. The most

Compute the mean
of the measured volume

Obtain power spectral
density of the volume

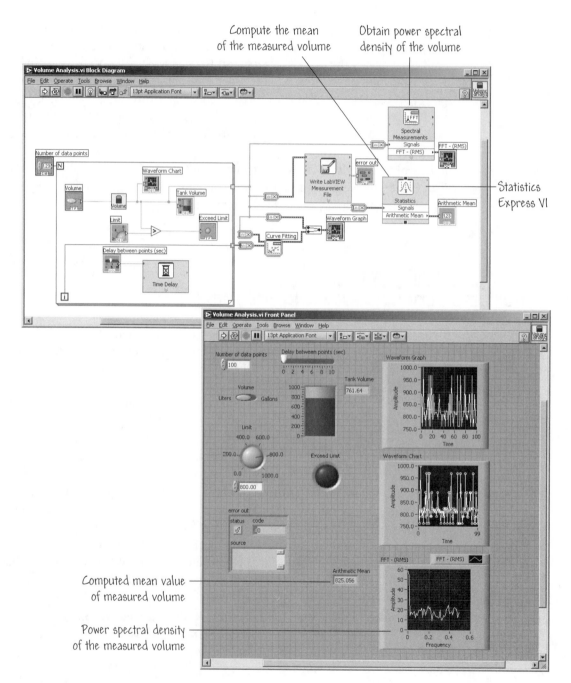

Statistics
Express VI

Computed mean value
of measured volume

Power spectral density
of the measured volume

FIGURE 11.48
Analyzing the volume data.

straightforward way to determine stopping distance is to bring the motorbike to the required speed before reaching a certain predetermined marked point, and then to apply the brake from that point until the motorbike stops. Using a distance-measuring instrument, the stopping distance can be determined. This is a tedious and time-consuming process and not practical during long trials over different terrain. Additionally, in this method, other parameters relating to braking, such as the brake-pad temperature and hydraulic pressure, cannot be obtained.

An on-road brake test system was designed to measure the braking distance of a motorbike under different road conditions in real-time. This data was then used offline to compute the performance parameters of the disc brakes. The test required a dedicated onboard computer to run the system and to acquire the test data. The test rig had to be compact enough to fit on the motorbike. A test system based on a small form factor rugged FieldWorks computer with a National Instruments data acquisition card, signal conditioning modules, and application software written in LabVIEW was constructed to meet the requirements.

The following sensors are used in the data acquisition system:

1. **Rotation encoder**—Measures the rotational displacement and the wheel speed.

2. **Thermocouple**—Measures the pad temperature during braking.

3. **Pressure sensor**—Measures the brake hydraulic line pressure.

4. **Accelerometer**—Measures the instantaneous acceleration of the vehicle.

For user inputs and displays, a small-sized integrated keyboard and LCD display was interfaced with the FieldWorks computer through the RS-232 port. Using this display, the motorbike rider can send commands, set parameters, and receive feedback from the system, such as the braking distance. The computer, user-interface kit, and signal conditioning modules were all powered by the motorbike battery. The conditioned signals from the transducers connected to the National Instruments PCMCIA multipurpose high-speed data acquisition card installed in the FieldWorks computer. Three analog input channels and two counter/timers from the DAQ card were used for the data acquisition.

The rotational encoder and the accelerometer were both used to acquire the data used in calculating the braking distance. Two complementary and independent methods were used to calculate the stopping distance. The first method uses the pulses per revolution output of the rotational encoder to measure both the speed and the displacement. The accelerometer data was used as supplementary data.

The data acquired by the various sensors is stored in the onboard computer. At the end of the test, the onboard computer can be connected to an Ethernet network and the data can be downloaded onto another PC for offline analysis. A software application was developed for the offline data review and analysis. This application includes a review module that can be used to review the test

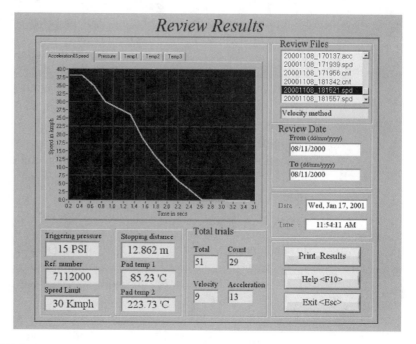

FIGURE 11.49
On-Road Brake Test Review Data Screen.

data and view graphs that plot various parameters against time, such as pressure, acceleration, speed, and brake-pad temperature.

With the user interface kit and integrated LCD display and keyboard, the user can enter various speeds at which to start braking, and the LabVIEW application monitors the motorbike speed and gives a signal to start braking at the desired speed range. When the bike comes to a stop, the display immediately shows the braking distance. During the braking, the brake-pad temperature and acceleration are also acquired using the high-speed data acquisition card.

The on-road brake test system measured braking distance to an accuracy of 1% over a distance of 30 m. The small form factor of the NI PCMCIA data acquisition cards and the FieldWorks 2000 embedded computer made it easy to mount it on the motorbike. With the user interface, the user can set up different tests during the road trials and also see the results without having to connect a laptop to the embedded computer. The whole system was rugged enough to handle accidental falls during testing.

For more information, contact:

V. Arunachalam or Gokul DassT. V.
Soliton Automation Pvt. Ltd.
info@solitonautomation.com

11.11 SUMMARY

LabVIEW provides a great computational environment for analysis of signals and systems. In this chapter, we presented some applications of the many VIs available for analysis of signals, systems, functions, and systems of equations. The material was intended to motivate you to look further into the capabilities of the G programming language in developing your own VIs for solving linear algebraic systems of equations, curve fitting, integrating ordinary differential equations, computing function zeroes, computing derivatives of functions, integrating functions, generating and analyzing signals, computing discrete Fourier transforms, and filtering signals.

KEY TERMS

Bandpass filter: A system that passes a specified band of frequencies.

Bandstop filter: A system that attenuates a specified band of frequencies.

Condition number: A quantity used in assessing the accuracy of solutions to systems of linear algebraic equations. The condition number can vary from 0 to ∞—a matrix with a condition number near 1 is closer to being nonsingular than a matrix with a very large condition number.

Determinant: A characteristic number associated with an $n \times n$ square matrix that is computed as a function of the minors and cofactors of the matrix. When the determinant is identically equal to zero, then we say that the matrix is singular.

Digital frequency: Computed as the analog frequency divided by the sampling frequency. Also known as the **normalized frequency**.

Discrete Fourier transform (DFT): A common practical algorithm for transforming sampled signals from the time domain into the frequency domain.

Even symmetric signal: A signal symmetric about the y-axis.

Fast Fourier transforms (FFT): The FFT is a fast algorithm for calculating the DFT when the number of samples (N) is a power of 2.

Frequency-domain representation: The representation of a signal in terms of its individual frequency components.

Impulse response: The output of a system (e.g., a filter) when the input is an impulse.

FIR filter: A finite impulse response filter in which the impulse response decays to zero in a finite time, and the output depends only on current and past values of the input signal.

Frequency response: The Fourier transform of the impulse response.

G Math Toolkit: A toolkit that enables you to interface real-world measurements to mathematical analysis algorithms.

High-pass filter: A system that passes high frequencies and attenuates low frequencies.

Homogeneous DEQ: A differential equation that has no input driving function (that is, $u(t) = 0$).

IIR filter: An infinite impulse response filter in which the impulse response continues indefinitely (in theory), and the output depends on current and past values of the input signal and on past values of the output.

Low-pass filter: A system that passes low frequencies and attenuates high frequencies.

Nyquist frequency: The highest frequency that a filter can process, according to the sampling theorem.

Odd symmetric signal: A signal symmetric about the origin.

Passband: The range of frequencies that is passed through a filter.

Sampling theorem: The statement that a continuous-time signal can be reconstructed from discrete, equally spaced samples if the sampling frequency is at least twice that of the highest frequency in the time signal.

Spectral leakage: A phenomenon that occurs when you sample a noninteger number of cycles, leading to "artificial" discontinuities in the signal that manifest themselves as very high frequencies in the DFT/FFT spectrum, appearing as if the energy at one frequency has "leaked out" into all the other frequencies.

Stopband: The range of frequencies that do not pass through a filter.

Trace: The sum of the diagonal elements of an $n \times n$ square matrix.

Windowing: A method used to reduce the amplitude of sampled signal discontinuities by multiplying the sampled original waveform by a finite-length window whose amplitude varies smoothly and graduates towards zero at the edges.

EXERCISES

E11.1 (a) Open a new VI and place a While Loop on the block diagram by going to **Functions≫Express≫Execution Control≫While Loop**.

(b) Within the While Loop, place the Simulate Signal Express VI on the block diagram by navigating to **Functions≫Express≫Input≫Simulate Signal**. The configuration window will open in LabVIEW by default, but if it does not, double-click on the Simulate Signal Express VI to view it.

(c) In the configuration window, alter the **Signal Type** (i.e., Sine, Square, Triangle, Sawtooth, DC) and the additional inputs associated with the signal (i.e., Frequency, Amplitude, Phase, etc). Notice that the **Result Preview** graph changes according to these controls.

(d) Add noise to the signal by adding a checkmark to the box **Add noise** and view how the various types of noise affect the signal.

(e) Feel free to experiment with the other options in the Configuration Window to create the simulated signal of your choice. Press the **OK** button to return to the block diagram of your VI.

(f) Right-click on the signal output of your Simulate Signal Express VI and select **Create≫Graph Indicator**.

(g) Run the VI and confirm that the signal that you configured in the Simulate Signal Express VI matches the signal displayed on the front panel of your VI.

(h) Stop the VI and go back into the Express VI **Configuration Window** to make changes to the simulated signal in order to observe other types of signals that can be generated.

E11.2 Open a new VI and go to **Help≫Find Examples**. . .. Click on the **Search** tab, type "express" into the string labeled **Enter keyword(s)**, and press the **Search** button.

(a) Open Limit Example.vi, which performs a limit test on a sine wave. Using the **Amplitude** control, you can modify the sine wave in order to make it pass or fail the test. If the signal goes over/under the limit, the integral of the area that is over the limit is displayed in the Area over/under limit indicator.

(b) Open Express Vibration Lab.vi, which simulates a vibratory system of two masses and springs. The parameters of the system, such as the mass and rigidity of the springs, can be changed. Refer to **Context Help** for more a detailed explanation of this example VI, and explore the block diagram to see an example of what simulation and analysis in LabVIEW 7.0 can accomplish with Express VIs!

E11.3 Navigate to Examples\Analysis\windxmpl.llb and open the Window Comparison.vi to observe the effects of the native LabVIEW window algorithms, including Hanning, Hamming, Triangle, etc. (prior to LabVIEW 7.0 Express VIs).

E11.4 Navigate to Examples\Analysis\fltrxmpl.llb and open the IIR Filter Design.vi to observe the effects of the native LabVIEW IIR filters, including Elliptic, Bessel, Butterworth, etc. (prior to LabVIEW 7.0 Express VIs).

PROBLEMS

P11.1 Create a VI that generates a **Sine Wave**, which is displayed on a Waveform Graph. The user should have the ability to change the **Frequency** and the **Amplitude** of this sine wave programmatically. A **Power Spectrum** analysis should be performed on the sine wave with a **Hanning** window and another with a **Flat Top** window, both displayed on the same graph on the front panel.

    Hint: Explore the Express VIs by navigating to **Functions≫Express≫Signal Analysis**.

P11.2 Continue your analysis on the sine wave that is being generated from your solution to Problem 11.1. Determine the **Positive Peak** value of the signal and the **Root Mean Square Value** of the signal, and display these numeric values on the front panel.

    Hint: Explore the Express VIs by navigating to **Functions≫Express≫Signal Analysis**.

P11.3 Continue your analysis on the sine wave that is being generated from your solution to Problem 11.2. Add **Uniform White Noise** with noise amplitude of 3 to the sine wave being generated. Filter the noisy sine wave through a **Lowpass Butterworth filter** (Order =4), and display the filtered sine wave to compare the affect the filter has on the noise.

    Hint: Explore the Express VIs by navigating to **Functions≫Express≫Signal Analysis**.

P11.4   Complete the crossword puzzle.

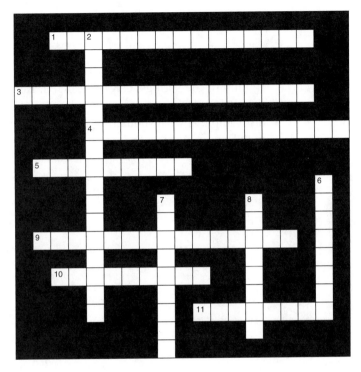

Across

1. A quantity used is assessing the accuracy of solutions to systems of linear algebraic equations.
3. The Fourier transform of the impulse response.
4. A phenomenon that occurs when you sample a noninteger number of cycles, leading to artificial discontinuities.
5. A filter in which the impulse response decays to zero in a finite time and the output depends only on current and past values of the input signal.
9. The output of a system when the input is an impulse.
10. A method used to reduce the amplitude of sampled signal discontinuities by multiplying the sampled original waveform by a finite-length window.
11. The range of frequencies that is passed through a filter.

Down

2. The highest frequency that a filter can process, according to the sampling theorem.
6. The range of frequencies that do not pass through a filter.
7. A filter in which the impulse response continues indefinitely.
8. A system that passes a specified band of frequencies.

CHAPTER 12

Other LabVIEW Applications

In this chapter we discuss several LabVIEW applications that may be of interest to you as you begin to master LabVIEW and want to further utilize the power of virtual instruments. The material presented here is intended to stimulate your curiosity about advanced LabVIEW programming and is not covered in the same detail as previous chapters. Did you know that LabVIEW can act in an event-driven nature, similar to some text-based programming environments? Did you know that you can actually control VIs over the Internet through a Web browser? Have you wondered how you can programmatically change properties of front panel objects? Have you ever wanted to create simple programs for your PDA? With LabVIEW, you can do all of this, from event-driven programming to creating programs for your PDA. These topics are introduced in this chapter, and you are pointed to other information sources from which you can continue to discover the many uses of LabVIEW.

GOALS

1. Learn about event-driven programming.
2. Read about viewing and controlling a VI remotely from a Web browser.
3. Learn about modifying the appearance of front panel objects using Property Nodes.

12.1   EVENT-DRIVEN PROGRAMMING

As you have learned throughout this book, LabVIEW is a data-flow programming environment. This means that the flow of data determines the execution order of block diagram elements. With event-driven programming, the user's direct interaction with objects on the front panel determines the execution of the program. Event-driven programming is a popular paradigm for managing user interfaces in many language environments such as CVI and Visual Basic because it allows those applications to sleep until something happens on the user interface, rather than having to repeatedly poll for such activity. This ability to sleep frees up time on the processor, allowing the program to run more efficiently.

Events are caused by actions the user performs. For example, clicking the mouse generates a mouse event, pressing a key on the keyboard generates a keyboard event, and so on. In an event-driven program, the program first waits for events to occur, responds to those events, and then returns to waiting for the next event. How the program responds depends on the code written for that specific event. The order in which an event-driven program executes depends on which events occur and on the order in which those events occur.

Event-driven programming is accomplished in LabVIEW using the Event Structure shown in Figure 12.1. The Event Structure can be found on your **Functions≫All Functions≫Structures** palette.

As with a Case Structure, you can add multiple cases to the Event Structure. You can then configure those cases to handle one or more events. When those events occur, LabVIEW executes the corresponding case. You should place the Event Structure inside a While Loop to handle a series of events until a terminating condition occurs.

On the Event Structure, there are three key elements—the **Event Selector**, the **Timeout terminal**, and the **Event data node**, as show in Figure 12.1. The **Timeout** terminal is used to specify the number of milliseconds an Event Structure waits until an event occurs. The default is −1, indicating never to timeout. If the timeout expires before an event occurs, LabVIEW will execute the code placed in the **Timeout** event.

The **Event Data Node** is similar in appearance to the Unbundle By Name function. This node is attached to the inside left and right borders of each event case. You can resize the node vertically, and you can set each element in the node to access an arbitrary Event Data Field, as you can with the Unbundle By Name function. The node displays data that is different in each case of the Event Structure, depending on which event(s) you configure that case to handle.

The event selector label at the top of the Event Structure displays which case is currently selected, including the events handled for that case. To add additional events for your Event Structure to handle, right click the event selector and select **Add Event Case**. . . .

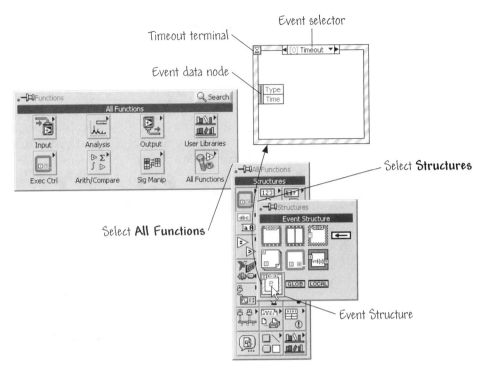

FIGURE 12.1
The Event Structure.

12.2   REMOTE FRONT PANELS

A new feature of the *LabVIEW 7 Express Student Edition* is the ability to have a LabVIEW VI viewed and controlled remotely from a Web browser. This technology allows you to create a VI on one machine, and in a few short steps be able to view and run that VI from a remote machine using only a Web browser such as Internet Explorer or Netscape.

Only a few steps are necessary to publish a VI on to the Web. First, open the VI you wish to control remotely, as illustrated in Figure 12.2. In Figure 12.2, we selected the Vibration Analysis.vi to configure the control remotely. Next, select **Web Publishing Tool**... from the **Tools** menu. The **Web Publishing Tool**, shown in Figure 12.2 will appear. This tool allows you to modify the look of the document you will see in the Web browser and gives the user the ability to modify Title, Header, and Footer text, as well as preview the document in a Web browser on his/her local machine. After modifying the text and previewing the document in a Web browser, click **Save to Disk**. This will save the .htm file to the www directory inside your LabVIEW 7 folder. Name and save the file and click **Ok**. This will bring up another window giving you the URL of the document for you to put in the Web browser. You can choose to connect at that time or click **OK**

Select
**Web Publishing
Tool**

You can
modify the
title header
and footer
text.

Click here to
save file as .htm
in the **window** directory
in Labview 7 folder.

Click here
when done.

Web
Publishing
Tool

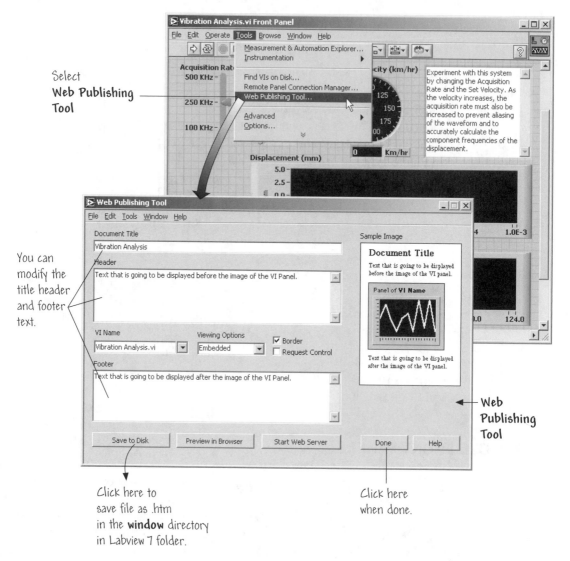

FIGURE 12.2
The **Web Publishing Tool**.

to return to the **Web Publishing Tool**. You are now finished setting up your VI
and can click **Done**.

Now that you have saved the .htm file on your local machine, you can view
and control that VI through the Web browser on a different machine.

*If you want clients/machines that do not have LabVIEW installed to be able
to view and control a front panel remotely using a Web browser, they must install
the LabVIEW Run-Time Engine. The LabVIEW Student Edition CD contains*

an installer for the LabVIEW Run-Time Engine, or it can be downloaded at www.ni.com/download.

To view the VI you published to the Web, open a Web browser and type in the URL you received after saving the VI with the **Web Publishing Tool**. You will see the front panel of the VI with the **Run** button and **Abort Execution** button grayed out. This means you do not have control over the VI. To gain control, right-click anywhere on the panel and select **Request Control of VI** as shown in Figure 12.3.

12.3 PROPERTY NODES

In some LabVIEW applications, you might want to programmatically modify the appearance of front panel objects in response to certain inputs. For example, if a user enters an invalid password, you might want an LED on your front panel to start blinking. Another example would be changing the color of a trace on a chart. When the data you are collecting is above a certain value, you might want to show a red trace instead of a green one, noting an alarm condition. These are just a few examples of when you might want to change the appearance and function of front panel objects programmatically. This functionality is achieved in LabVIEW using **Property Nodes**. You also can use **Property Nodes** to resize front panel objects, hide parts of the front panel, add cursors to graphs programmatically, and so on. **Property Nodes** in LabVIEW are very powerful and have many uses.

Create **Property Nodes** by right-clicking the object and selecting **Create≫ Property Node** from the shortcut menu, as illustrated in Figure 12.4. LabVIEW creates a **Property Node** on the block diagram that is implicitly linked to the front panel object. If the object has an owned label, the **Property Node** has the same label. You can create multiple **Property Nodes** for the same object.

When you create a **Property Node**, it initially has one terminal representing a property you can modify for the corresponding front panel object. Using this terminal on the **Property Node**, you can either set (write) the property or get (read) the current state of that property. For example, if you create a **Property Node** for a numeric control, it appears on the block diagram with the **Visible** property selected. A small arrow appears on the right side of that terminal, indicating that you are getting a value for that property. Set a property by right-clicking the terminal and selecting **Change To Write** from the shortcut menu. Wiring a Boolean FALSE to that terminal will cause the digital control to vanish from the front panel when the **Property Node** receives the data. Wiring a Boolean TRUE causes the control to reappear.

To add terminals to the node, right-click and select **Add Element** from the shortcut menu or use the **Positioning** tool to resize the node. You can then associate each **Property Node** terminal with a property from its shortcut menu.

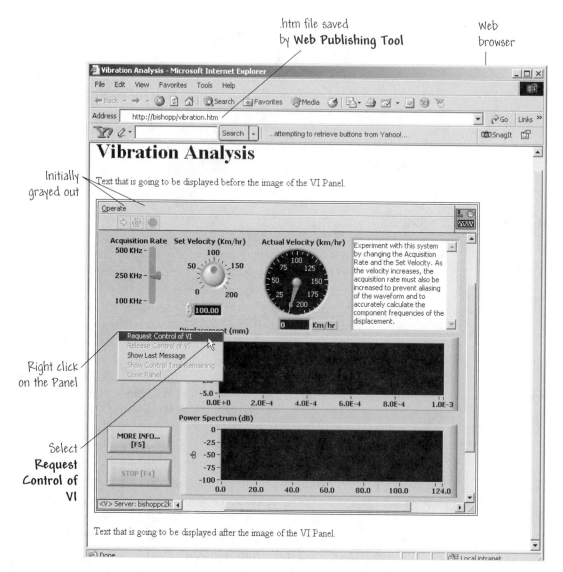

.htm file saved
by **Web Publishing Tool**

Web
browser

Initially
grayed out

Right click
on the Panel

Select
**Request
Control of
VI**

FIGURE 12.3
The Remote Front Panel.

To change the property for each terminal, simply click on the property and select a new one. Common properties that can be changed include:

- Visible
- Disable
- Key Focus
- Blink
- Position

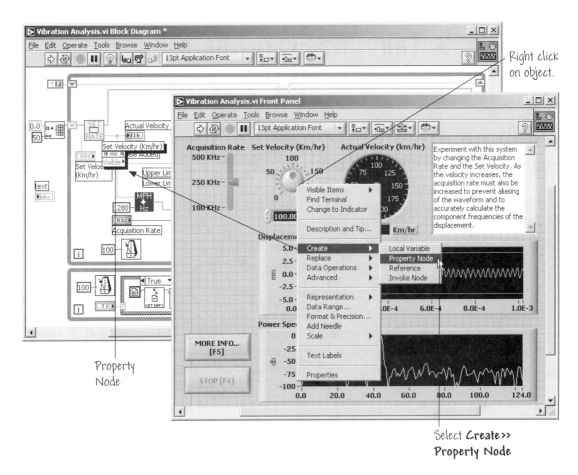

FIGURE 12.4
The **Property Node**.

- Bounds
- Value
- Color

12.4   LABVIEW EVERYWHERE

LabVIEW was first released in 1986 for use on the Macintosh Operating System. Years later, LabVIEW was released as a multi-platform environment capable of running on Macintosh, Windows, Sun, and eventually on Linux as well. Over the years, the vision behind LabVIEW has been to expand the reach of LabVIEW from a multi-platform tool to ever more embedded targets. This vision, referred to as **LabVIEW Everywhere**, means LabVIEW development could scale all the

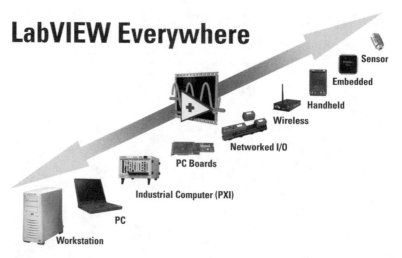

FIGURE 12.5
The LabVIEW Everywhere vision.

way from development on workstations and PCs to more embedded systems on chips. The LabVIEW Everywhere vision is depicted in Figure 12.5.

The first step in realizing this vision came with the introduction of LabVIEW Real-Time (RT) in 1999. LabVIEW RT allows users to program LabVIEW applications for deterministic, real-time performance for data acquisition and control systems. LabVIEW RT has grown to include targets such as industrialized computers, PCI DAQ boards, and rugged networked I/O systems.

With the introduction of LabVIEW 7 Express, National Instruments has taken another significant step in realizing the vision of LabVIEW Everywhere. LabVIEW 7 Express has expanded LabVIEW development to the PDA and FPGA. With the LabVIEW PDA Module, users can now develop a LabVIEW program on their personal computers and download it to their PDA running either the Palm or Pocket PC OS. Imagine the wide range of applications that can be accomplished using a simple development tool like LabVIEW to create programs for your PDA!

LabVIEW FPGA was also a significant technological advancement, extending LabVIEW development all the way down to the chip level of field-programmable gate arrays (FPGAs) that resides on special hardware. FPGA programming was previously reserved for advanced VHDL programmers and is known to be rather complicated. With LabVIEW, FPGA engineers can now achieve the benefits of FPGA technology, high-speed, deterministic execution, along with the simultaneous execution of parallel tasks, without having to learn VHDL.

As you can see from the examples above, LabVIEW's presence and reach is ever expanding. What began on the Macintosh has grown to include dozens of other targets as large as industrial PCs and as small as PDAs and FPGAs. It is clear that LabVIEW is closing in on its vision—that with one development tool you will be able to design from desktop to embedded, simply varying your target based on application timing, size, ruggedness, portability, and other requirements, without having to learn an entirely new development process. This vision simply extends the very purpose of LabVIEW, to increase the productivity of scientists and engineers. Stay tuned for further innovation and expansion of the LabVIEW platform from National Instruments.

12.5   RELAXED READING: VIRTUAL INSTRUMENTATION AND THE UNIVERSITY OF TENNESSEE FUTURE TRUCK

The University of Tennessee Future Truck entry for 2003, nicknamed *Evolution*, is based on a 2002 Ford Explorer SUV. Team Tennessee is converting *Evolution* into a highly efficient, low emission, hybrid electric vehicle. The goal is to increase on road fuel economy by 25% over the stock vehicle and achieve low emissions status, while maintaining the performance of the stock vehicle.

Team Tennessee is taking new directions on several aspects of the vehicle design. For example, the 2.7 liter V-4 engine used last year is being replaced with an efficient 2.3 liter Ford Duratec in-line four cylinder engine. The Unique Mobility SR-218H electric motor and CD40-400L inverter/controller from last year's configuration are being retained; however, the powertrain will be changed to a pre-transmission parallel configuration. The system voltage will be 300V, with the energy storage system configured as 25 12V Hawker Genesis 16Ah lead-acid batteries. The control system will be developed using LabVIEW RT software and implemented using National Instruments Compact FieldPoint hardware. The control strategy will be charge-sustaining with engine-off operation.

National Instruments Fieldpoint hardware and LabVIEW Real-Time software was used in the design process. The design team decided to make use of the recently available Compact Fieldpoint hardware. This decision was based on the fact that the compact version was more rugged than its predecessor while consuming less package space. A cFP-2020 network module is the heart of the *Evolution* control system. This module has a CompactFlash port for solid state data storage. LabVIEW RT was used as the programming environment.

The goal for the *Evolution* control system was to implement a robust control system that coordinates the interaction of the heat engine, traction motor, and high voltage battery pack to deliver the power demanded by the driver. The manner in which the control system carries out this function relies on several factors. The most fundamental of these is to translate the intent of the driver. The control system must interpret what the driver is trying to do, and to deliver what is expected up to the limitations of the entire system. The primary interface for the driver/vehicle is the accelerator and brake pedals. These inputs are transformed into control signals for the traction motor and heat engine. These two

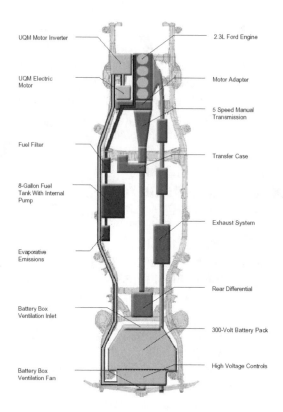

UQM Motor Inverter

2.3L Ford Engine

UQM Electric
Motor

Motor Adapter

5 Speed Manual
Transmission

Fuel Filter

Transfer Case

8-Gallon Fuel
Tank With Internal
Pump

Exhaust System

Evaporative
Emissions

Rear Differential

Battery Box
Ventilation Inlet

300-Volt Battery Pack

Battery Box
Ventilation Fan

High Voltage Controls

FIGURE 12.6
Overview of mechanical layout of *Evolution*.

primary motive forces work together to provide the necessary torque to satisfy the demands of the driver.

The Virtual Instrumentation applications employed in *Evolution's* development greatly enhanced the quality of the control system and reduced prototyping time. Perhaps the greatest benefit of the VI applications that were used with *Evolution* was the ability to run the control strategy in real-time, and to be able to make logic and calibrations corrections on the while the vehicle is operating. The VIs developed for this application were structured in such a way that any variable could be viewed in real-time. This allowed spot checks of the functionality of the controller, and the ability to quickly track down faulty logic or coding. The application of Virtual Instrumentation to the *Evolution* control system made development quick and painless.

Virtual Instrumentation was used to perform measurements during system and subsystem testing wherever possible due to ease of use and rapid turnaround. As an example, the hand test controller for the high voltage traction motor drive was replaced with a VI. The front panel of this VI is shown in Figure 12.7. This approach allowed more precise dynamometer testing of the motor through

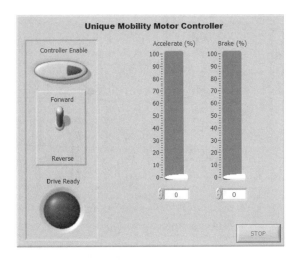

FIGURE 12.7
Traction motor control VI front panel.

precision inputs (0%, 25%, 50%, 100%). These values would have to be guessed or measured from the old hand unit.

In summary, a sophisticated control system was developed for the University of Tennessee Future Truck, *Evolution*. The control system was developed using extensive use of Virtual Instrumentation principles from base software development, to subcomponent testing, to the actual vehicle application. The use of Virtual Instrumentation streamlined the entire controls development process, making *Evolution* a successful hybrid electric vehicle.

For further information, contact:

David E. Smith
University of Tennessee, Knoxville
dsmith31@utk.edu

12.6 SUMMARY

In this chapter we discussed some LabVIEW applications to help you utilize the power of virtual instruments. The material presented here was intended to stimulate your curiosity about the topics of Event-Driven Programming, Remote Front Panels, Property Nodes, and more. It is beyond the scope of this introduction to LabVIEW to go into much detail on these advanced topics, but it is appropriate to point out to you the many interesting LabVIEW topics that remain for your study. Yes, you can use LabVIEW in conjunction with a variety of different industry standard tools and LabVIEW is designed to aid scientists and engineers in all their applications.

Event: Caused by actions the user performs.

Event Driven: The flow of data determines the execution of order of block diagram elements.

LabVIEW Everywhere: Vision to expand the reach of LabVIEW.

LabVIEW Real-time: Allows users to program LabVIEW for deterministic, real-time performance.

Property Node: Change appearance and function of front panel objects programmatically.

Remote Front Panel: A VI is viewed and controlled remotely from a Web browser.

Web Publishing Tool: Tool allows you to modify the document you will see on the Web browser.

EXERCISES

E12.1 Navigate to the NI Example Finder and Browse according to Task. Under the New Examples for LabVIEW 7.0 directory, select the Viewing and Controlling Front Panels Programmatically. There are two VIs titled **RemotePanelMethods-Client.vi** and **RemotePanelMethods-Server.vi**. Open both VIs and follow the directions on the **RemotePanelMethods-Client.vi**. This will give you an introduction to controlling a front panel remotely.

E12.2 Practice with Property Nodes. Navigate to the NI Example Finder and Browse according to Task. Under the New Examples for LabVIEW 7.0 directory, select the Viewing and Programmatically Controlling VIs. There are three example VIs available. Choose the one titled **Property Nodes.vi**. Run the VI and experiment with the various front panel objects to see how the properties can be varied programmatically.

E12.3 Navigate to the NI Example Finder and Browse according to Task. Under the New Examples for LabVIEW 7.0 directory, select the Registering Events Dynamically and Handling User Events. There are three VIs available. Open **Dynamically Monitor VI's.vi**. This VI uses both Dynamic Events and User Defined Events to monitor all VIs running with their Front Panels open. It dynamically registers the Close Panel event for each of them. This means that while this VI is running, if the user clicks on the X to close the Front Panel of any VI that is running with its Front Panel open, it will fire an event in this VI and bring up a dialog. This will give you an introduction to event-driven programming.

PROBLEMS

P12.1 Complete the crossword puzzle.

Across

4. Vision to expand the reach of LabVIEW.
5. Allows users to programLabVIEW for deterministic, real-time performance.
6. Change appearance and function of front panel objects programmatically.

Down

1. Tool allows you to modify the document you will see on the web browser.
2. The flow of data determines the execution order of block diagram elements.
3. A VI is viewed and controlled remotely from a web browser.
7. Caused by actions the user performs.

INDEX